Studies in Logic
Volume 82

Formal Logic
Classical Problems and Proofs

Volume 72
Fathoming Formal Logic: Volume II. Semantics and Proof Theory for Predicate Logic
Odysseus Makridis

Volume 73
Measuring Inconsistency in Information
John Grant and Maria Vanina Mrtinez, eds.

Volume 74
Dictionary of Argumentation. An Introduction to Argumentation Studies
Christian Plantin. With a Foreword by J. Anthony Blair

Volume 75
Theory of Effective Propositional Paraconsistent Logics
Arnon Avron, Ofer Arieli and Anna Zamansky

Volume 76
Argumentation and Inference. Proceedings of the 2nd European Conference on Argumentation. Volume I
Steve Oswald and Didier Maillat, eds.

Volume 77
Argumentation and Inference. Proceedings of the 2nd European Conference on Argumentation. Volume II
Steve Oswald and Didier Maillat, eds.

Volume 68
Logic and Philosophy of Logic. Recent Trends in Latin America and Spain
Max A. Freund, Max Fernández de Castro and Marco Ruffino, eds.

Volume 79
Games Iteration Numbers. A Philosophical Introduction to Computability Theory
Luca M. Possati

Volume 80
Logics of Proofs and Justifications
Roman Kuznets and Thomas Studer

Volume 81
Factual and Plausible Reasoning
David Billington

Volume 82
Formal Logic: Classical Problems and Proofs
Luis M. Augusto

Studies in Logic Series Editor
Dov Gabbay   dov.gabbay@kcl.ac.uk

Formal Logic
Classical Problems and Proofs

Luis M. Augusto

© Individual author and College Publications, 2019, 2020
All rights reserved.

ISBN 978-1-84890-317-3

College Publications
Scientific Director: Dov Gabbay
Managing Director: Jane Spurr

http://www.collegepublications.co.uk

All rights reserved. No part of this publication may be reproduced, stored in a retrieval system or transmitted in any form, or by any means, electronic, mechanical, photocopying, recording or otherwise without prior permission, in writing, from the publisher.

Contents

Preface		**xv**
Note to the 2nd printing		xx

I	**Formal Logic: Form, Meaning, and Consequences**		**1**
1	**Preliminary notions**		**3**
	1.1	Formal languages: Alphabets and grammars	3
	1.2	Logical languages: Form and meaning	7
		1.2.1 Object languages and metalanguages	7
		1.2.2 Logical sentences: From categorical propositions to set-theoretical expressions	8
	1.3	Logic and metalogic: Proofs and metaproofs	11
		1.3.1 Induction, mathematical and structural	12
		1.3.2 Proof by contradiction	13
	1.4	Logic and computation: Turing machines, decidability and tractability	15
2	**Logical form**		**33**
	2.1	Logical languages and well-formed formulas	33
		2.1.1 Alphabets, expressions, and formulas logical	33
		2.1.2 Orders	36
	2.2	Formalizing natural language	40
	2.3	Argument form	46
	2.4	Normal forms and substitutions for L1	50
		2.4.1 Literals and clauses	52
		2.4.2 Negation normal form	53
		2.4.3 Prenex normal form	54
		2.4.4 Skolem normal form	54
		2.4.5 Conjunctive and disjunctive normal forms	57
		2.4.6 Substitutions and unification for L1	62
3	**Logical meaning**		**77**
	3.1	Truth: Values, tables, and functions	77
	3.2	The Boolean foundations of logical bivalence	79

v

3.3 Compositionality and truth-functionality 83
3.4 Classical interpretations and valuations 88
3.5 Meaning and form 93

4 Logical consequences 99
4.1 Logical consequence: A central notion 99
    4.1.1 Consequences and systems logical, inferential, and deductive 99
    4.1.2 Syntactical consequence and proof theory 105
    4.1.3 Semantical consequence and model theory 111
    4.1.4 Adequateness of a deductive system 116
4.2 Logical theories and decidability 120
    4.2.1 Theories, subtheories, and extensions 121
    4.2.2 FOL theories and decidability 123
        4.2.2.1 Finite satisfiability and ground extensions 125
        4.2.2.2 Finite models and prefix classes 132

II The System CL and the Logic CL 139

5 The language of classical logic 141
5.1 Some preliminary remarks 141
5.2 L1 and classical subsets/extensions thereof 144
    5.2.1 The classical connectives 144
    5.2.2 The quantifiers of CFOL 148
5.3 Applications of L1 148
    5.3.1 Logical arguments: Categorical syllogisms 148
        5.3.1.1 Evaluating arguments with Euler diagrams 150
        5.3.1.2 Evaluating arguments with Venn diagrams 151
    5.3.2 Logic programming (I): **Prolog** 153
        5.3.2.1 The language (of) **Prolog** 155
        5.3.2.2 Increased expressiveness and ambivalent syntax 158
        5.3.2.3 Programs and substitutions 159
    5.3.3 Logic design: Logic circuits 162

6 Classical logical consequence 175
6.1 Classical $\heartsuit$-consequences 175
    6.1.1 Classical syntactical $\heartsuit$-consequences 176
    6.1.2 Classical semantical $\heartsuit$-consequences 178
6.2 Classical $\blacklozenge$-consequences 180

7 CL and extensions 183

	7.1	The logic CL	183
	7.2	The extension CL$^=$: CL with equality	186

8 Classical FO theories and the adequateness of CFOL — 193

III Classical Models — 201

9 Three formal semantics for classical logic — 203
    9.1 Tarskian semantics — 204
    9.2 Herbrand semantics — 206
    9.3 Algebraic semantics: Boolean algebras — 212

IV Classical Proofs I: Direct Proofs — 221

10 The validity problem, or VAL — 223
    10.1 The *Entscheidungsproblem* and Turing's negative answer — 223
    10.2 VAL and direct proofs — 227
    10.3 The complexity of VAL — 231

11 Hilbert-style systems — 237
    11.1 The axiom system $\mathcal{L}$ — 238
        11.1.1 The propositional system $\mathcal{L}$ — 238
        11.1.2 The FO system $\mathcal{L}$ — 245
    11.2 Further Hilbert-style systems — 246
        11.2.1 The class $\mathscr{H}$ — 246
        11.2.2 Other systems — 248

12 Gentzen systems — 253
    12.1 The natural deduction calculus $\mathcal{NK}$ — 253
        12.1.1 The propositional calculus $\mathcal{NK}$ — 254
        12.1.2 The FO predicate calculus $\mathcal{NK}$ — 265
        12.1.3 The extension $\mathcal{NK}^=$ for CL$^=$ — 270
    12.2 The sequent calculus $\mathcal{LK}$ — 271

V Classical Proofs II: Indirect Proofs — 285

13 The satisfiability problem, or SAT — 287
    13.1 SAT and refutation proofs — 288
        13.1.1 The different forms of SAT — 288
        13.1.2 Indirect proofs — 292

13.2 The complexity of *SAT* 295
13.3 Herbrand's Theorem and the SAT 300

14 The resolution calculus   **311**
14.1 The resolution principle 313
    14.1.1 The resolution principle for propositional logic . . 313
    14.1.2 The resolution principle for FOL 317
14.2 Resolution refinements 324
    14.2.1 Semantic resolution 325
    14.2.2 Linear resolution: Logic programming (II) 332
14.3 Paramodulation 346

15 The analytic tableaux calculus   **365**
15.1 Analytic tableaux as a propositional calculus 367
15.2 Analytic tableaux as a FO predicate calculus 376
    15.2.1 FOL tableaux without unification 378
    15.2.2 FOL tableaux with unification 380

Bibliography   **385**

Bibliographical references   **387**

Index   **395**

List of Figures

1.1.1	A syntactic, or derivation, tree.	6
1.2.1	Relations of inclusion and exclusion in diagrammatic representation. Source: Venn (1881). (Work in the public domain.)	8
1.4.1	Computer model of a Turing machine.	17
1.4.2	A Turing machine that computes the function $f(n,m) = n+m$ for $n,m \in \mathbb{N}^+$.	19
1.4.3	The hierarchy of complexity classes with corresponding tractability status.	21
1.4.4	The encodings $\langle M_T \rangle$ and $\langle M_T, z \rangle$.	24
1.4.5	State diagram of a Turing machine.	30
2.2.1	Formalizations for English by means of the language of classical propositional logic.	45
2.2.2	Formalizations for English by means of the language of classical FO logic.	47
2.3.1	Some classical formally correct arguments.	51
2.3.2	Two invalid argument forms.	52
2.4.1	Tseitin transformations for the connectives of L.	61
2.4.2	Unifying the pair $\langle P(a,x,h(g(z))), P(z,h(y),h(y)) \rangle$.	67
2.4.3	A FOL argument.	73
3.3.1	A truth table with $2^3 = 8$ rows.	84
3.3.2	Truth tables for the connective $\rightarrow$ in the 3-valued logics $Ł_3$, K_3^W, and Rn_3.	87
3.5.1	The properties of a Boolean algebra.	95
4.1.1	The complete lattice $\mathcal{S} = (2^A, \subseteq)$ for $A = \{a,b,c\}$.	103
4.1.2	Adequateness of a deductive system $L = (L, \Vdash)$.	119
5.1.1	Venn diagram of the set A.	143
5.2.1	Diagrammatic representations of the connectives of O_L.	145
5.2.2	Diagrammatic representations of the logical connectives $O_G = \{\uparrow^2, \downarrow^2, \nleftrightarrow^2\}$.	147
5.2.3	Euler diagrams for the classical quantifiers.	149
5.3.1	Euler diagrams of an invalid (1) and a valid (2) argument.	152

List of Figures

5.3.2   Venn diagram with eight minterms. 153
5.3.3   A Venn-diagram representation of argument A1. 154
5.3.4   A Venn-diagram representation of argument A2. 154
5.3.5   From a binary switch (i) to a series-parallel connection (v). 164
5.3.6   Logic gates and their graphical representations. 165
5.3.7   A logic circuit for the function $f(x_1, x_2, x_3) = (x_1 \wedge x_2) \vee (x_1 \wedge x_3)$. 166
5.3.8   A logic circuit for $f(x_1, x_2, x_3) = x_1 \wedge (x_2 \vee x_3)$. 166
5.3.11 Properties of XOR. 167
5.3.9   Two functionally equivalent logic circuits. 168
5.3.10 The De Morgan's laws and the NOR and NAND gates. . . 169
5.3.12 Logic circuits. 174

11.1.1 Proof of $\vdash_\mathcal{L} P \to P$. 239
11.1.2 Proof of an argument in $\mathcal{L}p$. 241
11.1.3 Proof in $\mathcal{L}q$ of a valid syllogism. 246

12.1.1 A proof of a propositional derivation in $\mathcal{NK}$. 256
12.1.2 A proof in $\mathcal{NK}$ of the distributivity property for $\wedge$. 259
12.1.3 Proof of $\vdash_{\mathcal{NK}} ((P \to Q) \wedge (P \to R)) \to (P \to (Q \wedge R))$. . . 260
12.1.4 Proof of an argument in (extended) $\mathcal{NK}$. 261
12.1.5 Proof of $\vdash_{\mathcal{NK}} \phi \leftrightarrow \neg\neg\phi$. 263
12.1.6 A proof with universal generalization. 266
12.1.7 An example of universal instantiation. 267
12.1.8 An example of existential generalization. 267
12.1.9 An example of existential instantiation. 268
12.1.10 A FO $\mathcal{NK}$ proof. 269
12.1.11 A proof in $\mathcal{NK}^=$. 271
12.2.1 Proof in $\mathcal{LK}$ of axiom $\mathscr{L}2$ of the axiom system $\mathscr{L}$. 276
12.2.2 Proof in $\mathcal{LK}$ of a FO theorem. 277

13.2.1 A tableau for the Turing machine M. 298
13.3.1 Closed semantic tree of $C = \{\mathcal{C}_1, \mathcal{C}_2, \mathcal{C}_3, \mathcal{C}_4, \mathcal{C}_5\}$ in Example 13.31. 303
13.3.2 A closed semantic tree. 304

14.1.1 A refutation tree. 315
14.1.2 A propositional argument as input in Prover9/Mace4. . . . 316
14.1.3 Output by Prover9: A valid propositional argument. . . . 316
14.1.4 Output by Prover 9: A valid formula. 318
14.1.5 Output by Mace4: A counter-model. 318
14.1.6 A resolution refutation-failure tree. 320
14.1.7 Input in Prover9/Mace4: A FO theory. 320

List of Figures

14.1.8	Output by Prover9.	321
14.1.9	Output of Prover9: A valid FO argument.	322
14.2.1	A PI-resolution tree.	328
14.2.2	Hyper-resolution of $\Xi = (\mathcal{C}_3; \mathcal{C}_1, \mathcal{C}_2)$.	331
14.2.3	A hyper-resolution deduction tree.	331
14.2.4	Theory of distributive lattices and commutativity of meet: Input in Prover9/Mace4.	332
14.2.5	Proof by Prover9 of the commutativity of meet in a distributive lattice.	333
14.2.6	A linear-resolution refutation tree.	334
14.2.7	Trace by SWI-Prolog.	338
14.2.8	A failed proof tree.	340
14.2.9	A successful reduction interpreted as a resolution proof.	342
14.2.10	A LI-resolution proof tree.	343
14.2.11	A SLD-resolution proof.	346
14.2.12	A complete proof tree.	347
14.2.13	SWI-Prolog answering a query and outputting traces for some "true" instantiations.	348
14.2.14	SWI-Prolog traces of a "true" and a "false" instantiation.	349
14.3.1	Theory of commutative groups: Input in Prover9/Mace4.	353
14.3.2	Output by Prover9.	354
15.1.1	Analytic tableaux expansion rules: $\alpha\beta$-classification.	369
15.1.2	A closed propositional tableau.	373
15.2.1	Analytic tableaux expansion rules: $\gamma\delta$-classification.	377
15.2.2	A closed FO tableau without unification.	379
15.2.3	A closed FO tableau with unification.	382

List of Algorithms

2.1	PNF transformation.	55
2.2	Skolemization.	56
2.3	Tseitin transformation.	60
2.4	The Robinson algorithm.	65
14.1	Binary resolution.	312
14.2	Reduction.	340
15.1	Analytic tableaux proof.	366

Preface

Often spoken of as the science of reasoning, *logic* can be *formal* or *informal*. While it is not unequivocal–there is significant overlap between both–, the use of these two adjectives allows us to distinguish between a largely mathematical from a substantially psychological approach, respectively, to logic. This might appear unwarranted to those well-acquainted with logic as an object language, but at the metalanguage and/or metalogical levels it becomes clear that formal logic has its foundations in mathematics, namely in what can be called abstract mathematics, whereas informal logic reposes on psychological theories of human reasoning. This book is an introduction to formal logic.

A second major distinction in contemporary logic segregates *classical logic* from the *non-classical logics*. These–note the plural–are typically rivals of the former–note the singular–, it being meant by this that they aim at replacing it in many contexts and/or applications. This rivalry notwithstanding, they are either extensions or restrictions of classical logic, which means that anyone advocating a non-classical logic should be well-versed in classical logic. This book is an introduction to classical logic.

While formal classical logic is certainly interesting per se, today its study is often associated to computer science with a plethora of computational implementations in view. This association of logic and computation can be roughly captured by the expression *computational logic*. This book is an introduction to computational logic.

Do we then need to specify that this book is an introduction to formal classical computational logic? Not really, because in it we take the adjectives *formal* and *computational* to be so intimately related that they can be often considered synonyms. This synonymy is more typically to be found between the expressions *formal language* and *computer language*, but we discuss here the language of classical logic as first and foremost a formal language, and hence the redundancy of the adjective *computational* in the title.

This book is thus an introduction to formal classical logic with its contemporary uses in mind, to wit, *logical problems* that are in fact *decision problems* that are in fact *computational problems* whose *proofs* are delegated to computer software. In effect, logic is—arguably—all about

Preface

proving, but proofs can be costly, often impossibly so, in terms of space and time, it being meant by this that proofs require storage space (i.e., a physical memory) and they take time to be computed; hence, monetary costs are also often associated to proofs, as space and time, as well as human work, cost money. Given these costs, unrealistic for human computers and undesirable for companies, today most proofs are delegated to (partly) automatic provers, namely the so-called *SAT solvers*. These are software based on the (Boolean) satisfiability problem, or *SAT*. This is the dual of the (Boolean) validity problem, or *VAL*, at the core of the conception of the digital computer via Hilbert's *Entscheidungsproblem* and the Universal Turing Machine.

These two problems, *VAL* and *SAT*, can be said to be the two classical problems that initiated the computational history of formal classical logic, a history that can be more immediately traced back to the *Entscheidungsproblem*, but that actually also requires digressions into the work of the likes of J. Venn, G. Frege, and A. Turing–if not Aristotle, too. In particular, we discuss the classical formal semantics conceived by, or originating in the work of, G. Boole, J. Herbrand, and A. Tarski. While this, as said, is an introduction to formal classical logic, we dispense with the adjective "classical" between "formal" and "logic" in the title, because this book has as its backbone these two semantical problems. The fragment "Formal logic: Classical problems" indicates that our introduction to formal logic is so via the classical problems, first and foremost *VAL* and *SAT*, but then also all the decision and computational problems that can be formulated in terms of these, namely with computer implementations in mind.

But, as stated above, logic is–arguably–all about proving. Without (adequate) proof systems at hand, these two problems and all the other problems formulated in their terms (let us call them all *classical problems* for the sake of simplicity) have no solution beyond propositional logic, given the undecidability of first-order logic (abbr.: FOL), a problem motivated by semantical structures known as *models* that, differently from proofs, which are finite by definition, may be infinite. Indeed, to say that *VAL* and *SAT* are formulated in semantical terms means that they are formulated in terms of *preservation of truth*: If all the, say, facts in a database are true, is a certain conclusion one wishes to draw therefrom always, or at least in some cases, also so? Given classical problems of very low complexity formulated in propositional logic, the semantical construct known as a truth table can provide a solution. But classical problems are more often than not highly complex, sometimes industrial-scale so, and they typically require a first- (or higher-) order language.

Fortunately, we have today a plethora of adequate proof systems for *VAL* and *SAT*. The Hilbert(-style) systems and the Gentzen systems, the latter divided into natural deduction and the sequent calculus, are proof systems to address *VAL*, and resolution and analytic tableaux are the two proof systems of election to find answers to classical problems formulated in terms of *SAT*. The comprehensive elaboration on these systems accounts for the expression "proofs" in our title, now complete as *Formal logic: Classical problems and proofs*. Although the first systems above are not algorithmic in nature, thus not providing efficient methods for classical problems, they are both historically and pedagogically relevant, and we accordingly discuss them in due detail. Resolution and analytic tableaux are at the root of many efficient SAT solvers, and we give equally full treatments of these calculi.

But there are more than these proofs. In the paragraph above we wrote "adequate" without brackets (compare with farther above), it being meant by this with respect to a proof system that one can prove in it every logical truth of the associated logic and nothing that is not a logical truth thereof. But these properties, known as completeness and soundness, require *metalogical proofs*—i.e. proofs at a level higher than the *logical proofs*. The same is true of the general undecidability of FOL, a result that is a celebrated answer to *VAL*. In turn, *VAL* and *SAT* have been proven to belong to specific classes of computational complexity—i.e. it has been shown how much they "cost"—, with these proofs constituting fundamental knowledge for the computational implementations of classical problems. Fulfilling our requirements of self-containment and comprehensiveness, we provide discussions of these celebrated proofs, as well as of the above-mentioned properties for all the proof systems we elaborate on in detail.

It is the moment now to convince the reader that ours is a truly original introduction to logic. Largely depending on the applications in view, logic can be approached today from three perspectives, to wit, mathematical, computational, or philosophical. Introductory textbooks to logic accordingly segregate their contents: Mathematical approaches typically concentrate on the mathematical properties of logical systems; computational approaches focus on computational implementations and automation of proofs; philosophical treatments greatly concentrate in argumentation. Gödel's (in)completeness and satellite results feature prominently in the first, as mathematical proof is a major concern of mathematical logic and it is unpalatable not to be able to prove a mathematical truth once one is discovered (or constructed, depending on one's philosophy of mathematics). The temporal and spatial costs of computational implementations, from the simple transformation of a formula into one acceptable by some software to the carrying out of a proof in it, are central topics in the second kind. Arguments, categorical syllogisms and fallacies included, occupy many of the pages of the third type. More technically, this can be reformulated as follows by invoking the four so-called *pillars of formal logic*: *Model theory* and *set theory* are major topics to be found in mathematical treatments of logic; *recursion*, or *computability, theory* features significantly in computational approaches; *proof theory* tends to be weighty in introductions to logic written for philosophy students. In particular, while the classical problems—VAL significantly less so than SAT—feature in introductory logic textbooks aimed at computer science students, they are largely or wholly absent from textbooks targeting a mathematical or philosophical studentship.

This segregation has constituted a successful recipe for a long time now, and possibly rightly so, but it does not reflect the current state of what can very generally be called formal logic. This book corrects this misguided state of affairs. Not focusing on the history of classical logic, this book nevertheless provides discussions and quotes central passages on its origins and development, namely from a philosophical perspective. Not being a book in mathematical logic, it takes formal logic from an essentially mathematical perspective. Biased towards a computational approach, with SAT and VAL as its backbone, this is thus an introduction to logic that covers essential aspects of the three branches of logic, to wit, philosophical, mathematical, and computational. More so, it gives practical applications of all these fields, namely in argumentation, theorem proving, logic programming, and even in logic design.

To be sure, the aim of reaching a large academic readership poses the risk of serving only a small one: The "traditional" tripartite segregation may in fact mirror some real distinctions, whether in skills or interests, in the different studentships. Moreover, the ambition of treating classical logic both at the object-language and at the metalanguage/metalogic levels while trying to keep the book in a "manageable" size may entail the suppression or obliteration of important contents of either of these components. To this we reply that no book stands alone, or is wholly self-contained; just as in any other field, certain treatments of logic have reached the status of standard works, and we refer to Hurley (2012), Mendelson (2015), and Boolos, Burgess, & Jeffrey (2007), for "classics" in philosophical, mathematical, and computational logic, respectively. Additionally, we hope the intersection of the above mentioned readerships is not empty. Our hope may in fact be a justified belief, as, for instance, linguists and computer scientists, to mention but these, may prove.

Be it as it may, we assume knowledge of, or at least familiarity with, mathematical concepts such as sets, functions, operations, and relations, providing solely definitions of less basic notions (e.g., Boolean algebra). In order to refresh their memory, or newly acquire such notions, mathematically literate readers can benefit from Bloch (2011) and the more mathematically reticent can do so from Makinson (2008). We also think that logic is a subject that requires both hands-on practice and reflection (or rumination), and we accordingly provide a vast selection of exercises ranging from the typical logic "drilling" exercise to commentary of relevant passages.

Finally: This book is in a large measure a selection, a restructuring, and an extension of contents first published in Augusto (2018). Main motivations for the present resulting text were the desire to improve, by reviewing and extending, the contents of the mentioned book, as well as the aim to provide a comprehensive stand-alone book on formal classical logic with the above-mentioned characteristics, in the belief that classical logic, particularly so in its formal version, is a subject both fascinating and–more and more–fundamental.

I wish to thank Dov M. Gabbay for accepting to publish this "extended remix," as well as Jane Spurr for her impeccable assistance as managing director of College Publications.

Madrid, Summer 2019

Luis M. S. Augusto

Preface

Note to the 2nd printing

The present 2nd printing corrects identified addenda and errata, has improved figures and a more uniform notation, and introduces a few notions that were either missing or not adequately defined in the original edition (e.g., *parameter, trivial quantification, free for*). A few paragraphs underwent minor changes, mostly in Chapter 2, namely in Sections 2.1.1-2 and 2.4.6. Concerning the notation, the major change was the decision to reserve the Greek letters in the metalanguage for utmost generality, with Backus-Naur definitions, as well as most axiom schemata and rules of inference, featuring the same letters from the Roman alphabet. Another minor change in notation was the replacement of the symbol $\Rightarrow$ by $\vdash$ in the rules of the sequent calculus $\mathcal{LK}$. All this done, the pagination is essentially the same as in the original edition.

Madrid, June 2020

Luis M. S. Augusto

Part I.

Formal Logic: Form, Meaning, and Consequences

1. Preliminary notions

Logic is the science of reasoning. Its main object is both *forms* and *instances* of reasoning; hence, it covers both the general and the particular, inasmuch as this conforms to the general. When these forms and instances of reasoning are formulated in a precisely defined formal language, we speak of *formal logic*. The notion of reasoning in formal logic is specific to this field: it entails the concept of *logical consequence*. This segregates formal logic with respect to psychology, more generally, and argumentation theory, more particularly. By *a formal logic*, we shall understand here a system of logical consequence over a formal language that is a logical language, and *formal logic* will be the study of any such system. In order to avoid equivocation with relation to these two notions we may speak synonymously of *metalogic* in the latter case. Being a formal language, a logical language can be used as a *computer language*, i.e. it can be used for computer implementations of reasoning. Adequately formalized reasoning instances are subject to both *decidability issues* arising in formal languages and *computational costs*; the Turing machine is the standard measure of both.

1.1. Formal languages: Alphabets and grammars

A logical language is first and foremost a formal language. Although we do not here elaborate on formal languages, our belief is that a summary understanding of this topic is relevant in the study of formal logic. We accordingly give the basics of formal languages and refer the reader to Augusto (2020b) for a comprehensive presentation.[1]

Definition 1.1. A *formal language* is a pair $\mathsf{F} = (\Sigma, G)$ where Σ is a set of symbols, called an *alphabet*, and G is a *formal grammar*, a four-tuple $G = (\Sigma_1, \Sigma_2, S, P)$ where $\Sigma_1 \cup \Sigma_2 = \Sigma$, $\Sigma_1 \cap \Sigma_2 = \emptyset$, S is the start symbol (a variable), and P is the set of *rules of production* (often just *productions*), or *rules of rewriting*.

---
[1] See also Augusto (2020a), Part III, for a briefer but still comprehensive presentation.

1. Preliminary notions

Example 1.2. Let $\Sigma_1 = \{a, b, c\}$, $\Sigma_2 = \{S, A, B\}$, and P be constituted by the following rules of rewriting:

$$P = \left\{ \begin{array}{l} S ::= aAB \\ A ::= Bba \\ B ::= bB \\ B ::= c \end{array} \right\}$$

The pair $\mathsf{F} = (\Sigma = \Sigma_1 \cup \Sigma_2, G)$, where $G = (\{a, b, c\}, \{S, A, B\}, S, P)$, is a formal language. The rules in P, in which "::=" denotes left-to-right replacement, generate the following *words* or *strings* over the alphabet Σ: abcbabc, abbcbabc, acbabbbbc, etc. For instance, starting with S, the start symbol, we can have the following word derivation:

$$S \Longrightarrow aAB \Longrightarrow a\,(Bba)\,B \Longrightarrow acbaB \Longrightarrow acba\,(bB) \Longrightarrow acbabc$$

In this Example 1.2, Σ_1 is the set of *terminal* symbols, reason why we can denote it by T; Σ_2 is the set of *variables*, and we thus may denote it by V. The language generated by a formal grammar is typically defined as the set of words, or strings, constituted solely by terminal symbols. The language of Example 1.2 is the set

$$L = \{ab^n cbab^m c \in T^* | n, m \geq 0\}$$

where T^* denotes all the strings, including the empty string (see the following Example), that can be built from the terminal alphabet T, and $n, m \geq 0$ denotes the number of occurrences of the symbols. The string above can be rewritten as $ab^0 cbab^1 c$.

Although a *natural language* like English is likely too complex to be fully captured as a formal language, large portions of it can in fact be so considered. The formal grammar of Example 1.3–a context-free grammar–can generate the sentence *Resourceful little ants work diligently*. In this grammar, the elements of V are those written as $\langle \cdot \rangle$, and those of T are the given English words in italics. Note in this grammar the symbol ϵ denoting the *empty word* or *string*, so called because it has no symbols.[2] For instance, in the third rule we may replace $\langle ARTICLE \rangle$ by the definite article (*the*) or–denoted by "|"– by the indefinite article (*a* or *an*), or we may replace it by the empty string (the *zero article*, in linguistics jargon).

---

[2] For instance, $ab\epsilon bd = ab^2 d$. As a legal string, the empty string may constitute a language, namely $L = \{\epsilon\}$, called the *language of the empty string*. As a matter of fact, we may also have the language with no strings whatsoever, i.e. $L = \emptyset$, called the *empty language*.

1.1. Formal languages: Alphabets and grammars

Given the complexity of this grammar, it is useful to depict the generation of a string by means of a *syntactic*, or *derivation*, *tree* (see Fig. 1.1.1), a downwards-growing tree whose root is the start symbol and the leaves are the terminal symbols, the terminal string being read from left to right.[3]

Example 1.3. Let the following be the production rules of a formal grammar G over $\Sigma = V \cup T$, where the symbols of Σ are either built from symbols of the Roman alphabet (e.g., the single symbol *the*) or built from these within $\langle \cdot \rangle$ (e.g., the start symbol $\langle SENTENCE \rangle$):

$$P = \left\{ \begin{array}{c} \langle SENTENCE \rangle ::= \langle NOUN\_PHR \rangle \langle VERB\_PHR \rangle \\ \langle NOUN\_PHR \rangle ::= \langle ARTICLE \rangle \langle ADJECTIVE \rangle \langle NOUN \rangle \\ \langle ARTICLE \rangle ::= \langle DEFINITE \rangle | \langle INDEFINITE \rangle | \epsilon \\ \langle ADJECTIVE \rangle ::= \langle ADJECTIVE \rangle \langle ADJECTIVE \rangle | \epsilon \\ \langle VERB\_PHR \rangle ::= \langle VERB \rangle \langle ADVERB \rangle \\ \langle DEFINITE \rangle ::= the \\ \langle INDEFINITE \rangle ::= a \,|\, an \\ \langle ADJECTIVE \rangle ::= resourceful \,|\, little \\ \langle NOUN \rangle ::= ants \\ \langle VERB \rangle ::= work \\ \langle ADVERB \rangle ::= diligently \end{array} \right\}$$

Figure 1.1.1 is the syntactic tree of the string

$$\epsilon resourcefullittleantsworkdiligently.$$

As said, this well-formed string corresponds to the English sentence *Resourceful little ants work diligently*.

The problem with the formal grammar of Example 1.3, as the North-American linguist N. Chomsky pointed out (Chomsky, 1957), is that it also generates meaningless, albeit syntactically correct, sentences, say, and famously, *Colorless green ideas sleep furiously*. This is an important remark, as a formal grammar is first and foremost a syntactical "device" to generate *well-formed* (also: *legal*) strings without regard to meaning. This is typically a secondary, though by all means not always dispensable, feature to be added to a formal language

▶ *Do Exercises 1.1-2.*

---
[3] This is actually frequently essential, as in the case of front-end compilation. See Augusto (2020b) for derivation trees for the context-free languages, the class of formal languages comprising the important class of most programming languages.

1. Preliminary notions

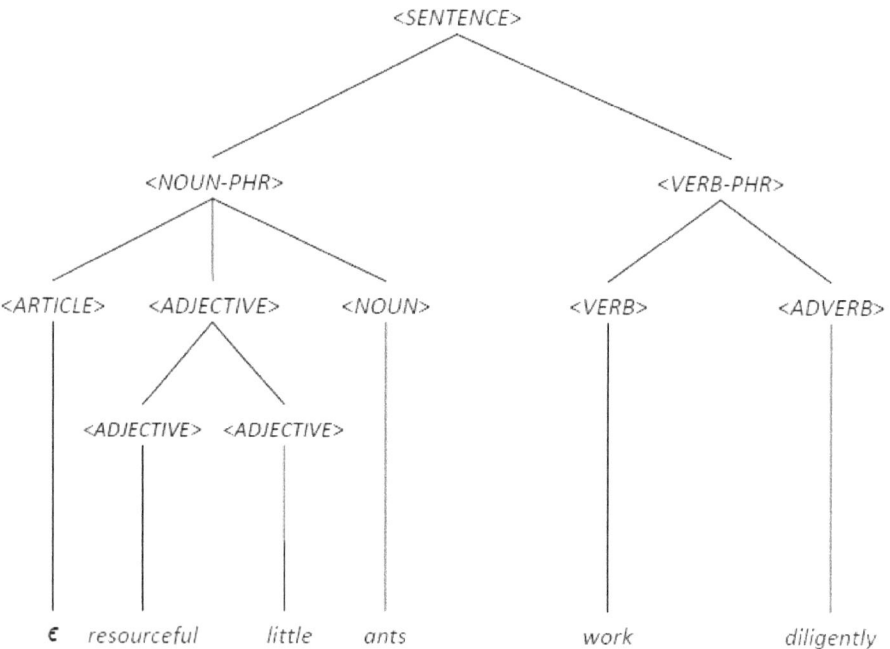

Figure 1.1.1.: A syntactic, or derivation, tree.

1.2. Logical languages: Form and meaning

1.2.1. Object languages and metalanguages

In the field of logic, it is important to distinguish an *object language* L, i.e. a logical language within which logical objects such as formulas and arguments are proved and/or interpreted, from a *metalanguage*, the language in which the study of an object language is conducted. In this book, the main object language is the first-order language of classical logic, denoted by L1, and the metalanguage is English supplemented with logical jargon (e.g., operator, quantifier) and symbols that are not part of the object language L1 (e.g., $\vdash$, $\Rightarrow$). Given the object language L1, μ^{L1} denotes the metalanguage of L1. In the metalanguage here adopted, arbitrary formulas are denoted by the Greek lowercase letters $\phi, \chi, \psi, ...$, with or without subscripts, and arbitrary sets of formulas are denoted by Greek uppercase letters (e.g., Φ, X, Ψ). This convention notwithstanding, where appropriate we shall also denote formulas in μ^{L1} by the Roman letters $A, B, ..., P, Q, R$, mostly. This practice has the pedagogical advantage of bridging in a clearer way the object language and the corresponding metalanguage.

Given an object language L, two aspects are of major importance: *form* and *meaning*. By *logical form*, we mean both what makes a logical expression *well-formed* (e.g., a well-formed formula) and what constitutes a *formal proof*, i.e. a proof that well-formed expressions follow logically from other well-formed expressions by virtue of form alone. Thus, logical form features as the central object of two components of logic related to *syntax*, to wit, the *grammar* of a logical language, and *proof theory*, the study of formal derivations and proofs, and of the diverse proof systems.

This tells us that logical statements and reasoning can be approached from a purely syntactical perspective, but in fact we more often than not also care for *meaning* in a logical language. This is typically provided by an *interpretation*, i.e. an assignment of meaning to the symbols and expressions of a logical language. This assignment of meaning is essentially an attribution of a distinguished element known as a *truth value* (i.e. a valuation) to a formula: we speak of a formula ϕ being *true* or *false* under an interpretation $\mathcal{I}$. This is the fundamental notion of *model theory*, the study of the interpretation of logical languages by means of set-theoretical structures. Given a logical language L, the set of (the classes of) these structures, called *models*, is called a *semantics* for L, but in a broader conception we can speak of semantics as the general study of meaning in a logical system.

1. Preliminary notions

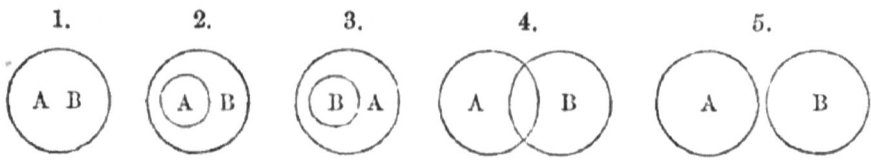

Figure 1.2.1.: Relations of inclusion and exclusion in diagrammatic representation. Source: Venn (1881). (Work in the public domain.)

1.2.2. Logical sentences: From categorical propositions to set-theoretical expressions

Before the rise of modern logic with G. Boole in the 19th century, simple logical sentences were of the categorical form

$$A \text{ is } B$$

of which form *Socrates is mortal* is a notable instance. This simplest of forms allowed for variations such as "No A is B" or "All As are not Bs." (See Fig. 1.2.1.) Essentially, we have the structure *Subject – Predicate*, which can actually be further specified in the form *Noun – Copula – Noun/Adjective*, where by "copula" we mean the verb "to be." Interestingly enough, this had been the case since Aristotle, and as late as in the last decades of the 18th century the celebrated philosopher I. Kant saw no need for any changes in the field of logic.[4]

G. Boole himself, who can be said to have given the first impulse to the birth of modern logic, still used this Aristotelian sentence structure in his algebraic study of logic:

> Every assertion that we make may be referred to one or the other of the two following kinds. Either it expresses a relation among *things*, or it expresses, or is equivalent to the expression of, a relation among *propositions*. ... The former

---

[4]In his celebrated *Critique of pure reason*, Kant wrote:

> [S]ince Aristotle it [logic] has not required to retrace a single step, unless, indeed, we care to count as improvements the removal of certain needless subtleties or the clearer exposition of its recognized teaching, features which concern the elegance rather than the certainty of the science. It is remarkable also that to the present day this logic has not been able to advance a single step, and is thus to all appearance a closed and completed body of doctrine. (Kant, 1787/1929)

class of propositions, relating to *things*, I call "Primary;" the latter class, relating to *propositions*, I call "Secondary." The distinction is in practice nearly but not quite co-extensive with the common logical distinction of propositions as categorical or hypothetical. For instance, the propositions, "The sun shines," "The earth is warmed," are primary; the proposition "If the sun shines the earth is warmed," is secondary. To say, "The sun shines," is to say, "The sun is that which shines," and it expresses a relation between two classes of things, viz., "the sun" and "things that shine." The secondary proposition, however, given above, expresses a relation of dependence between the two primary propositions, "The sun shines" and "The earth is warmed." (Boole, 1854)

In this perspective, we then say "Socrates is mortal" to express the relation between "Socrates" (A) and the class "things that are mortal" (B). While G. Boole was first to undertake–rather than just conceive the idea of, as G. W. von Leibniz did–a purely symbolic formalization of the statements of logic, it was the feat of J. Venn to have formalized these categorical relations in a more explicitly set-theoretical way by means of the now-called Venn diagrams, which graphically depict the relations of mutual inclusion or exclusion of classes of objects. The path was paved for the full presentation of logic in set-theoretical terms.[5]

Now, what if, instead of using this categorical form that closely mirrors statements in natural language, we write

$$mortal(socrates)$$

where *mortal* is a function and *socrates* its single argument, to express the assertion that Socrates is mortal? In particular, how can we still

---

[5]This is actually only part of what can be spoken of as the "mathematization" of logic, as G. Boole also undertook the presentation of what is now called classical logic in the language of algebra (Boole, 1847; 1854). In his own words:

> The design of the following treatise is to investigate the fundamental laws of those operations of the mind by which reasoning is performed; to give expression to them in the symbolical language of a Calculus, and upon this foundation to establish the science of Logic and construct its method ... There is not only a close analogy between the operations of the mind in general reasoning and its operations in the particular science of Algebra, but there is to a considerable extent an exact agreement in the laws by which the two classes of operations are conducted. (Boole, 1854)

This algebraic presentation is typically called *Boolean logic* and it features in this text in several Sections.

1. Preliminary notions

attribute meaning to a "statement" that appears so remote from the normal constructions of a natural language like English? This question was replied precisely at the same type that the above substitution was proposed, and it is in this that the true birth of *formal logic* can be placed. This was the achievement of the German mathematician G. Frege. In his own words, and as an invitation to adhere to what he called his *Begriffsschrift*, or, in English, *concept writing* or *ideography*:

> If it is one of the tasks of philosophy to break the domination of the word over the human spirit by laying bare the misconceptions that through the use of language often almost unavoidably arise concerning the relations between concepts and by freeing thought from that with which only the means of expression of ordinary language, constituted as they are, saddle it, then my ideography, further developed for these purposes, can become a useful tool for the philosopher. ... The mere invention of this ideography has, it seems to me, advanced logic. ... These deviations [that I was driven to make] from what is traditional find their justification in the fact that logic has hitherto always followed ordinary language and grammar too closely. In particular, I believe that the replacement of the concepts *subject* and *predicate* by *argument* and *function*, respectively, will stand the test of time. It is easy to see how regarding a content as a function of an argument leads to the formation of concepts. Furthermore, the demonstration of the connection between the meanings of the words *if, and, not, or, there is, some, all,* and so forth, deserves attention. (Frege, 1879)

The replacement proposed by Frege has stood the test of time, as will be evident in our formalization of classical logic.[6] Because we indeed think that all this deserves attention, we have gone to the great lengths of writing this book. In effect, much of what below we call *formal logic* is directly or indirectly related to this passage, even if we do not carry out here a historical approach.

▶ *Do Exercise 1.3.*

---

[6] Note, however, that today we write *Mortal*(*socrates*), where *Mortal* is a predicate symbol, instead of *mortal*(*socrates*), where *mortal* is a function symbol, to express the fact that Socrates is mortal. This development is mirrored in the fact that the term "predicate calculus" to refer to first-order logic (FOL) progressively replaced the original Fregean label "function calculus."

1.3. Logic and metalogic: Proofs and metaproofs

Although logic is all about *proving*, namely correctness and/or validity of reasoning instances or formulas, we must distinguish two notions of proving in logic, to wit, *proof* and *metaproof*. The distinction between these two notions follows directly from the segregation above between object language and metalanguage, i.e. a (logical) *proof* is a construct in an object language, whereas a (logical) *metaproof* is so of a metalanguage. While, as said, proofs fall on reasoning instances and formulas, metaproofs fall on properties of the logical constructs and systems in consideration. This latter focus constitutes the subject matter of *metalogic*, and so metaproofs carried out in a metalanguage are often spoken of as *metalogical proofs*. For instance, proofs of soundness and completeness with respect to some proof system or some semantics are metaproofs, or metalogical proofs. Other fundamental notions studied by metalogic are, for instance, logical consequence, logical satisfiability, logical validity, logical inference, decidability, etc. Importantly, metalogic can be considered a mathematical subject, being the study of logical systems by mathematical methods. This accounts for the largely mathematical character of metaproofs.[7]

Consider the following characterization of metalogic by A. Tarski, one of the most influential theoreticians in modern logic:[8]

> In contemporary methodology we investigate deductive theories in their entirety as well as individual sentences which constitute them; we consider the symbols and the expressions of which such sentences are composed, properties and sets of expressions and of sentences, relations holding among them (such as the relation of consequence), and even relations among expressions and the objects which the expressions "talk about" (such as the relation of designation); we

---

[7] To be more precise, it can be said that metalogic is a subfield of *metamathematics*, the study of (formal systems of) mathematics and its foundations, namely the subfield thereof restricted to logic.

[8] In this passage, the term *methodology* is synonymous with *metalogic*. In Tarski's own words:

> "*[M]ethodology*" means, essentially, "*the science of method*". Consequently, this expression is now often replaced by others–especially the terms "METALOGIC" and "METAMATHEMATICS"–, which mean about the same as "*the science of logic*" and "*the science of mathematics*". (Tarski, 1994)

1. Preliminary notions

establish general laws which govern these concepts. (Tarski, 1994)

In this passage, the set-theoretical viewpoint is explicit, and such terms as *properties* and *relations* should be taken as meaning *mathematical properties* and *mathematical relations*.

We elaborate at length on the notion of *proof* in Section 4.1.2 and this elaboration is then followed by the study of prominent proof systems in Parts IV and V. With respect to *metaproofs*, which are ubiquitous in this text, there are two main techniques to be found in them: *structural induction* and *proof by contradiction*. The former is based on the principle of *mathematical induction*, and we give a brief exposition of this central notion in mathematics. Although the latter is often present in proofs, in metaproofs it has a particular character, and thus we give here also a short elaboration of this technique in metaproofs. Other, less common, techniques in metaproofs, will be the object of short explanations where required.

1.3.1. Induction, mathematical and structural

Mathematical induction (abbr.: MI) proper is restricted to the set of natural numbers, but we can generalize it to any well-founded mathematical structure (e.g., formulas, graphs, trees), in which case we speak of *structural induction*.[9] In a proof by MI, we require two main steps: the *basis (step)* and the *induction step*. Say that we have a formula of n symbols; the idea is that if we can prove that this formula is well-formed for $n = 1$ (or some other $n \in \mathbb{N}^+$; the basis),[10] then we can prove (the induction step) that a formula formed with arbitrarily many n symbols is equally well formed by proving that if a formula formed with m symbols is well formed–the *induction hypothesis*–then a formula with $m + 1$ symbols is also well formed. It is this latter generalization that is of interest to formal logic, but it is useful to understand thoroughly MI proper.

Example 1.4. The following is an often-used example to introduce MI. We wish to prove that the following statement $P(n)$ holds for all positive natural numbers n.[11]

$$(P(n)) \quad \sum_{i=1}^{n} i = \frac{n(n+1)}{2}.$$

---

[9] Throughout this text, when we write "by induction" with respect to proofs we often mean "by structural induction."
[10] More rarely, $n = 0$ for the basis. Even more rarely, $n \in \mathbb{Z}$.
[11] $\sum_{i=1}^{n} i$ is an abbreviation for the *summation* $1 + 2 + ... + n$, $1 \leq i \leq n$.

1.3. Logic and metalogic: Proofs and metaproofs

Induction basis: We show that $P(1)$ holds. In effect, we have

$$1 = \frac{1(1+1)}{2}.$$

Induction step: We now show that if $P(m)$ holds, then $P(m+1)$ holds, too. We begin by assuming that $P(m)$ holds for arbitrary $m \in \mathbb{N}$ and then show that $P(m+1)$ also holds:

$$(P(m+1)) \quad \sum_{i=1}^{m+1} i = \sum_{i=1}^{m} i + (m+1) =$$

$$= \frac{m(m+1)}{2} + (m+1) = \frac{m^2 + 3m + 2}{2} = \frac{(m+1)(m+2)}{2} =$$

$$= \frac{(m+1)((m+1)+1)}{2}$$

and what was to be proven has been proven (abbreviated **QED**, for the Latin expression *"quod erat demonstrandum"*).

We formalize the above:

Definition 1.5. *(Principle of mathematical induction)* Let P be a proposition defined on the set $\mathbb{N}^+$. Then, for each $n \in \mathbb{N}^+$, $P(n)$ either holds or does not hold. Suppose that $P(1)$ holds and $P(m+1)$ holds whenever $P(m)$ holds. Then, $P(n)$ holds for arbitrary $n \in \mathbb{N}^+$.

1.3.2. Proof by contradiction

Proof by contradiction is a method of mathematical proof theoretically based on the following properties of Boolean expressions: For any (propositional) variable p, it holds that

$$(\text{LEM}) \quad p \vee \neg p = 1$$

and

$$(\text{LNC}) \quad p \wedge \neg p = 0.$$

In logical jargon, we say that for every proposition P, either P is true or its negation (i.e. $\neg P$) is, a principle known as *law of excluded middle* (LEM), and P cannot be both true and false, known as *law of non-contradiction* (LNC).

These two laws become the basis of a proof by contradiction in the following way: Of a given proposition P, we may assume that it is either true or false. Let us assume that P is false, so that $\neg P$ is true.

1. Preliminary notions

If by the end of our reasoning we have reached a proposition of the form $Q \wedge \neg Q$, for some proposition Q, then we have clearly–though indirectly[12]–reached a contradiction, and the only way out of it is to retract the truth of $\neg P$ and accept the truth of P instead.[13]

We formalize this:

Definition 1.6. Where P and Q denote propositions or assertions and the symbol $\neg$ denotes negation, a proof by contradiction has the following structure:

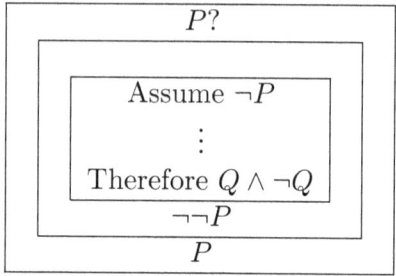

Example 1.7. This is a classic proof by contradiction of the irrationality of the number $\sqrt{2}$. Let there be given the proposition

The number $\sqrt{2}$ is irrational.

We assume, for the sake of contradiction, that this proposition is false, i.e. we assume that $\sqrt{2}$ is rational. If $\sqrt{2}$ is rational, then there are integers a and b, $b \neq 0$, for which

$$(2.1) \qquad \sqrt{2} = \frac{a}{b}.$$

---

[12] Reason why proofs by contradiction are a kind of *indirect proof*.

[13] Although proof by contradiction is often seen as equivalent to proof by *reductio ad absurdum*, it actually is a particular kind of the general form of proofs that involve an absurd, or impossible, conclusion. In fact, *reductio ad absurdum* is more properly used for argumentation, any form thereof resulting (supposedly) in absurdity qualifying as such. For example, the reasoning

If that's so, then pigs can fly.

fits into the broadest sense of *reductio ad absurdum*. This notwithstanding, it is also very frequent in mathematical reasoning to refer to a proof by contradiction as a proof by *reductio (ad absurdum)*. Also, it can be further distinguished between a "genuine" *reductio* proof and a proof of a negative proposition as in Example 1.7 (cf. Exercise 13.3.2). We approach this topic in Chapter 13.

We additionally assume that $\frac{a}{b}$ is reduced to its simplest terms, which entails that not both of a and b can be even. So, one or both of a and b must be odd. From 2.1, it follows that

$$(2.2) \qquad 2 = \left(\frac{a}{b}\right)^2 = \frac{a^2}{b^2}$$

or

$$a^2 = 2b^2$$

and the square of a is an even number. Hence, a itself is an even number. Let us now replace $a = 2k$ in 2.2; we have

$$2 = \frac{(2k)^2}{b^2} = \frac{4k^2}{b^2}$$

and hence

$$2b^2 = 4k^2$$

or equivalently

$$b^2 = 2k^2$$

and we have it that b is itself even. So we have it that both of a and b are even, but we know that not both of a and b can be even, so that we have reached a contradiction. We reject the statement that $\sqrt{2}$ is a rational number, as being false, and accept the irrationality of $\sqrt{2}$.
QED

▶ *Do Exercises 1.4-6.*

1.4. Logic and computation: Turing machines, decidability and tractability

The notion of the Turing machine is required for the (intuitive) definition of a *decision procedure* with respect to some adequately formalized instance of reasoning, but it is also required for the formal definition of *tractability*, or *actual solvability*, of a *computational problem*, i.e. an infinite collection of problem instances with a solution for every instance. It should be remarked that these two topics are relevant for formal logic only when this meets computability theory; this, however, is more and more the case, given not only the increasing automation of logical calculi but also the ever-growing use of logic-based computational models or implementations–of which, in this text, we mainly discuss three "classics," to wit, automated theorem proving (Part V), logic programming

1. Preliminary notions

(Sections 5.3.2 and 14.2.2) and logic design (Section 5.3.3). In any case, the two classical problems of formal logic, to wit, the validity problem (abbr.: *VAL*) and the satisfiability problem (*SAT*), which work as the backbone of this book (see Parts IV-V, mostly), are tightly associated to the Turing machine. Thus, we give here the essentials of Turing machines with a view to the decidability of logical theories and respective tractability, and refer the reader to Augusto (2020a, b) for fuller treatments of these topics and further, more specialized, bibliography. The essentials here provided suffice for a satisfactory grasp of the contents of Chapters 10 and 13, in which the validity and satisfiability problems, respectively, are elaborated on.

Definition 1.8. A *Turing machine* is a 7-tuple

$$M_T = (Q, \Gamma, \#, \Sigma, q_0, A, \delta)$$

where $Q = \{q_0, ..., q_n\}$ is a finite set of states, $\Gamma = \{\sigma_1, ..., \sigma_m\}$ is the *tape alphabet*, $\# \in \Gamma$ is the *blank symbol*, $\Sigma \subseteq (\Gamma - \#)$ is the *input alphabet*, $A \subseteq Q$ is the set of *accepting states*, and

$$\delta : (Q \times \Gamma) \longrightarrow (Q \times \Gamma \times \{L, R\})$$

where L, R denote direction (left and right, respectively), is the *transition function*.

1. A Turing machine is said to be *deterministic* if it prescribes a single action for each state; otherwise, it is *non-deterministic*.

Definition 1.9. The *computer model* for a Turing machine $M_T = (Q, \Gamma, \#, \Sigma, q_0, A, \delta)$ consists of a logic box programmed by δ equipped with a read-and-write head. This head reads an input string starting on the left of an input tape that is infinite to the right. When the machine reads a symbol a while in state q, it switches into state p, replaces symbol $a \in \Gamma$ by symbol $b \in \Gamma$, and moves one tape cell in direction $D = \{L, R\}$, so that we have

$$\delta(q, a) = (p, b, D).$$

Figure 1.4.1 shows the computer model of a Turing machine in which the tape is infinite to both the right and the left sides of the input string, a frequent variation to the model above of Definition 1.9. The current content on the tape is ...#11#1#... and the machine is–possibly back–in state q_0.

1.4. Logic and computation: Turing machines, decidability and tractability

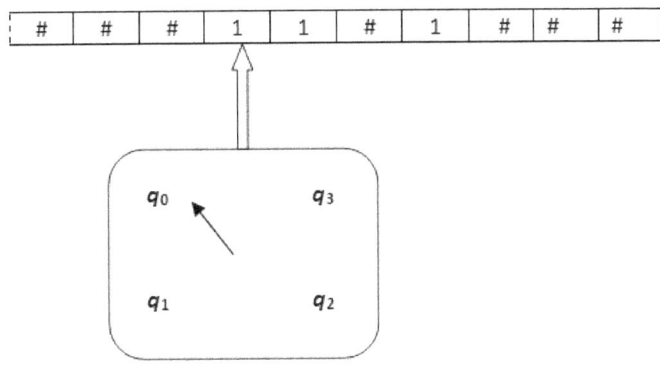

Figure 1.4.1.: Computer model of a Turing machine.

Figure 1.4.1 is also a graphical representation of a Turing machine *configuration*.

Definition 1.10. A *configuration* for a Turing machine M_T is a pair $C = (q, u\underline{a}v)$ indicating that M_T is in state q with the present tape string uav and reading the symbol a.

1. The M_T configuration $(q, u\underline{a}v)$ *yields* the configuration $(p, x\underline{b}y)$ in one step (or *move*), denoted by

    $$(q, u\underline{a}v) \vdash_T (p, x\underline{b}y)$$

    if and only if (abbr.: iff) the transition $\delta(q, a)$ changes configuration $(q, u\underline{a}v)$ to configuration $(p, x\underline{b}y)$.

2. A *starting configuration* for M_T is a configuration of the form $(q_0, \epsilon\underline{a}v)$ indicating that a is the symbol in the leftmost cell of the input tape.

3. An *accepting* (or *halting*)[14] configuration for M_T has the form $(q, u\underline{a}v)$, $q \in A$, indicating that uav is the output.

4. A *hanging configuration* for M_T is a configuration of the form $(q, \epsilon\underline{a}v)$ such that the transition function $\delta(q, a)$ instructs the machine to move left (i.e. off the tape).

---

[14] It is assumed that a Turing machine halts when it has accepted a string, because there are no more possible moves.

1. Preliminary notions

Example 1.11. In Figure 1.4.1, the Turing machine is in the configuration
$$C = q_0, \underline{1}1\#1$$

Definition 1.12. A *computation* for a Turing machine M_T on some input string w is any of three possibilities for $i = 1, ..., n$: (i) a finite sequence $C_0, C_1, ..., C_n$ of configurations, $C_0 = (q_0, \epsilon \# w)$, such that $C_{i-1} \vdash_T C_i$ and C_n is an accepting configuration; (ii) a finite sequence $C_0, C_1, ..., C_n$ of configurations, $C_0 = (q_0, \epsilon \# w)$, such that $C_{i-1} \vdash_T C_i$ and C_n is a hanging configuration; (iii) an infinite sequence $C_0, C_1, ...$ of configurations, $C_0 = (q_0, \epsilon \# w)$, such that $C_{i-1} \vdash_T C_i$.

Example 1.13. The following is the transition table for a Turing machine $M_T = (\{q_0, q_1, q_2\}, \{1, \#\}, \#, \{1\}, q_0, \{q_2\}, \delta)$ adding two positive integers (the integers are represented in unary notation $1^n = \underbrace{111...1}_{n}$; e.g. $1^3 = 111 = 3$):

	1	#
q_0	$q_0 1 R$	$q_1 1 R$
q_1	$q_1 1 R$	$q_2 \# L$
q_2	$q_2 \# R$	–

On input $11\#111\#\#...$, M_T carries out the following computation:

$$q_0\underline{1}1\#111\# \vdash q_0 1\underline{1}\#111\# \vdash q_0 11\underline{\#}111\# \vdash q_1 111\underline{1}11\# \vdash q_1 1111\underline{1}1\#$$

$$\vdash q_1 11111\underline{1}\# \vdash q_1 111111\underline{\#} \vdash q_2 11111\underline{1}\# \vdash q_2 11111\underline{\#}\#$$

$$\equiv q_0 11\#111\# \vdash^3 q_1 111111\underline{1}\# \vdash^4 q_2 11111\underline{\#}$$

The behavior of the machine is as follows: Given the input $11\#111\#\#...$, which corresponds to the two integers 2 and 3, in order to add both strings M_T has firstly to delete the blank space between them, replace it with 1, and then keep moving right until it finds the last symbol 1; finally, this is deleted, and the machine halts. The output string is 11111 followed by infinitely many blank symbols.

A graphical representation of a Turing machine is a useful visual aid. This representation, called *state diagram*, is essentially a directed graph in which the vertices are the states of the machine and the arcs are the

1.4. Logic and computation: Turing machines, decidability and tractability

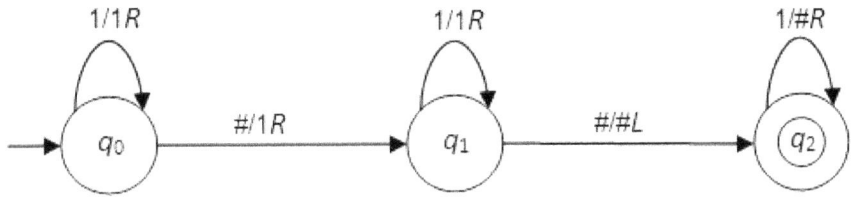

Figure 1.4.2.: A Turing machine that computes the function $f(n,m) = n + m$ for $n, m \in \mathbb{N}^+$.

transitions. A short single-vertex arc indicates the initial state and two concentric circles for a node represent a final state.[15]

Example 1.14. Figure 1.4.2 shows the state diagram of the Turing machine of Example 1.13.

Let now some decision problem be formulated as a language L.[16]

Definition 1.15. A Turing machine M_T *decides* a language L if M_T computes the characteristic function $\chi_L : \Sigma^* \longrightarrow \{0, 1\}$ defined as follows for some string x:

$$\chi_L(x) = \begin{cases} 1 & \text{if } x \in L \\ 0 & \text{otherwise} \end{cases}.$$

L is a *recursive language*, or a *decidable language*, if there is a Turing machine M_T that decides L.

---

[15] In several passages in this Section the mathematical notion of a graph is required. See Augusto (2020a, b), where further literature on discrete mathematics is given.

[16] We give here the *general* definition of a decision problem. Let there be given some domain of discourse (e.g., the set $\mathbb{N}$ of the natural numbers) whose elements we shall refer to as *instances*. Then, given some subset S of this domain

$$S = \{x \in \mathbb{N} | P(x)\}$$

where $P(x)$ stands for, say, "x is even," we ask whether there is an algorithm to decide for each instance x whether $x \in S$ or $x \notin S$ (see Examples 4.67-8). It is convenient, and common practice in the field of the theory of computation, to define a decision problem as a formal language, namely the language L of strings w over an alphabet of terminal symbols T such that every w has a specific property P:

$$L = \{w \in T^* | P(w)\}.$$

(See Section 1.1 for the specific terminology and notation.) Below, in Def. 4.63, we specify the notion of a decision problem in relation to a logical problem.

1. Preliminary notions

1. If M_T outputs 1 when $x \in L$ but fails to output 0 when $x \notin L$, then L is a *recursively enumerable*, or *semi-decidable*, *language*.

Because the Turing machine is central in the definition above, we speak of *Turing-decidability* for the case of decidable problems, and of *Turing-recognizability* for the semi-decidable problems. It should be further remarked that the latter class includes, but is not equal to, the former.

It is important to realize that the fact that a theory is decidable does not necessarily entail that it is *actually computable* or, in other words, *tractable*. In effect, a decidable problem is said to be tractable if a decision can be found in polynomial time, i.e. if given an input of size n, the running time of the algorithm, denoted by $T(n)$, is upper-bounded by a polynomial expression in n: given a constant k and a notation (*big O*) characterizing functions with respect to their (asymptotic) growth rates, we have $T(n) = \mathcal{O}(n^k)$.[17] In other words, a problem is tractable if it belongs to the complexity class **P**. A problem that is not tractable is said to be *intractable*, i.e. it can be solved, in theory and given a large but finite time window, but not in polynomial time. This leaves exponential time (see Fig. 1.4.3).[18] Let now **NP** be the complexity class of problems that can be *solved* by a non-deterministic Turing machine (whereas the problems in **P** can be solved by a deterministic Turing machine), or whose solution can be *checked* in polynomial time by a deterministic Turing machine. Then, if–an unsolved problem in computability theory–**P** $\neq$ **NP**,

---

[17] The notation $\mathcal{O}(\cdot)$ is called *big-O notation*. Consider the functions $f, g : \mathbb{N} \longrightarrow \mathbb{R}^+$, where $\mathbb{R}^+$ is the set of the positive real numbers. We say that $f(n)$ *is of order* $g(n)$, and write $f(n) = \mathcal{O}(g(n))$, if there exist $c, n_0 \in \mathbb{Z}^+$ such that, for $n \geq n_0$, c is a constant,
$$f(n) \leq cg(n).$$
When $f(n) = \mathcal{O}(g(n))$, we say that $g(n)$ is an (asymptotic) upper bound for $f(n)$. Clearly, $\mathcal{O}(g(n))$ is an approximation, or estimation, of $f(n)$. In practical terms, given a polynomial $p(x)$ of degree n we suppress the coefficient of the highest-order term in $p(x)$ and disregard all other terms to obtain $p(x) = \mathcal{O}(x^n)$. For example, given the polynomial $3x^3 + 7x^2 + x + 36$, we have
$$3x^3 + 7x^2 + x + 36 = \mathcal{O}(x^3).$$
Note, further, that $\mathcal{O}(\cdot)$ denotes a class of functions.

[18] Similar definitions can be made with respect to *space*, which equates with the amount of tape of a Turing machine required to solve a decision problem. Note the following further complexity classes in this hierarchy: **NL** (non-deterministic Turing-machine logarithmic space), **PSPACE** (deterministic Turing-machine polynomial space), **EXPTIME** (deterministic Turing-machine exponential time), and **EXPSPACE** (deterministic Turing-machine exponential space). This hierarchy is actually altogether a source of major open problems (denoted by $\stackrel{?}{=}$ in Fig. 1.4.3).

1.4. Logic and computation: Turing machines, decidability and tractability

then **NP**-complete problems, or the "hardest problems in **NP**," are also intractable.

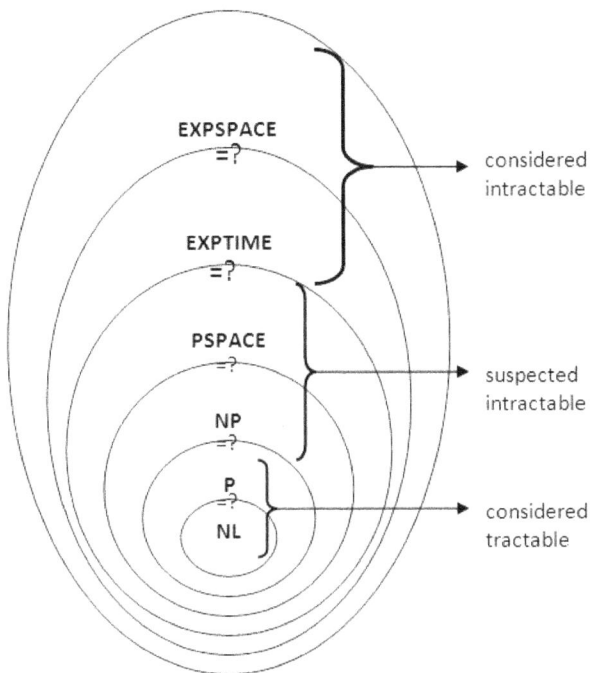

Figure 1.4.3.: The hierarchy of complexity classes with corresponding tractability status.

We shall actually require a formal definition of the **NP**-complete (also: **NPC**) class. This, in turn, requires some additional notions:

Definition 1.16. Given two languages A and B, we say that there is a *mapping reduction from A to B*, denoted $A \preceq B$, if there is a computable function $f : \Sigma^* \longrightarrow \Sigma^*$ such that, for every w, we have

$$(f_{red}) \qquad w \in A \quad \text{iff} \quad f(w) \in B.$$

The function f is the *reduction from A to B*.

1. If furthermore the condition

    $$(f_P) \qquad f \text{ can be computed in polynomial time}$$

    is satisfied, we say that A can be *polynomial-time reducible* to B and we write "$A \preceq_P B$".

1. Preliminary notions

Example 1.17. An undirected graph is a pair $\mathfrak{G} = (V, E)$ where $V = \{v_0, v_1, ..., v_n\}$ is a set of vertices and $E = \{e_0, e_1, ..., e_n\}$, $e_i = \{v_j, v_k\}$, is a set of edges such that all $v_j, v_k \in E$ are in V. By making use of this structure we illustrate a polynomial-time reduction. *The vertex cover problem* is as follows: Given an undirected graph $\mathfrak{G} = (V, E)$, a vertex cover is a set $V' \subseteq V$ such that if $(u, v) \in E$, then $u \in V'$ or $v \in V'$, or both. The size of a vertex cover is the number of vertices that are its members. *The clique problem*: Given an undirected graph $\mathfrak{G}$, is there a clique (i.e. a complete subgraph) of $\mathfrak{G}$ of size k? We want to find a vertex cover of (minimum) size k in $\mathfrak{G}$. We show in general traits how to reduce the Vertex Cover Problem to the Clique Problem. Let $(\mathfrak{G}, k)$ be an instance of the former, for $\mathfrak{G} = (V, E)$. We construct the complement of $\mathfrak{G}$, denoted by $\overline{\mathfrak{G}}$, such that $V_{\overline{\mathfrak{G}}} = V_{\mathfrak{G}}$, but the edge $(u, v) \in \overline{\mathfrak{G}}$ iff $(u, v) \notin \mathfrak{G}$. Where k is the integer parameter for the Vertex Cover Problem, we now define the integer parameter for the Clique Problem as $n - k$. Then–and the reduction consists in this–, $\overline{\mathfrak{G}}$ has a clique of size at least $n - k$ iff $\mathfrak{G}$ has a vertex cover of size at most k. Moreover, the construction for the Clique problem runs in polynomial time, so that this is a polynomial-time reduction.

We can now give a formal definition of the **NP**-complete complexity class.

Definition 1.18. A computational problem A is said to be **NP**-*complete*, or a member of the class **NPC** if (i) $A \in$ **NP**, and (ii) for every other computational problem $A_i \in$ **NP** it is the case that $A_i \preceq_P A$.

We conclude this discussion with the *universal Turing machine*, celebrated for at least two reasons: it was at the heart of the conception of the digital computer and was central in proving the negative result for the *Entscheidungsproblem*.[19] We next give the necessary notions to define this celebrated construct.

Definition 1.19. Let a Turing machine M_T be given; then, there is an *encoding function* $e : M_T \longrightarrow \{0, 1\}^*$ that is a unique description of M_T.[20]

In this Definition, it must be remarked that $\{0, 1\}^*$ is the set of all the strings that can be formed by concatenations of 0s and 1s, in other

---

[19] For the latter, see Chapter 10. For the former, see Introduction to Augusto (2020b).
[20] Turing (1936-7) referred to this encoding that, for practical ends, just is the program of a Turing machine, as the *Standard Description*. Of course, a Turing machine just is its program, or Standard Description.

1.4. Logic and computation: Turing machines, decidability and tractability

words, in binary code. In detail, given a Turing machine M_T, a generic move $\delta(q_i, \sigma_j) = (q_k, \sigma_l, D_m)$ is *encoded* by the binary string

$$\xi = 0^i 10^j 10^k 10^l 10^m$$

and a *binary code* for M_T is

$$111\ \text{code}_1\ 11\ \text{code}_2\ 11...11\ \text{code}_n\ 111 = \langle M_T \rangle$$

where every code_i has the form of ξ.

As said, this encoding is unique, i.e. every string ξ is interpreted as the code for at most one Turing machine, and a Turing machine has many codes.

Definition 1.20. A *universal Turing machine* is a machine M_{TU} such that, given an arbitrary Turing machine M_T and $z \in \Gamma^* \subset M_T$, and an encoding function e, upon receiving an input string of the form $\xi^* = e(M_T) e(z) = \langle M_T, z \rangle$

1. M_{TU} accepts ξ^* iff M_T accepts z.[21]

2. M_{TU} produces output $\langle y \rangle$ if M_T accepts z and produces output y.

Example 1.21. Let $q_{i,k}$, $i, k \geq 0$, be coded as $\langle q_0 \rangle = 0$, $\langle q_1 \rangle = 00$, etc. Let the following be the binary coding for the tape symbols: $\langle 1 \rangle = 1$, and $\langle \# \rangle = 0$. Finally, encode the directions $\langle L \rangle = 0$ and $\langle R \rangle = 00$. Then, Figure 1.4.4 shows the encoding of the Turing machine $M_T = (\{q_0, q_1, q_2\}, \{1, \#\}, \#, \{1\}, q_0, \{q_2\}, \delta)$ of Example 1.13, as well as the encoding of this Turing machine with the input string $z = 11\#111$.

▶ Do Exercises 1.7-12.

Exercises

Exercise 1.1. Let $\Sigma = \{0, 1\}$ be an alphabet. Give examples of strings (also: words) in the following languages over Σ:

1. $L = \{0^2 1^l 0^2 | l > 0\}$.

2. $L = \{(01)^n | n > 0\}$.

---
[21] Obviously, if $\Gamma \subseteq \{0, 1\}^*$, then for a string $z \in \Gamma^*$ we have $e(z) = \langle z \rangle = z$.

$\langle M_T \rangle = 1110100101001001101001001001001001001001001001001001000100010010010010010001001001000100100111$

$\langle M_T, z \rangle = 1110100101001001101001001001001001001001001001001001000100010010010010010001001001000100100111101111$

Figure 1.4.4.: The encodings $\langle M_T \rangle$ and $\langle M_T, z \rangle$.

1.4. Logic and computation: Turing machines, decidability and tractability

3. $L = \{1^l 01^m 01^n | l > 0, m, n \geq 0\}$.

Exercise 1.2. Find the language generated by the grammar $G = (V, T, S, P)$

1. with $V = \{S, A\}$, $T = \{a, b, c\}$, and the set of productions

$$P = \left\{ \begin{array}{l} S ::= aSb \\ Aab ::= c \\ aS ::= Aa \end{array} \right\}.$$

2. with $V = \{S, A, B\}$, $T = \{0, 1\}$, and the set of productions

$$P = \left\{ \begin{array}{l} S ::= 0B \\ A ::= 0B \\ B ::= 1A\,|\,1 \end{array} \right\}.$$

3. with $V = \{S, A, B\}$, $T = \{a, b\}$, and the set of productions

$$P = \left\{ \begin{array}{l} S ::= aA \\ A ::= aAB\,|\,a \\ B ::= b\,|\,\epsilon \end{array} \right\}.$$

4. with $V = \{S, A, B, C\}$, $T = \{0, 1\}$, and the set of productions

$$P = \left\{ \begin{array}{l} S ::= AS0\,|\,BS1\,|\,C \\ AC ::= 0C \\ A0 ::= 0A \\ A1 ::= 1A \\ BC ::= 1C \\ B0 ::= 0B \\ B1 ::= 1B \\ C ::= \epsilon \end{array} \right\}.$$

5. with $V = \{S, A, B\}$, $T = \{0, 1\}$, and the set of productions

$$P = \left\{ \begin{array}{l} S ::= 0AB \\ AB ::= 0\,|\,1 \\ B ::= AB \end{array} \right\}.$$

1. Preliminary notions

6. with $V = \{S, A, B\}$, $T = \{0, 1\}$, and the set of productions

$$P = \left\{ \begin{array}{c} S ::= 0AB \\ A0 ::= S0B \\ A1 ::= SB1 \\ B ::= SA \,|\, 01 \\ 1B ::= 0 \end{array} \right\}.$$

7. with $V = \{S, A, B\}$, $T = \{a, b, c\}$, and the set of productions

$$P = \left\{ \begin{array}{c} S ::= aAbc \,|\, abc \\ Ab ::= bA \\ Ac ::= Bbcc \\ aB ::= aaA \,|\, aa \\ bB ::= Bb \end{array} \right\}.$$

Exercise 1.3. Reflect on the following passages:

1. The sphere of logic is quite precisely delimited; its sole concern is to give an exhaustive exposition and a strict proof of the formal rules of all thought, whether it be *a priori* or empirical, whatever be its origin or its object, and whatever hindrances, accidental or natural, it may encounter in our minds. (Kant, 1787/1929)

2. [T]he universal and necessary laws of thought can only be concerned with its *form*, not in anywise with its *matter*. The science, therefore, which contains these necessary and universal laws is simply a science of the form of thought. And we can form a conception of the possibility of such a science, just as of a *universal grammar* which contains nothing beyond the mere form of language, without words, which belong to the matter of language. This science of the necessary laws of the understanding and the reason generally, or, which is the same thing, of the mere form of thought generally, we call *Logic*. (Kant, 1800/1885)

3. The rules of Logic, then, must not be derived from the *contingent*, but from the *necessary* use of the understanding, which, without any psychology, a man finds in himself. In Logic we do not want to know how the understanding is and thinks, and how it has hitherto proceeded in thinking, but how it ought to proceed in thinking. Its business is to teach us the correct use of reason, that is, the use which is consistent with itself. (Kant, 1800/1885)

1.4. Logic and computation: Turing machines, decidability and tractability

4. All the operations of Language, as an instrument of reasoning, may be conducted by a system of signs composed of the following elements, viz.: 1st. Literal symbols, as x, y, etc., representing things as subjects of our conceptions. 2nd. Signs of operation, as $+, -, \times$, standing for those operations of the mind by which the conceptions of things are combined or resolved so as to form new conceptions involving the same elements. 3rd. The sign of identity, $=$. And these symbols of Logic are in their use subject to definite laws, partly agreeing with and partly differing from the laws of the corresponding symbols in the science of Algebra. (Boole, 1854, Proposition I of Chapter II)

5. [A]ny system of propositions may be expressed by equations involving symbols x, y, z, which, whenever interpretation is possible, are subject to laws identical in form with the laws of a system of quantitative symbols, susceptible only of the values 0 and 1. But as the formal processes of reasoning depend only upon the laws of the symbols, and not upon the nature of their interpretation, we are permitted to treat the above symbols x, y, z, as if they were quantitative symbols of the kind above described. (Boole, 1854)

6. Now suppose that, instead of regarding the proposition as made up of a subject determined by a predicate, we regard it as assigning the relations, in the way of mutual inclusion and exclusion, of two classes to one another. It will hardly be disputed that every proposition *can* be so interpreted. Of course ... this interpretation may not be the most fundamental in a Psychological sense; but when, as here, we are concerned with logical methods merely, this does not matter. ... Now how many possible relations are there, in this respect of mutual inclusion and exclusion, of two classes to one another? Clearly only five. For the question here, as I apprehend it, is this:–Given one class as known and determined in respect of its extent, in how many various relations can another class also known and similarly determined, stand towards the first? Only in the following: It can coincide with the former, can include it, be included by it, partially include and partially exclude it, or entirely exclude it. In every recognized sense of the term these are distinct relations, and they seem to be the only such distinct relations that can possibly exist. These five possible arrangements would be represented diagrammatically as [in Figure 1.2.1]. (Venn, 1881)

Exercise 1.4. Reflect on the following paragraphs related to metalogic:

1. Preliminary notions

1. We here wish to emphasize that the theorems of this paper are *about* the logic of propositions but are *not included* therein. More particularly, whereas the propositions of 'Principia' [Whitehead & Russell (1910, 1912, 1913)] are *particular* assertions introduced for their interest and usefulness in later portions of the work, those of the present paper are about the set of *all* such possible assertions. Our most important theorem gives a uniform method for testing the truth of any proposition of the system; and by means of this theorem it becomes possible to exhibit certain general relations which exist between these propositions. These relations definitely show that the postulates of 'Principia' are capable of developing the complete system of the logic of propositions without ever introducing results extraneous to that system–a conclusion that could hardly have been arrived at by the particular processes used in that work. (Post, 1921)

2. [O]ne should observe the following: terms which denote expressions occurring in deductive theories, as well as terms denoting properties of these expressions, or relations among them, belong to the methodology of deductive sciences, but not to the domain of logic. This applies in particular to various terms ... such as "*variable*", "*sentential function*", "*quantifier*", "*consequence*", and many others. In order to make clearer to ourselves the difference between logical and methodological terms, let us consider such a pair of words as "*or*" and "*disjunction*". The word "*or*" belongs to sentential calculus–and therefore to logic–, although it is also used in all other sciences, and thus in particular in methodology. The word "*disjunction*", on the other hand, denotes a sentence which is constructed with the help of the word "*or*", and is a typical instance of a methodological term. (Tarski, 1994)

Exercise 1.5. Find proofs of the following statements using MI and identify the induction basis and the induction step in them:

1. $\sum_{i=0}^{n} 2^i = 2^{n+1} - 1$ (for all $n \in \mathbb{N}$).
2. $\sum_{i=1}^{n} i = n^2$ (for all $n \in \mathbb{N}^+$, n is odd).
3. $n < 2^n$ (for all $n \in \mathbb{N}$).
4. $n^3 - n$ is divisible by 3 (for $n \in \mathbb{N}^+$).
5. If $|S| = n$ for a set S and finite n, then S has 2^n subsets.

1.4. Logic and computation: Turing machines, decidability and tractability

6. For $n \geq 2$ and $A_i \in U$, U is some universal set, $\overline{\bigcap_{i=1}^{n} A_i} = \bigcup_{i=1}^{n} \overline{A_i}$.

Exercise 1.6. Find proofs of the following statements applying the method of proof by contradiction and identify the contradictions in them:

1. If $3n + 2$ is odd, then n is odd.

2. If A and B are sets, then $A \cap (B - A) = \emptyset$.

3. At least 4 of 22 chosen days fall on the same day of the week.

4. The sum of an irrational number and a rational number is an irrational number.

Exercise 1.7. Given the transition tables and the indicated tape strings, determine the computing behavior of each Turing machine and, if any, its output. Draw their state diagrams.

1. Tape string: 11##...; transition table:

	1	#
q_0	$q_0 1R$	$q_1 \# L$
q_1	$q_2 \# R$	$q_1 1R$
q_2	$q_2 \# R$	$q_3 1R$

2. Tape string: 111##; transition table:

	1	#
q_0	$q_1 \# R$	$q_5 \# L$
q_1	$q_1 1R$	$q_2 \# R$
q_2	$q_2 1R$	$q_3 1L$
q_3	$q_3 1L$	$q_4 \# L$
q_4	$q_4 1L$	$q_0 1R$
q_5	$q_5 1L$	$q_6 \# R$

3. Tape string: 1111##; transition table:

	1	#
q_0	$q_1 \# R$	$q_4 \# L$
q_1	$q_1 1R$	$q_2 \# L$
q_2	$q_3 \# L$	$q_4 \# L$
q_3	$q_3 1L$	$q_0 1R$
q_4	$q_4 1L$	$q_5 \# R$

4. Tape string: ...##0011##...; transition table:

	0	1	#
q_0	$q_1 1R$	$q_1 0R$	$q_1 0R$
q_1	$q_2 1L$	$q_1 0R$	$q_2 0L$

1. Preliminary notions

Exercise 1.8. Determine the language accepted by the Turing machine with the state diagram in Figure 1.4.5.

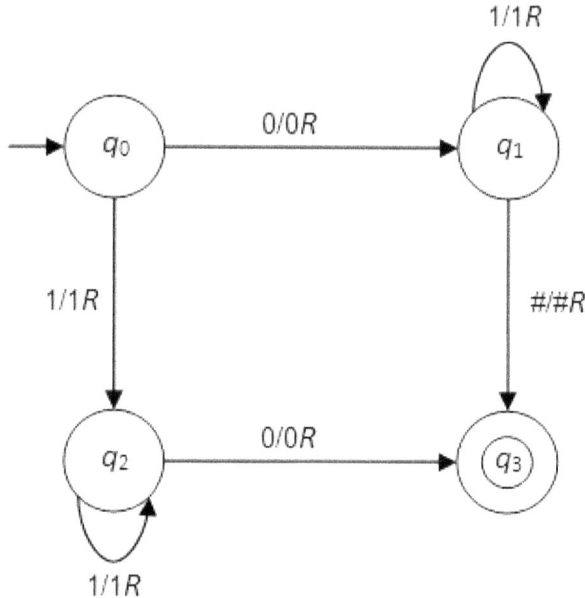

Figure 1.4.5.: State diagram of a Turing machine.

Exercise 1.9. Design Turing machines accepting the following languages:

1. $L = \{a^n b^n c^n | n \geq 1\}$
2. $L = \{a^m b^n | m < n\}$

Exercise 1.10. Design a Turing machine that runs forever.

Exercise 1.11. Give the encodings $\langle M_T \rangle$ and $\langle M_T, z \rangle$ of the Turing machines of Exercise 1.7 by applying the encoding of Example 1.21.

Exercise 1.12. Research into the following notable undecidable problems, formulate them in a set-theoretical manner, and sketch the proofs of their undecidability:

1. The halting problem.
2. The acceptance problem.

1.4. Logic and computation: Turing machines, decidability and tractability

3. The busy beaver problem.
4. Post's correspondence problem.
5. Hilbert's tenth problem.

2. Logical form

Given the two aforementioned components related to form (cf. Section 1.2), below we approach logical form firstly from the viewpoint of what makes a formal language a logical language, and then introduce some basic aspects of proof theory, namely arguments and argument form.[1]

2.1. Logical languages and well-formed formulas

2.1.1. Alphabets, expressions, and formulas logical

A *logical language* is a formal language that has the following specificity:

Definition 2.1. A formal language F whose alphabet Σ, denoted by Σ_F, contains logical constants is a logical language.

We shall henceforth denote a logical language by L.

Definition 2.2. For L a logical language, Σ_L consists of

1. an infinite supply of symbols for *variables*, as well as possibly of symbols for *predicates* and *functions* of arity $n \geq 0$ (i.e. n-place arguments), and

2. a finite number of (symbols for) *logical constants*, namely *operators* (also: *connectives*) $\heartsuit_1, ..., \heartsuit_r$ and *quantifiers* $\blacklozenge_1, ..., \blacklozenge_m$.

Furthermore, 0-place function symbols are called *(individual) constants*. *Brackets* are punctuation marks; in fact, in this text they are used mostly for readability. The variables and the *non-logical constants* (i.e. predicates and functions) of Definition 2.2 are typically symbols from the Roman alphabet, with or without subscripts (see below), and the logical constants are special conventional symbols.

---

[1] For many authors, the expression "logical form" is to be used solely for formal proofs. We account for our discrepancy with respect to this practice by the explicit emphasis taken in this book that a logical language is first and foremost a formal language.

2. Logical form

Example 2.3. Typically, arbitrary propositional variables are denoted by $p, q, r, s, ...$, arbitrary individual variables are so by $x, y, z, ...$, and arbitrary constants by $a, b, c, d, ...$; the symbols for arbitrary functions are $f, g, h, ...$, and typical symbols for arbitrary predicates are $P, Q, R, S, ...$. The symbols $\top$ and $\bot$, denoting *truth* and *falsity* (or *absurdity*), respectively, are logical constants. The symbol $\neg$, denoting the unary operator for *negation*, is a logical constant. The symbols $\wedge, \vee, \rightarrow$, denoting the binary operators for *conjunction*, *disjunction*, and the *conditional*, respectively, are logical constants. The symbols $\forall$ and $\exists$, known as *universal* and *existential quantifiers*, respectively, are logical constants.

The following definitions frame in a most general way the technical study of the object language to be approached in this text. For convenience, we introduce straightaway the important notion of *ground expression*.

Definition 2.4. For an alphabet Σ_L and a finite set of rules G_L of a logical language L, we define inductively the *expressions* of L as follows:

1. A *term* t is an expression built up from the variables and function–and therefore constant–symbols of Σ_L.

    a) A term that is not a variable or does not contain a variable is called a *ground term*.

2. An *atom* P is an expression of the form $P(t_1, ..., t_n)$ where P is a n-place predicate symbol, $n \geq 0$, and $t_1, ..., t_n$–the *arguments* of P (denoted by $arg\,(P)$)–are terms.

    a) An atom that contains only ground terms is a *ground atom*.

3. A *formula* A is an atom or a composition of atoms:

    a) An atom P is an *(atomic) formula*.
    b) If $\heartsuit_i^n$ is a n-place logical operator and $P_1, ..., P_n$ atoms, then $\heartsuit_i (P_1, ..., P_n)$ is a *(compound) formula*.
    c) $\blacklozenge_1 \otimes_1 ... \blacklozenge_n \otimes_n (A)$, for $1 \leq i \leq n$, where $\otimes_i$ is an individual variable, a function symbol, or a predicate symbol, and $\blacklozenge_i$ is a quantifier, is a *(quantified) formula*.

2.1. Logical languages and well-formed formulas

   d) A quantifier-free formula whose atoms are all ground atoms is a *ground formula*.

Some further fundamental specifications for formulas of a logical language L are given now.

Proposition 2.5. (Unicity of decomposition) *A formula $\phi \in \mathsf{L}$ can be formed in only one way:*

1. ϕ is an atom.

2. There is a *unique* formula ψ and a *unique* unary connective $\heartsuit_i^1$ such that $\phi = \heartsuit_i(\psi)$.

3. There is a *unique* pair of formulas (ϕ_1, ϕ_2) and a *unique* binary connective $\heartsuit_j^2$ such that $\phi = \heartsuit_j(\phi_1, \phi_2)$.

4. There is a *unique* quantifier $\blacklozenge_i$, a *unique* symbol $\otimes$, and a *unique* formula χ such that $\phi = \blacklozenge_i \otimes (\chi)$.

With respect to Proposition 2.5.3, in our object language we shall write $\psi \heartsuit \chi$ instead of $\heartsuit(\psi, \chi)$. The former is known as *infix notation*.

Definition 2.6. Given formulas $\phi, \psi, \chi \in \mathsf{L}$, some unary connective $\heartsuit_i^1$, and some binary connective $\heartsuit_j^2$, an *immediate sub-formula* is defined as follows:

1. Atomic formulas have no immediate sub-formulas.

2. $\phi = \heartsuit_i(\psi)$ has as immediate sub-formula only ψ.

3. $\phi = (\psi \heartsuit_j \chi)$ has as immediate sub-formulas only ψ and χ.

4. $\phi = \blacklozenge_i \otimes (\chi)$ has as immediate subformula only χ.

▶ *Do Exercises 2.1-2.*

2. Logical form

2.1.2. Orders

A logical language L *has order n*, or *is of order n*, denoted by Ln, according to the following definitions.

Definition 2.7. The *order of a formula* is the highest order of any of its predicates and quantifiers:

1. A predicate has order 1 if all its arguments are terms; otherwise, it has order $n+1$, for n the highest order of its argument that is not a term.

2. A quantifier has order 1 if it quantifies an individual variable; otherwise, it has order $n+1$, for n the order of the predicate (or function) quantified.

Predicates and functions can be considered as *sets*: for instance, $P(x) = P = \{x | x \in P\}$, and $f = \{(x,y) | x, y \in \mathbb{N}\}$ if f is the function $f(x) = y$ for all $x, y \in \mathbb{N}$. By writing the latter as $f(x,y)$, we are considering the function f as a predicate, and thus we have $f(x,y) = \{(x,y) | x, y \in f\}$. This is an important note, as a function seen as a predicate takes only terms for arguments, and thus any function name (e.g., f) is always of order 1.

Example 2.8. The predicates $P(a)$ and $R(f(x))$ are both of order 1, as both a and $f(x)$ are terms. $Q(P)$ and $S(g)$ are both of order 2, as P is a predicate and g is a function. The formula $P(Q) \wedge Q(S)$ is of order 3, because of the nesting $S \in Q, Q \in P$, i.e. $Q(S)$ is of order 2 and $P(Q)$ is of order 3. The quantified formulas $\forall x \forall y (P(x) \rightarrow Q(y))$ and $\forall x \exists y \exists z (R(x,y,z))$ are of order 1. $\exists f \forall x (P(f(x)))$ and $\forall y \exists Q (Q(y))$ are of order 2, as the order of a formula is the highest order of any of its predicates and quantifiers: in the first case, though $\forall x$ and $P(f(x))$ are of order 1, $\exists f$ is of order 2; with respect to the second formula, $\forall y$ and $Q(y)$ are of order 1 but $\exists Q$ is of order 2. $\forall x \exists Q \exists S (Q(x) \rightarrow S(Q))$ is of order 3, because $\forall x (Q(x))$ is of order 1, $\exists Q (Q(x))$ is of order 2, and $\exists S (S(Q))$ is of order 3.

Definition 2.9. A *n-th order logical language* Ln is a logical language whose formulas have order n or less.

We speak of a zeroth-order logical language when $n = 0$, of a first-order (second-order) logic for $n = 1$ ($n = 2$, respectively), and of a higher-order logic for $n > 2$. When the order is 0, i.e. when there are

2.1. Logical languages and well-formed formulas

only 0-ary predicates (i.e. propositional variables), we more commonly speak of a *propositional language*; when the order is 1, we have a *first-order* (abbreviated: *FO*) *predicate language* (often just *FO language*).

We begin by specifying a propositional language L0, and then augment it in order to obtain a FO language L1. For convenience, we do not consider distinct elements for languages of second or higher order, as our object languages will be essentially of zeroth or first order.

Definition 2.10. A *propositional language*, denoted by L0, is a pair (V, O) where $V = \{p, q, r, p_1, q_1, r_1, ...\}$ is a denumerable set of *propositional variables* and $O = \{\heartsuit_1^n, ..., \heartsuit_r^n\}$ is a finite set of r operators, called *connectives*, with arity $n \geq 1$ for finite n.[2]

We can also define a propositional language as the pair $\mathsf{L} = (F_{\mathsf{L}0}, O)$, where $F_{\mathsf{L}0}$ is the set of formulas of the propositional language L0 (often simply F if L0 is understood). In very general terms, a language L can be identified with its set of formulas F_L, because $F_\mathsf{L} \subseteq \mathsf{L}$.

Definition 2.11. A *well-formed formula* of $F \subseteq \mathsf{L}0$ is defined recursively as follows:

1. Every propositional atom is a well-formed formula, i.e., $V \subseteq F$.

2. A formula ϕ is a well-formed formula iff it has a finite number k of propositional variables (or atoms), i.e. iff $\phi = \phi(p_1, ..., p_k)$.

3. If $\phi_1, ..., \phi_n$ are well-formed formulas and $\heartsuit_i^n$ is a logical connective with arity n, then $\heartsuit_i(\phi_1, ..., \phi_n)$ is a well-formed formula.

4. Given a well-formed formula $\phi(p_1, ..., p_k)$ and a substitution σ, the formula $\sigma\phi = \phi(p_1/\sigma p_1, ..., p_k/\sigma p_k)$ obtained from ϕ by simultaneously substituting $p_1, ..., p_k$ by the propositional variables $\sigma p_1, ..., \sigma p_k$, is a well-formed formula.

We write "formula" to abbreviate "well-formed formula," as nothing else is here a formula. Definition 2.11.1 defines an *atomic propositional formula*; Definition 2.11.2-3 defines a *complex*, or *compound, propositional formula*. Definition 2.11.4 defines a propositional formula as *invariant under variable substitutions*. This latter property can be extended to *formula substitutions* (see Section 2.4.6).

---
[2] Actually, we often have $n = 0$, namely when we consider $\top$ and $\bot$ as operators.

2. Logical form

Definition 2.12. A *FO language* L1 is the language L0 augmented with the following sets:

1. $Q = \{\blacklozenge_1, ..., \blacklozenge_m\}$ of quantifiers.
2. $Vi = \{x, y, z, x_1, y_1, z_1, ...\}$ of nominal/individual variables standing for individual names.
3. $Cons = \{a, b, c, a_1, b_1, c_1, ...\}$ of individual constants.
4. $Pred = \{P, Q, R, P_1, Q_1, R_1, ...\}$ of predicates.
5. $Fun = \{f, g, h, f_1, g_1, h_1, ...\}$ of functions.

Definition 2.13. Given the FO language L1, the triple

$$\Upsilon = (Pred_{L1}, Fun_{L1}, ar)$$

where $ar : (Pred_{L1} \cup Fun_{L1}) \longrightarrow \mathbb{N}$ is the arity of the predicate and function symbols of L1, is called a *signature (for* L1*)*.[3]

Definition 2.14. The set $F \subseteq $ L1 is defined as above for L0; atomic formulas are treated similarly as propositional variables and the following additional conditions hold:

1. A formula $\phi \in F$ is well-formed iff it has a finite number of terms, i.e. iff $\phi = \phi(t_1, ..., t_n)$.
2. If ϕ belongs to F and x is a variable, then $\blacklozenge_i x (\phi) \in F$.
3. Given a substitution σ, we have

    a) $\sigma c = c$, for $c \in Cons$;

    b) $\sigma (\diamond (t_1, ..., t_n)) = \diamond (\sigma t_1, ..., \sigma t_n)$, for $\diamond \in (\Upsilon - Cons)$.

The following definition synthesizes the above in Backus-Naur form (in which "::=" denotes left-to-right replacement):

Definition 2.15. Over the signature $\Upsilon = (Pred_{L1}, Fun_{L1}, Cons_{L1})$ for L1, the expressions of L1 are inductively defined as follows:

---
[3] A 0-ary function symbol is considered a constant. As a matter of fact, an alternative equivalent definition of a signature for L1 is $\Upsilon_{L1} := Pred_{L1} \cup Fun_{L1} \cup Cons_{L1}$.

2.1. Logical languages and well-formed formulas

$$\text{Terms} \qquad t \qquad ::= \quad x \,|\, a \,|\, f(t_1, ..., t_n)$$

$$\text{Atoms} \quad P\,(Q, ...) \quad ::= \quad p \,|\, P(t_1, ..., t_n)$$

$$\text{Formulas} \quad A\,(B, ...) \quad ::= \quad P \,|\, \heartsuit_l^1 A \,|\, A \heartsuit_j^2 B \,|\, \blacklozenge_i x\,(A)$$

We contemplate here no higher arity than 2 for the connectives of a logical language. This is solely for the sake of simplicity, given that we shall be working in this text with no higher arity than 2 for connectives.

Definition 2.16. In the formula $\psi = \blacklozenge_i x\,(\phi)$, (ϕ) is the *scope* of the quantifier $\blacklozenge_i$ and we say that $\blacklozenge_i$ *binds* (or *quantifies*) a variable x if x is in the scope of $\blacklozenge_i$. If $x \notin Vi\,(\phi)$, then ϕ is said to be *trivially quantified*.

1. Every occurrence of a variable x in the scope of a quantifier $\blacklozenge_i$ is said to be *bound*; *free* otherwise. A variable x may be both bound and free in ψ, often denoted by $\psi' = \blacklozenge_i x \phi\,(x)$.

2. A formula is said to be *closed*, or a *sentence*, if it has no free variables; otherwise, it is *open*.

3. Suppose now that ψ is the formula with free variables $\phi\,(x_1, ..., x_n)$. Then:[4]

    a) the *universal closure* of ψ is the formula $\psi' = \forall x_1 ... \forall x_n\,(\phi)$;

    b) the *existential closure* of ψ is the formula $\psi' = \exists x_1 ... \exists x_n\,(\phi)$.

Example 2.17. The following are examples of binding and substitutions in quantified formulas:

- The formula $\forall x\,(R\,(x, y))$ is open, as the variable y is not bound by any quantifier, but the formula $\forall x\,(R\,(x, a))$ is closed, as the only variable x in $R\,(x, a)$ is bound by $\forall$.

- The existential closure of $\forall x\,(R\,(x, y))$ is $\exists y \forall x\,(R\,(x, y))$.

---

[4] We shall also often write $\psi' = \blacklozenge_i x_1 ... \blacklozenge_i x_n \phi\,(x_1, ..., x_n)$, namely to remove trivially quantified formulas from consideration. This does not clash with item 1, as every formula $\psi = \phi\,(x)$ can be turned into a formula $\psi' = \blacklozenge_i \phi\,(x)$, and vice-versa, as we shall see.

2. Logical form

- Let there be given the formula $\exists z\,((S(x,z) \wedge R(y,z)) \to Q(z))$. Its universal closure is $\forall x \forall y \exists z\,((S(x,z) \wedge R(y,z)) \to Q(z))$.

- In the closed formula $\forall x\,(P(x) \wedge \exists x\,(Q(x)))$, the scope of $\forall$ is $P(x) \wedge \exists x\,(Q(x))$, but $\forall$ only binds the variable x in $P(x)$, as the variable x in $\exists x\,(Q(x))$ is already bound by $\exists$. In order to disambiguate it suffices to rename x, resulting in the formula $\forall y\,(P(y) \wedge \exists x\,(Q(x)))$. This allows the equivalence

$$\forall y\,(P(y) \wedge \exists x\,(Q(x))) \equiv \forall y \exists x\,(P(y) \wedge Q(x)).$$

Another way of variable renaming is by subscripting, so that we obtain the formula $\forall x_1\,(P(x_1) \wedge \exists x_2\,(Q(x_2)))$.

From the above, it should be evident that every logical formula can be read in a unique way. This *unique readability* can be represented by means of a *formula tree*, a downwards-growing labeled tree whose root is the main connective or the main quantifier and the leaves are labeled with propositional atoms or atomic terms (see Example 2.19).

Proposition 2.18. *The following rewritings (denoted by $\Longrightarrow$ or $\Longleftarrow$) are permissible for the quantifiers $\blacklozenge_{i,j} = \forall, \exists$:*

1. *Quantifier reversal:* $\blacklozenge_i x \blacklozenge_j y\,(P(x,y)) \overset{\Longrightarrow}{\Longleftarrow} \blacklozenge_j y \blacklozenge_i x\,(P(x,y))$ if $i = j$.

2. *Negation distribution:* $\neg \blacklozenge_i x\,(P(x)) \overset{\Longrightarrow}{\Longleftarrow} \blacklozenge_j x\,(\neg(P(x)))$ if $i \neq j$.

3. *Existential distribution:* $\exists y \forall x\,(P(x,y)) \Longrightarrow \forall x \exists y\,(P(x,y))$.[5]

Proof: Left as an exercise.

▶ *Do Exercises 2.3-6.*

2.2. Formalizing natural language

Although a natural language like English or Sioux is a formal language in a certain sense, its grammar and lexicon are too complex to allow an unequivocal interpretation of its utterances. This is especially so in the

---

[5]But the reverse rewriting is not permissible.

2.2. Formalizing natural language

context of scientific discourse, and *very* especially so in mathematics, a field in which ambiguity is to be avoided at all costs. Mathematics, just like any other field of knowledge, uses a natural language to convey linguistically most of its facts and conjectures. But any terms in a natural language come with both a long history and regional, often strictly local, usages. For these reasons, it is often the case that students have to revise their vocabulary when they are introduced to mathematical terms such as *continuous*, *discrete*, *natural*, *real*, etc.

One of the main objectives in conceiving a formal language is that of *formalizing* natural language, i.e. giving an unequivocal form to its utterances. We formalize utterances in a natural language by means of a logical language when we are interested in unequivocally analyzing some *reasoning instance*, such as a mathematical proof. But we have to start by formalizing individual utterances.

In an important sense, by formalizing an utterance in a natural language via a logical language, we are abstracting it by means of its logical form. In particular, we typically only care to formalize *propositions* (also *assertions*, or *statements*) i.e. utterances that convey factual information.[6] In our classical formalization via L1, utterances such as questions, orders, commands, interjections, or ejaculations are not considered formalizable.

Technically, formalizing consists in identifying propositions in some natural language, identifying their simpler components (propositions or predicates), and assigning to each of these a propositional variable or a predicate/function of a formal language Ln, for $n \geq 0$.[7] In the latter case, we very likely have to quantify the (sub-)formulas if there are individual variables involved.[8] This done, we decide whether the corresponding formulas are atomic or complex.

A propositional language, albeit exhibiting some advantages over a FO language, as will be seen, is typically not *expressive* enough to formalize most interesting instances of reasoning, reason why one must know how to formalize statements in a FO language, or even in a language of higher order. For instance, any reasoning instance involving equality requires at least a first-order language to be formalized properly (see Figure 2.2.1).

The expressiveness of a FO or higher-order language resides to a large extent in its allowing for complex statements about both specific and unspecific entities, often in the same atom (see Example 2.19.d below). In particular, in a FO language a *domain*, or *universe* **U**, whose mem-

---

[6] Below we shall specify that these are utterances that can be *true* or *false*.
[7] For graphical convenience, we shall use the symbol for identity for this assignment (see Example 2.19).
[8] Constants are typically not quantifiable, at least not by means of $Q = \{\forall, \exists\}$.

2. Logical form

bers are classes of entities, is required, so that we know what (class) it is specifically that we are reasoning about; furthermore, a *domain of discourse* $\mathscr{D} \subseteq \mathbf{U}$ with individual entities is needed, in order that we can reason about particular entities in that domain. For instance, the universe may be *human beings*, and the universe of discourse may be *female students*, or even just *Peter*. For practical ends, however, we may identify both the universe and the domain of discourse, and then specify a function assigning to variables in the domain particular entities in it; this is actually the only thing to do when the domain is infinite and we want to reason about an individual entity in the domain (see next Section).

Example 2.19. Let there be given the propositions (a) *Fritz is a cat.*; (b) *Fritz likes fish but it doesn't like pasta.*; (c) *All cats are finicky.*; (d) *Fritz is finicky about everything*. In the case of propositions (a) and (b), we may remain in the terrain of a propositional language; propositions (c) and (d) require a full FO language.[9]

- We can assign to proposition (a) the propositional variable c, and we thus formalize (a) simply as the atomic formula

$$C$$

with the formula tree

$$C$$
$$\bullet$$

with only the root node, which coincides with the atom C.

- Proposition (b) contains two segregable propositions, to wit, *Fritz likes fish* and *Fritz doesn't like pasta*. Assign to the first the propositional variable f. The second proposition requires more care, as it is a negation; in this case, we assign a propositional variable to the affirmative version, i.e. *Fritz likes pasta* $= p$, so that *Fritz doesn't like pasta* $= \neg p$. The presence of the English connector *but* indicates a complex formula, and indeed we have the formula

$$F \wedge \neg P$$

---

[9] We use italics here for convenience only, namely to segregate these propositions with respect to the main text. We shall keep to this usage whenever it proves convenient to do so.

with the formula tree

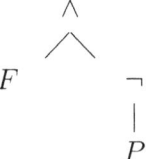

- Proposition (c) contains a quantifier (*all*), and we thus enter the terrain of a FO language *proper*. This means that we no longer assign propositional variables to atomic assertions, but we assign predicate variables to atomic assertions stating relations. Importantly, we have to identify whether these relations are stated for unspecified individuals or for specific individuals: if the first, then we are employing individual variables; if the second, we employ individual constants. In the case of proposition (c), we have two unary predicates stated for unspecified individuals, to wit, *x is a cat* and *x is finicky*, which we formalize as $C(x)$ and $F(x)$, respectively. Then, note that this is a *universal* assertion, i.e. it asserts something about *all* cats. Thus, we are employing the universal quantifier, denoted by $\forall$. We now have to reformulate proposition (c) in a way that allows for formalization. In fact, proposition (c) means "For all x, if x is a cat, then x is finicky," which abstracts further into the FO formula

$$\forall x \, (C(x) \rightarrow F(x))$$

with the formula tree

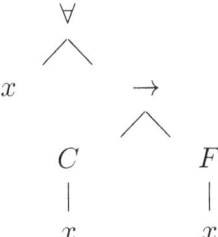

- Clearly, given a domain of discourse $\mathscr{D} = \{fritz\}$, we can now write $C(fritz)$ and $F(fritz)$ to denote that Fritz is a cat and Fritz is finicky. Note how both $C(fritz)$ and $F(fritz)$ are quantifier-free atoms. In effect, in a FO language quantifiers bind solely individual variables. The FO formalization of proposition (d) is

2. Logical form

the formula

$$\forall x\, (F(fritz, x))$$

with the formula tree

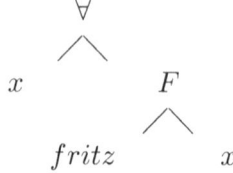

and where we now have the binary predicate $F(y, x)$ denoting "y is finicky about x." Note that if we do not specify an individual y, then we can have the formula

$$\exists y \forall x\, (F(y, x))$$

with formula tree

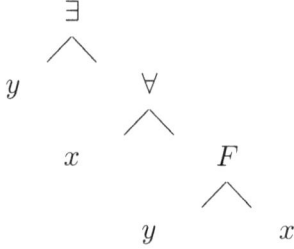

formalizing the assertion that there is someone who is finicky about everything. Note also that if we want to specify both individuals denoted by the variables y and x in this formula, we then need to specify two universes, namely $\mathcal{D}_1$ and $\mathcal{D}_2$, as, respectively, the domains of people and, say, food items.

Figure 2.2.1 in Example 2.20 provides some common formalizations for English by means of the logical constants of a propositional language. Figure 2.2.2 in Example 2.21 does the same for English statements involving quantification and equality. As a matter of fact, the logical language employed in both Examples is that of classical logic, i.e. L1.

Example 2.20. Figure 2.2.1 shows some common formalizations for English in a propositional language. We consider the set of connectives $O = \{\neg^1, \wedge^2, \vee^2, \to^2, \leftrightarrow^2\}$. The connectives $\wedge^2$ and $\vee^2$ are commutative.

Example 2.21. Figure 2.2.2 shows some formalizations for English in

2.2. Formalizing natural language

English	Formalization
not p – it is not the case that p	$\neg P$
p and q – both p and q – p but q – p; q	$P \wedge Q$
neither p nor q	$\neg P \wedge \neg Q$
p or else q	$(P \vee Q) \wedge (\neg P \vee \neg Q)$
p or q – either p or q	$P \vee Q$
there is only one option: p or q	$(P \wedge \neg Q) \vee (\neg P \wedge Q)$
if p then q – if p, q – q only if p – q provided that p p, so q – not p unless q – p is sufficient for q q is necessary for p	$P \rightarrow Q$
p unless q	$\neg Q \rightarrow P$
p if and only if (iff) q p is necessary and sufficient for q	$P \leftrightarrow Q$

Figure 2.2.1.: Formalizations for English by means of the language of classical propositional logic.

2. Logical form

the FO language of classical logic. We consider the set of quantifiers $Q = \{\forall, \exists\}$ and we augment L1 with a symbol for equality $(=)$.[10]

▶ *Do Exercises 2.7-8.*

2.3. Argument form

Although there may be diverse perspectives on logic and correspondingly different applications thereof, *formal logic* is essentially a means for *reasoning formally*. Indeed, reasoning can be, and is more often than not, carried out *informally*, i.e. without the aid of a logical language. Above we saw how form is fundamental for the definition of both logical languages and their constituents; now we concentrate on form from the viewpoint of reasoning.

Example 2.22. Let there be given the following statements in English:

> *Susan doesn't like dogs but she likes cats and rabbits. She also likes birds, and she has a problem if she also keeps cats. She only keeps dogs and rabbits. Hence, she doesn't have a problem.*

This constitutes a reasoning instance, as attested by the presence of the English connector *hence* in the final statement. That is, anyone uttering or writing the above is not just saying or writing a collection of statements in English, but is concluding something from the previously uttered or written statements. Instead of *hence*, we could have any of the following English connectors: *therefore, thus, so, as a result, consequently, accordingly, in conclusion*, etc. This clearly distinguishes an *argument* from other statements such as explanations, warnings, suggestions and pieces of advice, opinions and beliefs, and illustrations.

Whether in the reasoning instance of Example 2.22 the conclusion is correct or not, that is not easy to determine by analyzing the statements in English. We require a means to decide *formally* whether our reasoning is correct. The best way we have to do it is precisely by making use of a logical language: We first *formalize* the reasoning instance above, and then *formally* analyze it in the sense that we determine whether the conclusion follows logically from the previous statements *by virtue of form alone*.

We begin by defining formally *argument*.

---

[10] This introduction of the symbol "=" is not without issues in classical FO logic. See Section 7.2 below.

English	Formalization
all/every x – everything	$\forall x$
not all/every x – not everything	$\neg\forall x$
all R are S	$\forall x\,(R(x) \to S(x))$
no R are S	$\forall x\,(R(x) \to \neg S(x))$
there is a x – some x – something	$\exists x$
there isn't any x – nothing	$\neg\exists x$
some R is/are S	$\exists x\,(R(x) \land S(x))$
some R is/are not S	$\exists x\,(R(x) \land \neg S(x))$
only a is R	$R(a) \land \forall x\,(R(x) \to (x = a))$
no R except a is S	$R(a) \lor S(a) \land \forall x\,[(R(x) \lor S(x)) \to (x = a)]$
all R except a is S	$R(a) \land \neg S(a) \land \forall x\,[(R(x) \land (x \neq a)) \to S(x)]$
there is at most one R	$\forall x \forall y\,[(R(x) \lor R(y)) \to (x = y)]$
there is exactly one R	$\exists x \exists y\,[(R(x) \lor R(y)) \to (x = y)]$

Figure 2.2.2.: Formalizations for English by means of the language of classical FO logic.

2. Logical form

Definition 2.23. In very general terms, given a logical language L and formulas $A_1, ..., A_n, B \in F_L$, a *(logical) argument* is a construction of the form

$$(\alpha) \quad \frac{A_1 \wedge ... \wedge A_n}{B}$$

where $A_1, ..., A_n$ are *premises* and B is the *conclusion*. The following are alternative ways to represent α:

1.
$$\{A_1, ..., A_n\}/B$$
(where the set brackets can be omitted as an abbreviation)

2.
$$\begin{array}{c} A_1 \\ \vdots \\ A_n \\ \hline B \end{array}$$

3.
$$\begin{array}{c} A_1 \\ \vdots \\ A_n \\ \therefore B \end{array}$$

If for α we have $n = 1$, then α is *not* an argument.[11] Nevertheless, we can still speak of a reasoning of the form α with $n = 1$ that is formally correct–say, p/p–as a *(derived) rule*.

Example 2.24. Let there be given the logical language $\mathsf{L0} = (F, O)$ where $O = \{\neg^1, \wedge^2, \vee^2, \rightarrow^2\}$. In order to formalize the reasoning instance of Example 2.22 we begin by assigning a propositional variable to the different atomic statements to be found in it. For example:

Susan likes dogs $= d$
Susan likes cats $= c$
Susan likes rabbits $= r$
Susan likes birds $= b$
Susan has/keeps cats $= k$

---

[11] This prescribes that a simple conditional statement, i.e. a statement of the form "if p then q" (cf. Figure 2.2.1), is *not* an argument. This is an important note, as the connective $\rightarrow$ plays a rather special role in formalized arguments, as will be seen. In any case, this condition can be relaxed without (too) much ado.

Susan has/keeps birds = a
Susan has/keeps dogs = e
Susan has/keeps rabbits = f
Susan has a problem = p

According to Definition 2.15, propositional atoms are formulas, and we can use either the atoms above or the corresponding formulas in uppercase letters. We choose the latter. In a reasoning instance, each statement segregated by means of a period is a single premise. We opt for using the argument form specified in Definition 2.23.2, and thus every premise is to be written in a different line. By using the connectives in O and the propositional atomic formulas above, we have the following formal argument:

1. $\neg D \wedge C \wedge R$
2. $B \wedge ((K \wedge A) \rightarrow P)$
3. $E \wedge F \wedge \neg K \wedge \neg A$
4. $\neg P$

We more often than not require formal analysis for more complex instances of reasoning, but the principle is the same as applied to Example 2.22. However, as stated above, a propositional language is typically not expressive enough for more complex instances of reasoning. In particular, mathematical proofs are typically carried out in a FO language–sometimes in a language even of higher order. Example 2.25 shows a very simple mathematical proof that, however simple, cannot be formulated in a propositional language. Below, more sophisticated examples are provided.

Example 2.25. We reason that alternate interior angles formed by a diagonal of a trapezoid are equal. In order to formalize this reasoning we require a FO language. We let the alternate angles be abd and cdb. We further let $T(x, y, u, v)$ mean that $xyuv$ is a trapezoid with upper-left vertex x, upper-right vertex y, lower-right vertex u, and lower-left vertex v. We denote the property that the line segment xy is parallel to the line segment uv by $P(x, y, u, v)$. We represent the equality of the angles xyz and uvw as $E(x, y, z, u, v, w)$. Given the specific trapezoid $T(a, b, c, d)$, we reason that $E(a, b, d, c, d, b)$ will follow as a conclusion. We have the following premises and conclusion:

1. $\forall x \forall y \forall u \forall v (T(x, y, u, v) \rightarrow P(x, y, u, v))$
2. $\forall x \forall y \forall u \forall v (P(x, y, u, v) \rightarrow E(x, y, v, u, v, y))$
3. $T(a, b, c, d)$
4. $E(a, b, d, c, d, b)$

2. Logical form

In this short Section, we elaborated on argument form solely, omitting any discussion on how to verify the correctness or validity of the reasoning instances. This requires the fundamental notion of *logical consequence*, to be approached below. However, we can already provide a few classical arguments that are correct by virtue of form alone (see Example 2.27). Below, the reader can verify both their *correctness* and *validity*.[12] In Example 2.27 below, note the use of Greek lowercase letters to convey the fact that these argument forms are schemata.

Definition 2.26. Given an object language L, a *schema* ς (plural: *schemata*) is a formula in which one or more variables of μ^L, known as *metalinguistical variables*, occur. These metalinguistical variables can be replaced by an expression of L, in order to form an *instance* of the schema ς.

As is evident, a schema ς represents *infinitely* many formulas of an object language.

Example 2.27. The arguments in Figure 2.3.1 are formally correct in classical logic. For convenience, we use Definition 2.23.1 and we replace the symbol / by the symbol $\vdash \in \mu^L$.[13] The order of the premises is irrelevant.

In the classical formalization of logic, there is also the reverse case: there are arguments that are not correct in virtue of the form alone.

Example 2.28. The two argument forms in Figure 2.3.2 determine always incorrect reasoning instances. It is important to remark that they can be easily confused with the valid argument forms MP and MT (see Fig. 2.3.1).

▶ *Do Exercises 2.9-13.*

2.4. Normal forms and substitutions for L1

When given a logical language L1, it is often required that the formulas be in a specific *normal form*. For instance, the resolution calculus

---
[12] As a matter of fact, arguments are more often than not classified as valid or invalid, rather than correct or incorrect.
[13] See Section 4.1 below for this replacement. Note that $\phi, \psi \vdash \chi$ abbreviates $\{\phi, \psi\} \vdash \{\chi\}$ or, equivalently, $\phi \wedge \psi \vdash \chi$.

2.4. Normal forms and substitutions for L1

Argument form	Name	Abbr.
$\{\phi \to \psi, \phi\} \vdash \psi$	Modus ponens	MP
$\{\phi \to \psi, \neg\psi\} \vdash \neg\phi$	Modus tollens	MT
$\{\phi \to \psi, \psi \to \chi\} \vdash \phi \to \chi$	Hypothetical syllogism	HS
$\{\phi \vee \psi, \neg\phi\} \vdash \psi$	Modus tollendo ponens	TP
$\{\phi \to \psi, \phi \to \chi\} \vdash \phi \to (\psi \wedge \chi)$	Conditional product	CP
$\{(\phi \to \psi) \wedge (\chi \to \xi), \phi \vee \chi\} \vdash \psi \vee \xi$	Constructive dilemma	CD
$\{(\phi \to \psi) \wedge (\chi \to \xi), \neg\psi \vee \neg\xi\} \vdash \neg\phi \vee \neg\chi$	Destructive dilemma	DD
$\{\phi, \neg\phi\} \vdash \psi$	Ex contradictione quodlibet	ECQ

Figure 2.3.1.: Some classical formally correct arguments.

2. Logical form

Argument form	Name	Abbr.
$\{\phi \to \psi, \psi\} \vdash \phi$	Affirming the consequent	AC
$\{\phi \to \psi, \neg\phi\} \vdash \neg\psi$	Denying the antecedent	DA

Figure 2.3.2.: Two invalid argument forms.

(Chapter 14) requires the formulas of L1 to be in conjunctive normal form. Additionally, in order to test for validity or satisfiability of FO theories or formulas, it is often essential to carry out so-called *ground substitutions*.[14] In this Section, we elaborate on these additional aspects of logical form for a FO language.

2.4.1. Literals and clauses

Definition 2.29. We define a *literal*, denoted by L, to be an atom (e.g., P) or the negation of an atom (e.g., $\neg P$). We say that the literals P and $\neg P$ are *complementary*.

We shall often indicate a *negative literal*, i.e. a literal that is the negation of an atom, by $\neg L$.

Definition 2.30. A *clause* $\mathcal{C}$ is a finite disjunction of literals, i.e. $\mathcal{C} = L_1 \vee ... \vee L_n = \|L_1, ..., L_n\|$.

1. A one-literal clause is a *unit* clause.

2. $\mathcal{C}$ is a *Horn clause* if it contains at most one positive literal.

    a) A Horn clause with exactly one positive literal is a *definite clause*.

    b) $\mathcal{C}$ is a *dual-Horn clause* if it has at most one negative literal.

3. The *empty clause* $\| \|$, denoted by $\square$, is a clause that contains no literals.

4. A clause is called *ground* if no individual variables occur in it.

---

[14]See Chapter 3 for the definitions of validity and satisfiability.

2.4. Normal forms and substitutions for L1

5. A clause is said to be *positive* (*negative*) if it has no negative (positive, respectively) literals.

We shall further denote a finite set of clauses $\{\|\cdot\|_1, ..., \|\cdot\|_n\}$ by C.[15]

2.4.2. Negation normal form

Definition 2.31. A formula $A \in F_{L1}$ is said to be in *negation normal form* (NNF) iff the negation connective $\neg$ is applied only to atoms and the only other connectives are conjunction and disjunction.

Proposition 2.32. *Any formula $A \in F_{L1}$ can be transformed into NNF by applying the following rewriting rules:*

1. *Inter-definitions:*

   a) $(\rightarrow_{df})$ $A \rightarrow B \implies \neg A \vee B$

   b) $(\leftrightarrow_{df})$ $A \leftrightarrow B \implies (A \rightarrow B) \wedge (B \rightarrow A)$

2. *De Morgan's laws:*

   a) $(\text{DeM}_\vee)$ $\neg(A \vee B) \implies \neg A \wedge \neg B$

   b) $(\text{DeM}_\wedge)$ $\neg(A \wedge B) \implies \neg A \vee \neg B$

3. *Double negation, or involution, law (DN):* $\neg\neg A \implies A$

4. *Quantifier duality:*

   a) $(\text{QN}_\exists)$ $\neg \exists x\,(A) \implies \forall x\,(\neg A)$

   b) $(\text{QN}_\forall)$ $\neg \forall x\,(A) \implies \exists x\,(\neg A)$

Proof: Left as an exercise.

---

[15] Note that $\|\cdot\|$ just is another way to represent the *set* of literals of a clause C. It is a convenient way, as the common practice of representing a set of clauses as $C = \{\{\cdot\}, ..., \{\cdot\}\}$ can be confusing.

2. Logical form

2.4.3. Prenex normal form

Definition 2.33. A formula $A \in F_{L1}$ is said to be in *prenex normal form* (PNF) iff if it is in the form

$$(\dagger) \qquad \blacklozenge_1 x_1 ... \blacklozenge_n x_n (M)$$

where every $\blacklozenge_i x_i$, $i = 1, ..., n$ is either $\forall x_i$ or $\exists x_i$, and M is a formula containing no quantifiers. $\blacklozenge_1 x_1, ..., \blacklozenge_n x_n$ is the *prefix* and (M) is the *matrix* of A.

Proposition 2.34. *Every formula $A \in$ L1 can be transformed into a PNF by an algorithmic process constituted of the recursively applicable steps in Algorithm 2.1.*

Proof: Left as an exercise. (Hint: Read Chapter 3 and base your proof on the notion of logical equivalence, denoted by $\equiv$. In particular, consider Propositions 3.31-2.)[16]

2.4.4. Skolem normal form

Definition 2.35. Let a formula A be in the PNF $\dagger$. Then, the procedure in Algorithm 2.2 is called *Skolemization*.

Example 2.36. From the formula

$$A = \exists x \forall y \forall z \forall u \exists v \left((P(x) \vee Q(y)) \wedge R(z, u, v) \right)$$

by applying Algorithm 2.2 above we obtain the formula

$$A_{Sk} = \forall y \forall z \forall u \left((P(a) \vee Q(y)) \wedge R(z, u, f(y, z, u)) \right)$$

Definition 2.37. The formula A_{Sk} in Example 2.36 is in *Skolem normal form* (SNF). The constant a and the function $f(y, z, u)$ used to replace the existential variables of A are called *Skolem constant* and *Skolem function*, respectively.

The universal quantifiers can simply be removed from A_{Sk}. In effect, we have (see Exercise 3.13)

$$\forall y \forall z \forall u \left((P(a) \vee Q(y)) \wedge R(z, u, f(y, z, u)) \right)$$

---

[16] Although we can define *logical equivalence* syntactically as a relation between logical forms that are in some sense the same, the semantical definition is more intuitive. In particular, we can combine both into the definition of logical equivalence as the relation between logical forms that express the same meaning, where "meaning" is to be understood in the sense of a valuation or interpretation (see Chapter 3).

2.4. Normal forms and substitutions for L1

Algorithm 2.1 PNF transformation.

- **Input:** Some FO formula A
- **Output:** Formula A' in PNF such that $A' \equiv A$

Steps:

1. Remove the connective $\leftrightarrow$ by applying $\leftrightarrow_{df}$.

2. Push the connective $\neg$ inwards by applying $\text{DeM}_{\wedge(\vee)}$ and $\text{QN}_{\exists(\forall)}$.

3. Rename the bound variables so that each variable occurs only once.

4. Push the quantifiers outwards by means of the following rewriting rules *when x does not appear as a free variable in B*:

   a) $\forall x \, (A(x)) \wedge B \Longrightarrow \forall x \, (A(x) \wedge B)$

   b) $\forall x \, (A(x)) \vee B \Longrightarrow \forall x \, (A(x) \vee B)$

   c) $\exists x \, (A(x)) \wedge B \Longrightarrow \exists x \, (A(x) \wedge B)$

   d) $\exists x \, (A(x)) \vee B \Longrightarrow \exists x \, (A(x) \vee B)$

   e) $\forall x \, (A(x)) \to B \Longrightarrow \exists x \, (A(x) \to B)$

   f) $\exists x \, (A(x)) \to B \Longrightarrow \forall x \, (A(x) \to B)$

   g) $B \to \forall x \, (A(x)) \Longrightarrow \forall x \, (B \to A(x))$

   h) $B \to \exists x \, (A(x)) \Longrightarrow \exists x \, (B \to A(x))$

5. If x appears as a free variable in B, then rename the bound variable x in $\blacklozenge x \, (A(x))$ as y, obtaining the equivalent $\blacklozenge y \, (A[x/y]) = \blacklozenge y \, (A(y))$.

55

2. Logical form

Algorithm 2.2 Skolemization.

- **Input:** Some formula A in the PNF

$$\blacklozenge_1 x_1 ... \blacklozenge_n x_n (M)$$

with $\blacklozenge_r$, $1 \leq r \leq n$, an existential quantifier

- **Output:** A skolemized formula A_{Sk} such that $A_{Sk} \equiv_{sat} A$

Steps:

1. If there is no universal quantifier before $\blacklozenge_r$, we choose a new constant c that does not occur in M, replace all x_r in M with c, and delete $\blacklozenge_r x_r$ from the prefix.

2. If before $\blacklozenge_r$ there are the universal quantifiers $\blacklozenge_{s_1} ... \blacklozenge_{s_m}$, i.e., $1 \leq s_1 < s_2 < ... < s_m < r$, we select a new m-place function symbol f which does not occur in M, replace all $\blacklozenge_r$ in M by $f(x_{s_1}, ..., x_{s_m})$ and delete $\blacklozenge_r x_r$ from the prefix.

$$\equiv$$
$$(P(a) \vee Q(y)) \wedge R(z, u, f(y, z, u))$$

In order to carry out some proof procedures (e.g., resolution) a formula in SNF must be represented as a set of clauses (understood to be universally quantified).[17] For instance, A_{Sk} must be represented by:

$$C_{A_{Sk}} = \{P(a) \vee Q(y), R(z, u, f(y, z, u))\} =$$
$$= \{\|P(a), Q(y)\|, \|R(z, u, f(y, z, u))\|\}$$

The process of transforming a formula A into a formula A_{Sk} does not guarantee that $A \equiv A_{Sk}$. It particular, it does not necessarily preserve validity, denoted by $\equiv_\top$, but solely satisfiability (again, cf. Exercise 3.13).[18] We say that the formula A_{Sk} is satisfiable iff A is satisfiable, and we call this relation equisatisfiability.

Definition 2.38. Two formulas ϕ and ψ are *satisfiability-equivalent*, or *equisatisfiable*, denoted by $\phi \equiv_{sat} \psi$, iff they are both satisfiable or they are both unsatisfiable.

2.4.5. Conjunctive and disjunctive normal forms

As said above, for many applications of classical logic it is often required that the formulas be in a conjunctive or disjunctive normal form. Although both forms are important, we focus here on the transformation of a formula into a conjunctive normal form, a practice that is sanctioned by the duality of these two forms.

Definition 2.39. A formula $A \in$ L1 is said to be in a *conjunctive normal form* (CNF) iff A has the form $A = C_1 \wedge ... \wedge C_n$, $n \geq 1$, where each of $C_1, ..., C_n$ is a clause, i.e.,

$$A = \bigwedge_{i=1}^{n} \left(\bigvee_{j=1}^{m_i} L_{i,j} \right)$$

and is in NNF.

---

[17] Thus, it helps if the matrix M of A is already in a conjunctive normal form, a topic we elaborate on in the next Section.

[18] Preservation of satisfiability is, however, all that is required in the case of refutation-based proof procedures, such as resolution and analytic tableaux.

2. Logical form

Definition 2.40. A formula $A \in \mathsf{L1}$ is said to be in a *disjunctive normal form* (DNF) iff A has the form $A = \mathcal{E}_1 \vee ... \vee \mathcal{E}_n$, $n \geq 1$, where each of $\mathcal{E}_1, ..., \mathcal{E}_n$ is a conjunction of literals, i.e.,

$$A = \bigvee_{i=1}^{n} \left(\bigwedge_{j=1}^{m_i} L_{i,j} \right)$$

and is in NNF.

Proposition 2.41. *Let $\{A_1, ..., A_n\}$ be a finite set of formulas. Then*

$$\neg \left(\bigwedge_{i=1}^{n} A_i \right) \equiv \left(\bigvee_{i=1}^{n} \neg A_i \right)$$

and

$$\neg \left(\bigvee_{i=1}^{n} A_i \right) \equiv \left(\bigwedge_{i=1}^{n} \neg A_i \right).$$

Proof: (Sketch) We have it that $\neg(A) \equiv (\neg A)$. Then, obviously the proposition is proved for $n = 1$ by the equivalence $\neg \left(\bigwedge_{i=1}^{1} A_i \right) \equiv \left(\bigvee_{i=1}^{1} \neg A_i \right)$. The proof then follows by induction on n. **QED**

Proposition 2.42. *Let A be a formula in CNF and B be a formula in DNF. Then $\neg A$ is equivalent to a formula in DNF, and $\neg B$ is equivalent to a formula in CNF.*

Proof: If A is in CNF, then A is the formula $\bigwedge_{i=1}^{n} \left(\bigvee_{j=1}^{m_i} L_{i,j} \right)$. By Proposition 2.41, we have

$$\neg A = \neg \bigwedge_{i=1}^{n} \left(\bigvee_{j=1}^{m_i} L_{i,j} \right) \equiv \bigvee_{i=1}^{n} \neg \left(\bigvee_{j=1}^{m_i} L_{i,j} \right) \equiv \bigvee_{i=1}^{n} \left(\bigwedge_{j=1}^{m_i} \neg L_{i,j} \right).$$

The proof runs similarly for $\neg B$ being equivalent to a formula in CNF. **QED**

Proposition 2.43. *Any formula can be transformed into a conjunctive/disjunctive normal form in three simple main steps, by applying $\rightarrow_{def}$ and/or $\leftrightarrow_{def}$, together with DN, $\text{DeM}_{\wedge(\vee)}$ and the distributive laws*

$$(\text{D}_\wedge) \qquad A \wedge (B \vee C) \equiv (A \wedge B) \vee (A \wedge C)$$

2.4. Normal forms and substitutions for L1

$$(D_\vee) \quad A \vee (B \wedge C) \equiv (A \vee B) \wedge (A \vee C)$$

Proof: Left as an exercise.

Theorem 2.44. *Every formula A is equivalent to any formula A_1 in DNF and some formula A_2 in CNF.*

Proof: Follows immediately from the above. The proof is by induction on the complexity of A. **QED**

Example 2.45. We show the transformation into CNF of the formula $A = (p \vee (q \wedge \neg r)) \to s$ by following Proposition 2.43:[19]

		Rules
1	$(p \vee (q \wedge \neg r)) \to s$	
2	$\neg (p \vee (q \wedge \neg r)) \vee s$	$\to_{def}$
3	$(\neg p \wedge \neg (q \wedge \neg r)) \vee s$	$DM_\vee$
4	$\left(\neg p \wedge \underline{(\neg q \vee \neg \neg r)}\right) \vee s$	$DM_\wedge$
5	$(\neg p \wedge (\neg q \vee \underline{r})) \vee s$	DN
6	$(\neg p \vee s) \wedge ((\neg q \vee r) \vee s)$	$D_\vee$
7	$((\neg p \vee s) \wedge (\neg q \vee r)) \vee ((\neg p \vee s) \wedge s)$	$D_\wedge$
8	$((\neg p \vee s) \vee (\neg p \vee s)) \wedge ((\neg p \vee s) \vee s) \wedge ((\neg q \vee r) \vee (\neg p \vee s)) \wedge ((\neg q \vee r) \vee s)$	$D_\vee$
9	$(\neg p \vee s) \wedge ((\neg p \vee s) \vee s) \wedge ((\neg q \vee r) \vee (\neg p \vee s)) \wedge ((\neg q \vee r) \vee s)$	$\phi \vee \phi \equiv \phi$
10	$(\neg p \vee s) \wedge \left(\neg p \vee (s \vee s)\right) \wedge ((\neg q \vee r) \vee (\neg p \vee s)) \wedge ((\neg q \vee r) \vee s)$	Assoc.
11	$(\neg p \vee s) \wedge (\neg p \vee \underline{s}) \wedge ((\neg q \vee r) \vee (\neg p \vee s)) \wedge ((\neg q \vee r) \vee s)$	$\phi \vee \phi \equiv \phi$
12	$(\neg p \vee s) \wedge ((\neg q \vee r) \vee (\neg p \vee s)) \wedge ((\neg q \vee r) \vee s)$	$\phi \wedge \phi \equiv \phi$
13	$(\neg p \vee s) \wedge \left((\neg p \vee s) \vee (\neg q \vee r)\right) \wedge ((\neg q \vee r) \vee s)$	Commut.
14	$(\neg p \vee s) \wedge ((\neg q \vee r) \vee s)$	Absorpt.
15	$(\neg p \vee s) \wedge \left(\neg q \vee r \vee s\right)$	Assoc.

Although the procedure in Example 2.45 guarantees logical equivalence of a formula A with its DNF or CNF, it can lead to an exponential growth (a.k.a. "explosion") of the formula A. In effect, in Example 2.45,

---

[19]If a rule applies only to part of a formula, we indicate this by underlining the corresponding part.

2. Logical form

Algorithm 2.3 Tseitin transformation.

- **Input:** Some formula A
- **Output:** Formula A' in CNF such that $A' \equiv_{sat} A$

Steps:

1. Introduce a new atom, say p_1, not occurring anywhere else in the formula to abbreviate the innermost sub-formula and conjoin the abbreviated formula with the definition of p_1.

2. Proceed as in 1., introducing new atoms $p_2, p_3, ..., p_n$ as required in the direction of the main connective in A.

3. Put each of the conjuncts in CNF by applying the rules in Figure 2.4.1.

the CNF of A is the formula obtained in step 8; from this step on, simplification rules were applied that are in fact well-known properties of Boolean expressions (associativity, commutativity, and absorption; cf. Fig. 3.5.1). If one wishes to restrict this to a linear growth, then one may have to give up on equivalence in favor of equisatisfiability. This can be done by applying the Tseitin algorithm for CNF formulas (Algorithm 2.3).

Remark 2.46. Note in Figure 2.4.1 the following equivalences:

$$(\circledast) \qquad p \rightarrow (q \wedge r) \equiv (p \wedge q) \rightarrow (p \wedge r)$$

$$(\odot) \qquad ((p \wedge q) \vee (\neg p \wedge \neg q)) \rightarrow r \equiv ((p \wedge q) \rightarrow r) \wedge ((\neg p \wedge \neg q) \rightarrow r)$$

Example 2.47. Let $A = (p \vee (q \wedge \neg r)) \rightarrow s$. Then we have the following definitions:

$p_1 \leftrightarrow (q \wedge \neg r)$
$p_2 \leftrightarrow (p \vee p_1)$
$p_3 \leftrightarrow (p_2 \rightarrow s) \equiv p_3 \leftrightarrow (\neg p_2 \vee s) \equiv p_3 \leftrightarrow (p_2 \wedge \neg s)$

We have the conjuncts

2.4. Normal forms and substitutions for L1

(♡)			Rule
$x \leftrightarrow (\neg p)$	$\equiv$	$(x \to \neg p) \land (\neg p \to x)$	$\leftrightarrow_{df}$
$x \leftrightarrow (p \land q)$	$\equiv$	$(x \to p) \land (x \to q) \land ((p \land q) \to x)$	$\leftrightarrow_{df}, \circledast$
	$\equiv$	$(\neg x \lor p) \land (\neg x \lor q) \land (\neg (p \land q) \lor x)$	$\to_{df}$
	$\equiv$	$(\neg x \lor p) \land (\neg x \lor q) \land (\neg p \lor \neg q \lor x)$	DeM$_\land$
$x \leftrightarrow (p \lor q)$	$\equiv$	$(p \to x) \land (q \to x) \land (x \to (p \lor q))$	$\leftrightarrow_{df}, \circledast$
	$\equiv$	$(\neg p \lor x) \land (\neg q \lor x) \land (\neg x \lor p \lor q)$	$\to_{df}$
$x \leftrightarrow (p \leftrightarrow q)$	$\equiv$	$(x \to (p \leftrightarrow q)) \land ((p \leftrightarrow q) \to x)$	$\leftrightarrow_{df}$
	$\equiv$	$(x \to ((p \to q) \land (q \to p))) \land ((p \leftrightarrow q) \to x)$	$\leftrightarrow_{df}$
	$\equiv$	$(\neg x \lor ((\neg p \lor q) \land (\neg q \lor p))) \land (((p \land q) \lor (\neg p \land \neg q)) \to x)$	$\to_{df}$ $\leftrightarrow_{df}$
	$\equiv$	$(\neg x \lor (\neg(\neg p \lor q) \lor \neg(\neg q \lor p))) \land ((p \land q) \to x) \land ((\neg p \land \neg q) \to x)$	DeM$_\land$ $\odot$
	$\equiv$	$(\neg x \lor ((p \land \neg q) \lor (q \land \neg p))) \land ((\neg p \lor \neg q) \to x) \land ((p \lor q) \to x)$	DeM$_\lor$ DeM$_\land$
	$\equiv$	$(\neg x \lor (\neg(p \land \neg q) \land \neg(q \land \neg p))) \land (\neg(\neg p \lor \neg q) \lor x) \land (\neg(p \lor q) \lor x)$	DeM$_\lor$ $\to_{df}$
	$\equiv$	$\neg x \lor ((\neg p \lor q) \land (\neg q \lor p)) \land ((p \land q) \lor x) \land ((\neg p \land \neg q) \lor x)$	DeM$_\land$ DeM$_\lor$
	$\equiv$	$(\neg x \lor \neg p \lor q) \land (\neg x \lor \neg q \lor p) \land (\neg p \lor \neg q \lor x) \land (q \lor p \lor x)$	D$_\lor$ DeM$_\land$

Figure 2.4.1.: Tseitin transformations for the connectives of L.

2. Logical form

$$(p_1 \leftrightarrow (q \wedge \neg r)) \wedge (p_2 \leftrightarrow (p \vee p_1)) \wedge (p_3 \leftrightarrow (p_2 \wedge \neg s)) \wedge p_3$$

By putting the conjuncts into CNF (see Fig. 2.4.1), we obtain the formula

$$(\neg p_1 \vee q) \wedge (\neg p_1 \vee r) \wedge (\neg q \vee \neg r \vee p_1) \wedge$$
$$(\neg p_2 \vee p \vee p_1) \wedge (p_2 \vee \neg p) \wedge (p_2 \vee \neg p_1) \wedge$$
$$(\neg p_3 \vee p_2) \wedge (\neg p_3 \vee \neg s) \wedge (\neg p_2 \vee s \vee p_3) \wedge p_3$$

Although the conversion of formula A of Example 2.47 into CNF by the Tseitin algorithm yields a longer formula than the method of Proposition 2.43 (see Example 2.45, Step 8), its application may pay off in many other cases.[20]

2.4.6. Substitutions and unification for L1

The existence of individual variables in the alphabet of L1 requires special methods that allow deduction over it. These are mostly *substitutions* and *unification*.

Definition 2.48. Given a set $Vi = \{x_1, ..., x_n\}$ of variables and a set $T = \{t_1, ..., t_n\}$ of terms, a *substitution* is a mapping $\sigma : Vi \longrightarrow T$ such that $\sigma x_i = t_i$ almost everywhere, and where typically $t_i \neq x_i$. We represent a substitution σ as a finite set of expressions of the form $x_i \mapsto t_i$, where no two different terms substitute the same variable, i.e.

$$\sigma = \{x_1 \mapsto t_1, ..., x_n \mapsto t_n\}.$$

We define the domain of a substitution σ as $dom(\sigma) = \{x | \sigma x \neq x\}$ and its range as $rg(\sigma) = \{\sigma x | x \in dom(\sigma)\}$.

1. We say that σ is a *ground substitution* when $Vi(rg(\sigma)) = \emptyset$.

2. For $\sigma = \emptyset$, we speak of the *empty substitution*, denoted by ϵ.

3. A substitution σ such that $rg(\sigma) \subseteq Vi$ is a *(variable) renaming*.

---

[20] But note that this formula can be further simplified; for instance, the laws of associativity of $\vee$ and of absorption for $\wedge$ (cf. Fig. 3.5.1) allow the removal of the clause $(\neg p_2 \vee s \vee p_3)$. Below, we give other methods for obtaining the CNFs/DNFs of formulas. See Example 3.34.

2.4. Normal forms and substitutions for L1

The following remark completes and specifies Definition 2.14:

Remark 2.49. Let A, B be formulas. For the unary connective $\neg$ and the binary connectives $\rightarrow, \leftrightarrow$ and $\heartsuit = \wedge, \vee$, as well as for the quantifiers $\blacklozenge = \forall, \exists$, the following facts concerning a substitution σ hold:

1. $\sigma(\neg A) = \neg(\sigma A)$.

2. $\sigma\left(A \genfrac{}{}{0pt}{}{\rightarrow}{\leftrightarrow} B\right) = \sigma A \genfrac{}{}{0pt}{}{\rightarrow}{\leftrightarrow} \sigma B$.

3. $\sigma(A_1 \heartsuit ... \heartsuit A_n) = \sigma A_1 \heartsuit ... \heartsuit \sigma A_n$.

4. Let $A = \blacklozenge x A(x)$. Then, t is said to be *substitutable* (or *free*) for x in A, and we write $\sigma(A) = A_t^x = A[x/t]$, if (i) $Vi(t) = \emptyset$, (ii) $t \neq y$ or $y \notin Vi(t)$ if y is bound in A, or (iii) $t = x$.

Note with respect to Remark 2.49.4 that x must be free in A. For instance, $[\forall x\,(P(x)) \wedge Q(x)]_a^x = \forall x\,(P(x)) \wedge Q(a)$. But $[\forall x\,(R(x))]_x^x = R(x)$.

Definition 2.50. For a substitution $\theta = \{x_1 \mapsto t_1, ..., x_n \mapsto t_n\}$ and an expression E, $E\theta = \theta E$ is an expression obtained from E by replacing simultaneously each occurrence of the variable x_i, $1 \leq i \leq n$, in E by the term t_i. We say that $E\theta$ is an *instance* of E. If $Vi(E\theta) = \emptyset$, then $E\theta$ is called a *ground instance*.

Example 2.51. Let $\theta = \{x \mapsto a, y \mapsto f(b), z \mapsto c\}$, $E = P(x, y, z)$. Then $E\theta = P(a, f(b), c)$.

Definition 2.52. Given the two substitutions $\theta = \{x_1 \mapsto t_1, ..., x_n \mapsto t_n\}$ and $\lambda = \{y_1 \mapsto u_1, ..., y_n \mapsto u_n\}$, their *composition*, denoted by $\theta \circ \lambda$, is obtained from the set $\{x_1 \mapsto t_1\lambda, ..., x_n \mapsto t_n\lambda, y_1 \mapsto u_1, ..., y_n \mapsto u_n\}$ by deleting any element $x_j \mapsto t_j\lambda$ for which $t_j\lambda = x_j$, and any element $y_i \mapsto u_i$ such that y_i is among $\{x_1, ..., x_n\}$.

Example 2.53. Given the two substitutions $\theta = \{x \mapsto f(y), y \mapsto z\}$ and $\lambda = \{x \mapsto a, y \mapsto b, z \mapsto y\}$, we have $\theta \circ \lambda = \{x \mapsto f(b), z \mapsto y\}$.

Definition 2.54. We say that a set of expressions $E = \{E_1, ..., E_n\}$ is *unifiable* by a substitution σ (called the *unifier* for E) if $E_i\sigma = E_j\sigma$ for all $E_i, E_j \in E$. A unifier σ for the set E is a *most general unifier* (MGU) iff for each unifier θ for the set there is a substitution λ such that $\theta = \sigma \circ \lambda$. (Equivalently, let us say that λ is more general than θ, denoted $\lambda \leq_s \theta$, if there is a substitution σ such that $\sigma \circ \lambda = \theta$. Then, σ is the MGU of the set E if for every other unifier λ it is the case that $\sigma \leq_s \lambda$.)

2. Logical form

Example 2.55. For substitutions $\theta = \{x \mapsto z, y \mapsto z, u \mapsto f(z,z)\}$, $\lambda = \{y \mapsto z\}$, and $\sigma = \{x \mapsto y, u \mapsto f(y,z)\}$, σ is the MGU, because $\theta = \sigma \circ \lambda$.

Example 2.56. $\theta = \{x \mapsto f(b), y \mapsto b, z \mapsto u\}$ is a unifier of the expressions $E_1 = f(x, b, g(z))$ and $E_2 = f(f(y), y, g(u))$, because $E_1\theta = E_2\theta = f(f(b), b, g(u))$.

Definition 2.57. Let E_1 and E_2 be two expressions. Then, $E_1 \leq_{ss} E_2$ if there is a substitution σ such that $E_1\sigma \subseteq E_2$, and we say that E_1 subsumes E_2.

Definition 2.58. For a non-empty set of expressions E, the *disagreement set* $D(E)$ of E is the set of sub-expressions of E obtained by locating the leftmost symbol at which not all expressions in E have exactly the same symbol, and then extracting from each expression in E the sub-expression that begins with the symbol occupying that position.

Example 2.59. The disagreement set of

$$E = \{P(x, f(g(z))), P(x, h(k(x,y))), P(x, a)\}$$

is

$$D(E) = \{f(g(z)), h(k(x,y)), a\}.$$

Definition 2.60. The *unification problem* is that of finding a MGU of two given terms.

This is a purely mechanical procedure for which there is more than one algorithm. We give here the Robinson algorithm (Algorithm 2.4).

Remark 2.61. The application of *unification rules* is an alternative procedure more efficient than the Robinson algorithm.

Definition 2.62. Let σ and θ stand for substitutions and denote failure in the application of a rule by $\bot$. For P and Q pairs of expressions $\{\langle E_1, F_1 \rangle, ..., \langle E_n, F_n \rangle\}$, the unification rules for P and Q have the general form

$$P; \sigma \implies Q; \theta$$

or

$$P; \sigma \implies \bot.$$

The (successive) application of the unification rules ends either in success, denoted by $\emptyset$, or in failure. The order of application of the rules is non-deterministic.

2.4. Normal forms and substitutions for L1

Algorithm 2.4 The Robinson algorithm.

Given a set *Lit* of literals as input:

```
Let σ := ε
while |σ (Lit)| > 1 {
  select a disagreement pair d in σ (Lit);
  if d does not contain any variable then {
    stop and return "not unifiable";
  } else {
    let d = (x, t), x is a variable;
    if x occurs in t, then "occurs check" {
      stop and return "not unifiable";
    } else {
      let σ := σ ○ (x ↦ t);
    }
  }
}
return σ;
```

Definition 2.63. The *unification (inference) rules* are as follows:

1. *Trivial:*
$$\{\langle s, s \rangle\} \cup P'; \sigma \Longrightarrow P'; \sigma$$

2. *Decomposition:*
$$\{\langle f(s_1, ..., s_n), f(t_1, ..., t_n) \rangle\} \cup P'; \sigma \Longrightarrow \{\langle s_1, t_1 \rangle, ..., \langle s_n, t_n \rangle\} \cup P'; \sigma$$
   if $f(s_1, ..., s_n) \neq f(t_1, ..., t_n)$.

3. *Orient:*
$$\{\langle t, x \rangle\} \cup P'; \sigma \Longrightarrow \{\langle x, t \rangle\} \cup P'; \sigma$$
   if t is not a variable.

4. *Variable elimination:*
$$\{\langle x, t \rangle\} \cup P'; \sigma \Longrightarrow P'\theta; \sigma\theta$$
   given that x does not occur in t and $\theta = \{x \to t\}$.

5. *Symbol clash:*
$$\{\langle f(s_1, ..., s_n), g(t_1, ..., t_m) \rangle\} \cup P'; \sigma \Longrightarrow \bot$$

2. Logical form

if $f \neq g$.

6. *Occurs check:*
$$\{\langle x, t\rangle\} \cup P'; \sigma \Longrightarrow \bot$$
if x occurs in t but $x \neq t$.

Example 2.64. We provide examples of the application of each of the above rules:

1. For the pair of expressions $\langle P(a), P(a)\rangle$ we have

$$\{\langle P(a), P(a)\rangle\}; \epsilon \Longrightarrow_{Tr}$$
$$\emptyset; \epsilon$$

2. Given the pair of expressions $P(f(a), g(x))$ and $P(y, z)$ we have

$$\{\langle P(f(a), g(x)), P(y, z)\rangle\}; \epsilon \Longrightarrow_{Dec}$$
$$\{\langle f(a), y\rangle, \langle g(x), z\rangle\}; \epsilon$$

3. Applying this rule to the example immediately above, we have

$$\{\langle f(a), y\rangle, \langle g(x), z\rangle\}; \epsilon \Longrightarrow_{Or}$$
$$\{\langle \underline{y, f(a)}\rangle, \langle g(x), z\rangle\}; \epsilon$$

4. Given the pair of expressions $P(x, f(a))$ and $P(g(y), x)$ we have

$$\{\langle P(x, f(a)), P(g(y), x)\rangle\}; \epsilon \Longrightarrow_{VE}$$
$$\{\langle g(y), f(a)\rangle\}; \{x \to f(a)\}$$

5. Applying this rule to the result obtained immediately above, we have

$$\{\langle g(y), f(a)\rangle\}; \epsilon \Longrightarrow_{SCl}$$
$$\bot$$

6. For the pair of expressions $P(x)$ and $P(f(x))$ we have

$$\{\langle x, f(x)\rangle\}; \epsilon \Longrightarrow_{OCh}$$
$$\bot$$

Example 2.65. The pair $P(a, x, h(g(z)))$ and $P(z, h(y), h(y))$ is unifiable (see Fig. 2.4.2).

2.4. Normal forms and substitutions for L1

$\{\langle P(a,x,h(g(z))), P(z,h(y),h(y))\rangle\}; \epsilon \Longrightarrow_{Dec}$

$\{\langle a,z\rangle, \langle x,h(y)\rangle, \langle h(g(z)), h(y)\rangle\}; \epsilon \Longrightarrow_{Or}$

$\{\langle z,a\rangle, \langle x,h(y)\rangle, \langle h(g(z)), h(y)\rangle\}; \epsilon \Longrightarrow_{VE}$

$\{\langle x,h(y)\rangle, \langle h(g(a)), h(y)\rangle\}; \{z \mapsto a\} \Longrightarrow_{VE}$

$\{\langle h(g(a)), h(y)\rangle\}; \{z \mapsto a, x \mapsto h(y)\} \Longrightarrow_{Dec}$

$\{\langle g(a), y\rangle\}; \{z \mapsto a, x \mapsto h(y)\} \Longrightarrow_{Or}$

$\{\langle y, g(a)\rangle\}; \{z \mapsto a, x \mapsto h(y)\} \Longrightarrow_{VE}$

$\emptyset; \{z \mapsto a, x \mapsto h(y), y \mapsto g(a)\}$

Figure 2.4.2.: Unifying the pair $\langle P(a,x,h(g(z))), P(z,h(y),h(y))\rangle$.

▶ *Do Exercises 2.14-19.*

Exercises

Exercise 2.1. Let there be given the set $O = \{\neg^1, \wedge^2, \vee^2, \rightarrow^2\}$ of logical connectives (with superscripts indicating arity) and the set of quantifiers $Q = \{\forall, \exists\}$. Identify which of the following expressions are terms, atoms, formulas, or none of these:

1. b
2. $R(x,y)$
3. Q
4. $\neg x$
5. $P \vee f(x)$
6. $\exists (P(z))$
7. $g(a,b)$
8. $\forall Q (Q(P))$
9. f
10. $\forall x (P(f(x)), \exists y (Q(y))))$

2. Logical form

11. $h(a, f(a))$

12. $\forall z \forall w (S(z) \to T(w))$

13. $A \land \neg (B \lor C)$

Exercise 2.2. Let the sets O and Q be as in Exercise 2.1 above. Identify (i) the immediate sub-formulas and (ii) all the other sub-formulas of the following formulas:

1. $\neg (P \land \neg Q)$
2. $R(x) \to (P(x) \lor (Q(a) \land S(b)))$
3. $\neg S \lor \neg (T \to \neg R)$
4. $P(y) \lor (\neg R(x) \lor R(x))$
5. $(R \lor \neg (P \to R)) \land (Q \land (S \to \neg T))$
6. $\exists P (P(a,b) \to \forall Q (Q(b,a)))$
7. $\forall x \exists y (P(x,y) \land \forall z (R(z,a)))$
8. $\forall f \exists x (S(f(x)))$

Exercise 2.3. Determine the order of the following formulas:

1. $\exists g \forall x (P(g(x)))$
2. $\exists g \forall x (P(g(x)), f(x)))$
3. $\exists P \exists Q \forall y (P(y) \land \neg Q(y))$
4. $R(f) \to Q(f(x))$

Exercise 2.4. Construct the formula trees of the formulas in Exercise 2.2 above.

Exercise 2.5. Identify the closed and the open FO formulas. For the open formulas, give their universal closure.

1. $\forall x \exists y (P(x,y))$
2. $\forall z \exists x (R(z) \to S(z,x)) \land T(z)$

2.4. Normal forms and substitutions for L1

3. $\exists z \left(P\left(x,y \right) \right) \vee \neg R\left(z \right)$

4. $\forall x \forall y \left(P\left(x \right) \to \neg Q\left(y \right) \right) \wedge \forall z R\left(z,x \right)$

Exercise 2.6. Prove Proposition 2.18.

Exercise 2.7. Formalize the following statements in L if they are formalizable. If so, decide whether they can be formalized in the propositional language L0, or require L1 or higher. If the latter, clearly define the universe **U** and the domain of discourse $\mathscr{D}$.

1. Give the employees a bonus and they'll all be happy.

2. Give me that book, will you?

3. Any formula is a sub-formula of itself.

4. Cats are finicky.

5. Invariably, kids are truants, but some kids are studious.

6. All Vikings, both male and female, were warriors, but for the Greeks they were barbarians.

7. If something is cheap, then it is not of good quality.

8. No cat is both finicky and serendipitous.

9. There is a function that grows faster than the exponential function.

10. Radioactivity is lethal, because it messes with your genes.

11. Karim is a camel if it's not a dromedary.

12. Bob loves Samantha but she doesn't love him back.

13. Gosh, you're such a liar.

14. A superstar told Mary and John that they were Oscar winners.

15. Only warm clothes make good clothes.

16. There is something that Ron likes that everybody else with some commonsense does not like.

17. Not everything that shines is made of gold.

18. There is at least a white raven and some swans are not white.

2. Logical form

19. Anytime, anyone can fool a superb logician but a superb logician can never fool anyone.

20. Alas, the task was too difficult.

21. Some teacher introduced a nurse who was also a philosopher to an astronaut who was also a farmer.

22. Every integer is even or odd.

23. A sequence converges iff it is bounded.

24. There is a set of natural numbers that does not contain the number 7.

25. Are you going out today?

26. Come on!

27. Hurry up, or we'll be late.

28. For every non-empty set there is a set that is a proper subset thereof and that is the empty set.

29. A necessary condition for hypothermia is cold weather. However, it is not a sufficient condition.

30. There are only so many happy marriages.

Exercise 2.8. Translate the following sentences in L1 into English by paying attention to the domains and the defined predicates:

1. Consider the domain $\mathscr{D} \subseteq \mathbf{U}$ to be the set of all the people and $L = \{(x, y) \,|\, x \text{ loves } y\}$:

$$(\forall x \exists y L\,(x, y) \land \neg \exists x \forall y L\,(x, y)) \lor (\exists x \forall y L\,(x, y) \land \exists x \forall y \neg L\,(x, y))$$

2. Let the domain $\mathscr{D} \subseteq \mathbf{U}$ be the set of all the people and of all the times, and consider $F = \{(x, y, t) \,|\, x \text{ can fool } y \text{ at time } t\}$:

$$\forall x \,(\exists y \forall t\,(F\,(x, y, t)) \land \forall y \exists t\,(F\,(x, y, t)) \land \exists y \exists t\,(\neg F\,(x, y, t)))$$

Exercise 2.9. Identify the form of the following arguments:

2.4. Normal forms and substitutions for L1

1. $\{\neg P \to \neg Q, \neg Q \to R\} \vdash \neg P \to R$

2. $\{(P \to \neg Q) \land (\neg R \to S), P \lor \neg R\} \vdash \neg Q \lor S$

3. $\{\neg(P \land R) \to \neg Q, Q\} \vdash P \land R$

4. $\{(P \to Q) \to (R \lor T), R \lor T\} \vdash P \to Q$

5. $\{((T \lor P) \to (Q \land \neg O)) \land (S \to (R \leftrightarrow U)), (T \lor P) \lor S\}$
   $\vdash (Q \land \neg O) \lor (R \leftrightarrow U)$

6. $\{P \lor Q, (P \lor Q) \to (R \land \neg S)\} \vdash R \land \neg S$

7. $\{P \lor Q, (P \to R) \land (Q \to S)\} \vdash R \lor S$

8. $\{(R \land S) \to (P \lor (Q \land T)), \neg(R \land S)\} \vdash \neg((P \lor (Q \land T)))$

9. $\{\neg P, ((R \lor S) \land (T \to O)) \to P\} \vdash \neg((R \lor S) \land (T \to O))$

10. $\{\neg(Q \land P) \lor ((R \lor S) \land (T \to O)), Q \land P\} \vdash ((R \lor S) \land (T \to O))$

Exercise 2.10. Formalize the following reasoning instances.

1. There is a set $\mathbb{N}$ of natural numbers that does not include the number 6. Therefore, for any numbers $x, y \in \mathbb{N}$, $x, y \neq 6$, it is never the case that $f(x, y) = 6$ for any function $f : \mathbb{N}^2 \longrightarrow \mathbb{N}$.

2. Let $x \in \mathbb{Z}$. If x is even, then it is divisible by 2, and if x is even, then it is not odd. Therefore, if x is even then it is divisible by 2 and it is not odd.

3. All cats are finicky about some food. Fritz is a cat. Hence, Fritz if finicky about fish sticks.

4. A group $\mathcal{G}$ is a set G with an operation $\star$ such that (1) for all elements x, y, z of G we have $(x \star y) \star z$ is equal to $x \star (y \star z)$, (2) for every x of G there is an element e in G such that $x \star e$, $e \star x$, x are all equal, and (3) for every x of G there is an element x^{-1} in G such that $x \star x^{-1}$, $x^{-1} \star x$, e are all equal. If $\mathcal{G} = (G, \star)$ is a group, then for every x, y, z of G the equality of $x \star z$ and $y \star z$ implies that x and y are identical.

5. The same as above, but considering a group $\mathcal{G} = (G, f)$ where $f : G^2 \longrightarrow G$.

2. Logical form

6. An idempotent semi-ring is a structure $\mathcal{S} = (S, \star, +, 1, 0)$ such that $(S, +, 0)$ is a commutative monoid with idempotent addition, $(S, \star, 1)$ is a monoid, multiplication distributes over addition from the left and right, and 0 is a right and left annihilator of multiplication. If the relation $\leq$ defined as $x \leq y \leftrightarrow x + y = x$ for all x, y in S is a partial order, then for all $x, y, z \in S$ we have

$$x + y \leq z \leftrightarrow (x < z \land y < z)$$

and every idempotent semi-ring is also a semi-lattice $(S, \leq)$ with addition as join.

7. If everybody is a lover (i.e. loves somebody) and everybody loves every lover, then everybody loves everybody.

8. A person who is big is assumed to be strong, unless there is reason to believe they are weak. A person who is small is assumed to be weak, unless there is reason to believe they are strong. A person who is weak but muscular is assumed to be strong, except if there is some reason to believe the opposite. Mary is small and not muscular, but we have no reason to believe she is weak. Therefore, she is strong.

9. Cats are typically finicky. Nevertheless, finicky cats can eat what they would otherwise reject if they are truly hungry. But super-finicky cats would rather starve to death than eat what they typically reject. Fritz is a super-finicky cat. As a consequence, Fritz risks starving to death.

10. *Russell's paradox*: A barber shaves those men who do not shave themselves. If he shaves himself, then he does not shave himself. (Hint: Make the premise a bi-conditional proposition.)

Exercise 2.11. What change(s) would it entail adding the following statement to the reasoning of Examples 2.22-4: *But cats are not best friends either with birds or with rabbits.*

Exercise 2.12. Let the following premises be given:

$$\forall x \forall y \, (P(x, y) \to Q(x, y))$$

$$\forall x \forall y \, (R(x, y) \to Q(x, y))$$

$$\forall x \forall y \, (P(x, y) \lor R(x, y))$$

2.4. Normal forms and substitutions for L1

Without carrying out a formal proof, indicate which of the following two formulas is a conclusion of these premises: (i) $\forall x \exists y \, (Q(x,y))$; (ii) $\exists y \forall x \, (Q(x,y))$. Explain your reasoning.

Exercise 2.13. For convenience (for instance, in logic programming), predicates can be formalized in a more "denotational" manner. Reconstruct the original argument from the FOL formalization in Figure 2.4.3.

$(Native\_of\_c(x) \land Weapon(y) \land Sells(x,y,z) \land Hostile(z)) \rightarrow Criminal(x)$
$Missile(m1)$
$Owns(daffy, m1)$
$Enemy(daffy, c)$
$(Missile(x) \land Owns(daffy, x)) \rightarrow Sells(west, x, daffy)$
$Native\_of\_c(west)$
$Missile(x) \rightarrow Weapon(x)$
$Enemy(x, c) \rightarrow Hostile(x)$
$\therefore Criminal(west)$

Figure 2.4.3.: A FOL argument.

Exercise 2.14. Transform each of the following formulas into a set of ground clauses:

1. $(x_1 \land x_2) \leftrightarrow \neg (x_2 \lor (\neg x_1 \land x_2))$
2. $(p_1 \land p_2) \rightarrow (\neg (p_3 \land p_4) \rightarrow p_1)$
3. $\exists y \forall x \, (P(x,y) \rightarrow Q(x))$
4. $\exists y \forall x \, (P(x,y)) \rightarrow Q(x)$
5. $\forall x \forall y \, (P(x,y) \lor \exists z Q(x,y,z))$
6. $\forall w \exists u \exists v \, ((\neg P(w,u) \land Q(w,v)) \rightarrow R(w,u,v))$
7. $\forall y \neg \exists z \, (P(y) \rightarrow Q(y,z)) \land \forall x \forall u \exists w \, (R(x,w) \leftrightarrow S(x,u,w))$
8. $\forall x \, ((P(x) \leftrightarrow Q(x)) \land \forall z \neg \exists y \, (R(x,z) \rightarrow S(x,y,z)))$

Exercise 2.15. Rewrite the following formulas in DNF and CNF clearly showing every step of the transformation:

2. Logical form

1. $x_1 \vee (x_2 \wedge (x_1 \vee x_3) \vee (x_2 \wedge \neg x_1))$
2. $\forall x \neg \exists y \, [(P(x) \to S(y)) \wedge \forall z \, (Q(x, y, z) \leftrightarrow R(x, z))]$
3. $\forall x ((P(x) \leftrightarrow Q(x)) \wedge \forall z \neg \exists y (R(x, z) \to S(x, y, z)))$
4. $\forall y \neg \exists z \, (P(y) \to Q(y, z)) \wedge \forall x \forall u \exists w \, (R(x, w) \leftrightarrow S(x, u, w))$

Exercise 2.16. How many clauses has the CNF of the formula ϕ by applying the Tseitin transformation?

$$\phi = \neg \, (p_1 \wedge (p_2 \vee ... \vee p_n))$$

Exercise 2.17. For the pairs of expressions E and F, determine whether σ is (i) a unifier and (ii) a MGU of E and F:

1. $E = P(a, f(y), z) \, ; \, F = S(x, f(f(b)), b)$

   $\sigma = \{x \mapsto a, y \mapsto f(b), z \to b\}$

2. $E = P(x, h(a, z), f(x)) \, ; \, F = P(g(g(v)), y, f(w))$

   $\sigma = \{x \mapsto g(g(v)), y \mapsto h(a, z), w \mapsto x\}$

3. $E = Q(f(x), g(y)) \, ; \, F = Q(z, g(v))$

   $\sigma = \{x \mapsto a, z \mapsto f(a), y \mapsto v\}$

4. $E = Knows(john, x) \, ; \, F = Knows(john, jane)$

   $\sigma = \{x \mapsto jane\}$

Exercise 2.18. Determine whether you can unify the following pairs of atoms:

1. $Q(x, y, z)$ and $Q(u, h(v, v), u)$

2.4. Normal forms and substitutions for L1

2. $P(x, f(x))$ and $P(y, y)$

3. $P(f(x))$ and $P(a)$

4. $P(x, y)$ and $P(y, f(y))$

5. $P(a, y)$ and $P(x, f(x), b)$

6. $S(f(a), g(x))$ and $S(u, u)$

7. $R(a, x, h(g(z)))$ and $R(z, h(y), h(y))$

8. $P(f(a), g(x))$ and $P(x, x)$

9. $R(a, x, f(x))$ and $R(a, y, y)$

Exercise 2.19. Prove (or complete the proof of) the Propositions of Section 2.4.

3. Logical meaning

Meaning with respect to a logical language can be approached from different perspectives. Within the fields of the philosophy of logic and logical philosophy, natural language usually plays a fundamental role, and meaning is object of "knotty" philosophical discussions. We provide here an introductory formal treatment of meaning in a logical language L1 that is largely algebraic. The importance of this algebraic, namely Boolean, foundation for a formal approach to classical logic notwithstanding, it does not capture all the nuances of the subject of meaning in the classical formalization; both Tarskian and Herbrand semantics, discussed in other Sections below, in particular in Part III, contribute with fundamental insights to a fuller grasp of this central topic. These three semantics for classical logic–Tarskian, Herbrand, and algebraic (Boolean) semantics–are what we call *formal semantics* in this book.[1]

Regardless of the perspective taken, *valuations* and *interpretations* are two central components of logical meaning. In the present Chapter, we proceed in a bottom-up way and give some basic algebraic foundations for classical valuations and interpretations.

3.1. Truth: Values, tables, and functions

Definition 3.1. Given a set of formulas $F \subseteq$ L1 and a set $W_n = \{v_0, v_1, ..., v_{n-1}\}$ of *distinguished elements* known as *(logical) truth values*, a *valuation* is a function $val : F \longrightarrow W_n$. Let $\phi \in F$ be an atomic formula; then, we write $val(\phi) = v_i$.

In other words, a valuation is an assignment of a truth value to the formulas of L1. Given a truth-value set $W_n = \{v_0, v_1, ..., v_{n-1}\}$, typically, $v_0 = 0 = $ f (false) and $v_{n-1} = 1 = $ t (true).

Definition 3.2. In a logical language L with a truth-value set W_n, with each operator $\heartsuit_i^k$ with arity k, $k = 0, 1, 2$, and each quantifier $\blacklozenge_j$ there can be associated

1. a *truth table* $\widetilde{\heartsuit}_i^k : W_n^k \longrightarrow W_n$, k is the arity of $\heartsuit_i$, and

---
[1] See Chapter 9 for a clarification of the expression "formal semantics."

3. Logical meaning

2. a *truth (or distribution) function* $\widetilde{\blacklozenge}_j : \left(2^{W_n} - \emptyset\right) \longrightarrow W_n$.[2]

In this text, we consider mostly the set of *classical* truth values, to wit, $W_2 = \{0, 1\}$. As such, we henceforth often omit the subscript in W when we consider W_2.

Truth tables and truth functions are extensions of valuations; in particular, by means of truth tables the function *val* is extended to all propositional formulas and by means of truth functions *val* is extended to all quantified formulas.

Example 3.3. For the sake of simplicity, we begin by considering the atomic propositions $P, Q \in F_{L0}$ and the set of truth values W. An important feature of propositional atoms is that they have a fixed truth value, in the sense that their valuation, contrarily to predicates and functions (see below), does not depend on truth-value assignments to their arguments (because they have none). Thus, given two atoms $P, Q \in F_{L0}$, there can be four cases: (i) both P and Q are valuated to `true`, (ii) P is valuated to `true` and Q is valuated to `false`; (iii) P is valuated to `false` and Q is valuated to `true`; (iv) both P and Q are valuated to `false`. We represent these four cases in tabular form as follows:

P	Q
1	1
1	0
0	1
0	0

Let now there be given the connective $\to^2 \in O_L$. Recall Proposition 2.5 and Definition 2.6: This is a binary connective, so that we have formulas $\to^2 (\phi_1, \phi_2)$, here written $\phi_1 \to \phi_2$, where $\phi_1, \phi_2 \in \mu^{L0}$ are either atomic or complex formulas. We now give the truth table $\widetilde{\to^2}$ for the formula $P \to Q$:

P	Q	$P \to Q$
1	1	1
1	0	0
0	1	1
0	0	1

---

[2] Given a set A, 2^A denotes the *power set* of A. For example, let $A = \{a, b, c\}$; then
$$2^A = \{\emptyset, \{a\}, \{b\}, \{c\}, \{a, b\}, \{b, c\}, \{a, c\}, \{a, b, c\}\}.$$

As said above, non-mathematical introductions to logic often provide lengthy explanations on the "meaning" of these four valuations of Example 3.3, namely by appealing to reasoning in natural language; in particular, it is often claimed that it is not acceptable that falsity can follow from truth (second line of the table), and from falsity anything can follow (lines three and four), a principle known as *explosion*. We ask the reader to take the truth table above as a simple convention for the time being. In the next Sections, the algebraic basis of this convention will become evident.

3.2. The Boolean foundations of logical bivalence

Before we address the issue of truth-functionality, which will clarify the inter-relation between meaning and truth tables, we discuss the property of *logical bivalence*, and we do so from an algebraic perspective.[3] The *bivalent* nature of classical logic is evident from Example 3.3: Any formula of classical logic is *either true or false*. Although a valuation can be, and often is, a matter of convention, as stated above,[4] there is an *algebraic basis for logical bivalence*, Boolean algebras providing a mathematical foundation for the truth tables of the connectives ¬, ∧, ∨. In this algebraic perspective, these connectives are said to be primitive, in the sense that → and ↔ are defined via them.

Boolean algebras are named after G. Boole, who first established a formal correspondence between human thought and mathematical logic. The importance of his work for modern logic is impressive: in Section 5.3.3 we elaborate on Boolean algebras from the viewpoint of practical applications in electronics, and in Section 9.3 we discuss their foundational role in the semantics of classical logic. In this Section, we concen-

---

[3] An *(abstract) algebra* is a pair

$$\mathfrak{A} = (A, O) = (A, \{o_i | i \in I\})$$

where A is a non-empty set, called the *carrier* or *universe* of $\mathfrak{A}$, and $O = \{o_1, o_2, ..., o_n\}$ is a set of n operations on A, with $i \in I = \{1, 2, ..., n\}$, I is called the *index set* of $\mathfrak{A}$. An algebra $\mathfrak{B} = (B, \{f_k | k \in K\})$ is said to be *similar* to the algebra $\mathfrak{A}$ iff $I = K$ and $o_i^m = f_i^m$ for each $i \in I$ and m the arity of o_i, f_i. An algebra $\mathfrak{C} = (C, G) = (C, \{g_j | j \in J\})$ is said to be a *reduct* of the algebra $\mathfrak{A} = (A, O)$ iff $G \subsetneq O$; otherwise, if $G \supsetneq O$, then $\mathfrak{C}$ is called an *expansion* of $\mathfrak{A}$.

[4] This holds for every truth table for every single connective of a given logical language. The conventional character of truth tables for single connectives is clearly easier to see in the many-valued logics, there being different truth tables for the same connectives in the different logics (cf. Fig. 3.3.2 below).

3. Logical meaning

trate on Boolean functions and Boolean expressions.

Definition 3.4. A *Boolean algebra* is a 6-tuple $\mathfrak{B} = (B,', \wedge, \vee, 0, 1)$ where B is a non-empty set, $', \wedge, \vee$ are the operations of complementation, meet, and join, respectively, over B, and 0 and 1 are distinguished elements (nullary operations) such that[5]

1. for the unary operation of complementation ($'$): $0' = 1$ and $1' = 0$
2. for the binary operations meet ($\wedge$) and join ($\vee$):

$\wedge$	1	0
1	1	0
0	0	0

$\vee$	1	0
1	1	1
0	1	0

Further properties of a Boolean algebra are given in Figure 3.5.1 (cf. Exercise 3.1).

By means of a homomorphism,[6] we can have the above directly applied to the logical operations of negation ($\neg$), conjunction ($\wedge$), and disjunction ($\vee$). The above can then be extended to the logical connectives $\rightarrow$ and $\leftrightarrow$ via inter-definitions (cf. Prop. 2.32.1).

---

[5] More correctly, a Boolean algebra is a 6-tuple $\mathfrak{B} = (B,', \cdot, +, 0, 1)$, and $\mathfrak{B} = (B,', \wedge, \vee, 0, 1)$ is a *Boolean lattice*, but it can be shown that every Boolean lattice is a Boolean algebra (and vice-versa).

[6] A *homomorphism* is a structure-preserving mapping $h : A \longrightarrow B$ between two similar algebras $\mathfrak{A} = (A, \{o_1, ..., o_n\})$ and $\mathfrak{B} = (B, \{f_1, ..., f_n\})$, denoted by $hom(\mathfrak{A}, \mathfrak{B})$, such that for each $i \in I$ and for arbitrary elements $a_1, ..., a_{m(i)} \in A$ we have

$$h\left(o_i\left(a_1, ..., a_{m(i)}\right)\right) = f_i\left(h\left(a_1\right), ..., h\left(a_{m(i)}\right)\right).$$

In particular, a *Boolean homomorphism* $h : \mathfrak{A} \rightarrow \mathfrak{B}$ preserves the operations $', \wedge, \vee, 0, 1$ between two similar Boolean algebras $\mathfrak{A}$ and $\mathfrak{B}$. Let now $\mathfrak{B}$ be a Boolean algebra and $\mathfrak{L} = (F, \neg, \wedge, \vee, 0, 1)$ be an *algebra of (propositional) formulas* similar to it. Then, the Boolean homomorphism $h : \mathfrak{B} \longrightarrow \mathfrak{L}$ satisfies the following properties for $p, q \in F$ whenever $p, q \in B$:

$$h(p) = p$$
$$h(p \wedge q) = h(p) \wedge h(q) = p \wedge q$$
$$h(p \vee q) = h(p) \vee h(q) = p \vee q$$
$$h(p') = \neg(h(p)) = \neg p$$
$$h(0) = 0$$
$$h(1) = 1$$

See below and also Section 9.3 for further relations between Boolean algebras and formal classical logic.

3.2. The Boolean foundations of logical bivalence

It is this algebraic foundation of classical logic that sanctions the expression *Boolean logic* as synonymous with *bivalent logic*. As a matter of fact, it is not rare to see classical logic formalized in purely Boolean terms. We show how.

Definition 3.5. Let $\mathfrak{L} = (F, ', \wedge, \vee, 0, 1)$ be an algebra of (propositional) formulas, in which F is a set of formulas. A variable $x_i \in (F_0 = V) \subseteq F$ in a finite sequence $x_1, ..., x_n$ is said to be a *Boolean variable* if there is a *Boolean function of degree n*

$$f : \{0,1\}^n \longrightarrow \{0,1\}$$

such that we have $f(x_i) = 0$ or $f(x_i) = 1$, spoken of as *truth-value assignments*. Additionally, we have

1.
$$[f(x_i) = \neg x_i] = 1 - f(x_i)$$

2.
$$\left[f(x_1, ..., x_n) = \bigwedge_{i=1}^{n} x_i \right] = min(f(x_1), ..., f(x_n))$$

3.
$$\left[f(x_1, ..., x_n) = \bigvee_{i=1}^{n} x_i \right] = max(f(x_1), ..., f(x_n))$$

We often abbreviate and write simply $x_i = 1$ or $x_i = 0$, as well as

$$\neg x_i = 1 - x_i$$

$$\bigwedge_{i=1}^{n} x_i = min(x_1, ..., x_n)$$

$$\bigvee_{i=1}^{n} x_i = max(x_1, ..., x_n)$$

Definition 3.6. A *Boolean expression* in the individual variables $x_1, ..., x_n$ is an expression defined inductively in the following way:

1. the constants $0, 1$, and all variables x_i are Boolean expressions in $x_1, ..., x_n$;

3. Logical meaning

2. if ϕ and ψ are Boolean expressions in the variables $x_1, ..., x_n$, then so are ϕ', $(\phi \wedge \psi)$, and $(\phi \vee \psi)$.

For consistency, we shall write $\neg \phi$ instead of ϕ'.

Example 3.7. The following are examples of Boolean expressions:

- $\phi(p) = \neg p$
- $\omega(p, q, r) = (p \vee q) \wedge (\neg p \vee r)$
- $\chi(x, w, y, z) = (x \wedge \neg w) \vee \neg y \vee z$
- $\psi(x_1, x_2, x_3) = \neg x_1 \wedge (\neg x_2 \vee x_1) \wedge (\neg x_2 \vee \neg x_3)$

Definition 3.8. An expression of the form

$$\phi(p_1, ..., p_n)$$

where the p_i are propositional Boolean variables is called a *propositional Boolean formula*.

Example 3.9. All the Boolean expressions of Example 3.7 are propositional Boolean formulas. We evaluate

$$\psi(x_1, x_2, x_3) = \neg x_1 \wedge (\neg x_2 \vee x_1) \wedge (\neg x_2 \vee \neg x_3)$$

for the assignments $x_1 = 1, x_2 = 0, x_3 = 1$, according to Definitions 3.4-5:

$$\underbrace{\neg 1}_{0} \wedge \left(\underbrace{\neg 0 \vee 1}_{1} \right) \wedge \left(\underbrace{\neg 0 \vee \neg 1}_{1} \right) =$$

$$= 0 \wedge \left(\underbrace{1 \vee 1}_{1} \right) \wedge \left(\underbrace{1 \vee 0}_{1} \right) =$$

$$= \underbrace{(0 \wedge 1)}_{0} \wedge 1 \left(\text{or } 0 \wedge \underbrace{(1 \wedge 1)}_{1} \right) =$$

$$= 0$$

Definition 3.10. An expression of the form

$$\blacklozenge_1 x_1 ... \blacklozenge_n x_n (\phi(x_1, ..., x_n))$$

where the x_i are individual Boolean variables and the $\blacklozenge_i \in \{\forall, \exists\}$ is called a *quantified Boolean formula*.

Example 3.11. $\forall x \exists y ((x \wedge \neg y) \vee \neg x)$ is a quantified Boolean formula.

The evaluation of a quantified Boolean formula is not so simple a matter as for a propositional Boolean formula. As a matter of fact, this holds for any formula with quantifiers and bound variables. In Section 3.4, we elaborate on valuations and interpretations for FO formulas. Further below, we shall see there are ways to "propositionalize" quantified formulas.

▶ *Do Exercises 3.1-2.*

3.3. Compositionality and truth-functionality

Why 0 and 1 can be, and often are, interchangeable with `false` and `true`, respectively, that seems to be a philosophical hard nut to crack, but we take the latter as distinguished elements simpliciter in the above mathematical sense. By this, we mean that they are not to be confused with the English words *true* and *false*, and we emphasize this by writing them in a distinct `font` from that used for English words.[7]

Proposition 3.12. *Let ϕ be any formula such that $\phi \in F_{L0}$; in order to interpret L0, ϕ is provided with a* meaning. *The meaning of ϕ is its* semantical correlate. *Let G be the range of all the semantical correlates; a mapping $s : F_{L0} \to G$ requires two conditions (Frege, 1892):*

1. *With each $\phi \in F_{L0}$, exactly one semantical correlate is associated (i.e., s is a function).*

2. *Any two formulas χ, ψ such that $\chi, \psi \in F_{L0}$ are interchangeable in any propositional context $\phi \in F_{L0}$ whenever $s(\chi) = s(\psi)$ or, in other words, when for any $\phi \in F_{L0}$, $p \in V_{L0}$ we have it that*

$$s(\phi[p/\chi]) = s(\phi[p/\psi]) \quad \text{iff} \quad s(\chi) = s(\psi)$$

---
[7]This can be seen as a more informal sense of "distinguished." This said, we do not wish to remove completely from `true` and `false` the meanings they have in English, namely for the sake of an *intuitive semantics*, often a desideratum for a logical system. Tarski (1935), without dissolving the philosophical problem (much on the contrary!), provides an interesting definition of *truth* as a property of a formal*ized* language. See Section 9.1 below.

3. Logical meaning

where $\phi[p/\varsigma]$ stands for the formula that results from ϕ after the substitution ς instead of p.

Proposition 3.12 is known as the *Fregean axiom* and 3.12.2 is often referred to as the *principle of extensionality*. Generalization of Proposition 3.12 to a FO language follows naturally, though not without restrictions, as will be seen below.

In particular, Proposition 3.12.2, also referred to as *principle of the compositionality of meaning*, states that the meaning of a proposition is a function of the meaning of its components.

Example 3.13. Note how the valuation $val(\psi(x_1, x_2, x_3))$ in Example 3.9 is a function of the valuation of the components x_1, x_2, x_3 of the formula $\psi(x_1, x_2, x_3)$. A truth table just is the exhaustive evaluation of all the possible valuations for a formula $\heartsuit(\phi_1, ..., \phi_n)$ where $\heartsuit$ is the main connective. Figure 3.3.1 shows the truth table for $\psi(x_1, x_2, x_3)$. Note how the valuation of Example 3.9 corresponds to the third row of this table. Note also that there can only be a single main connective in a given formula; in the example given, we apply the associative property of $\wedge$ (cf. Fig. 3.4.1) and obtain the formula

$$\psi(x_1, x_2, x_3) = \underbrace{(\neg x_1 \wedge (\neg x_2 \vee x_1))}_{\phi_1} \overset{\heartsuit}{\wedge} \underbrace{(\neg x_2 \vee \neg x_3)}_{\phi_2}$$

where $\heartsuit$ indicates the main connective and ϕ_1, ϕ_2 are the immediate subformulae of ψ. The digits 1 to 4 below the binary connectives indicate the order by which they were evaluated by means of the corresponding truth tables $\widetilde{\heartsuit}_i$.

x_1	x_2	x_3	$(\neg x_1$	$\wedge$	$(\neg x_2$	$\vee$	$x_1))$	$\wedge$	$(\neg x_2$	$\vee$	$\neg x_3)$
1	1	1	0	0	0	1	1	0	0	0	0
1	1	0	0	0	0	1	1	0	0	1	1
1	0	1	0	0	1	1	1	0	1	1	0
1	0	0	0	0	1	1	1	0	1	1	1
0	1	1	1	0	0	0	0	0	0	0	0
0	1	0	1	0	0	0	0	0	0	1	1
0	0	1	1	1	1	1	0	1	1	1	0
0	0	0	1	1	1	1	0	1	1	1	1
				2		1		4		3	

Figure 3.3.1.: A truth table with $2^3 = 8$ rows.

Compositionality of meaning just is the logical property known as

3.3. Compositionality and truth-functionality

truth-functionality, and this can find a mathematical foundation in an algebraic interpretation of Proposition 3.12.

Definition 3.14. Let $n \geq 2 \in \mathbb{N}$; we denote by G_n the set $\{0, 1, ..., n-1\}$ and by $\mathfrak{A}_n$ any algebra of the form $(G_n, f_1, ..., f_m)$. In particular, we denote the set of all m-ary mappings defined on G_n with values in the same set by

$$Z_n^m = \{f | f : G_n^m \longrightarrow G_n\}, \quad m \geq 0, m \text{ finite}$$

and we denote the set of all mappings defined on G_n with values in Z_n^m by

$$Z_n = \bigcup_{m \in \omega} Z_n^m$$

for ω the set of all natural numbers.

Example 3.15. Let $G_2 = \{0, 1\}$. Then, we have the functions $f_1 : G_2^1 \longrightarrow G_2$ and $f_2 : G_2^2 \longrightarrow G_2$ for unary and binary mappings respectively defined on G_2. The functions f_1 and f_2 are *Boolean functions*, i.e. functions defined on $G_2 = W$. Given the set $O = \{\neg^1, \wedge^2, \vee^2, \rightarrow^2, \leftrightarrow^2\}$, it is evident that $\neg \in f_1$ and $\wedge, \vee, \rightarrow, \leftrightarrow \in f_2$. We may also consider $\top$ and $\bot$ as 0-ary functions, i.e. as degenerate operations. This allows a *characterization of classical logic* as the set

$$Z_2 = \bigcup_{i=0}^{2} Z_2^i = Z_2^0 \cup Z_2^1 \cup Z_2^2.$$

Definition 3.16. Let now $\mathfrak{L} = (F, o_1, ..., o_m)$ be an algebra of formulas freely generated by the set of generators $V \subseteq F_{\text{L0}}$.[8] Then $\mathfrak{L}$ is a propositional language. Given an algebra $\mathfrak{A} = (G_n, f_1, ..., f_m)$ similar to $\mathfrak{L}$, it is easy to see that a homomorphism $hom\,(\mathfrak{L}, \mathfrak{A})$ can give rise to a function $h : V_\mathfrak{L} \longrightarrow G_\mathfrak{A}$, which in turn can be extended to the function $h : T_\mathfrak{L} \longrightarrow G_\mathfrak{A}$ such that for each operation o_i with arity m and given terms $t_i \in T$ we have:

$$h\,(o_{i_\mathfrak{L}}(t_1, ..., t_m)) = f_{i_\mathfrak{A}}(h\,(t_1), ..., h\,(t_m))$$

---

[8] Recall from above in this Chapter the definition of an algebra (cf. footnote 3). The structure $\mathfrak{A} = (A, \{o_i | i \in I\})$ is an *absolutely free* algebra if there is a set $Z \subseteq A$ of generators of $\mathfrak{A}$ such that every mapping $s : Z \longrightarrow Y$ of a similar algebra $\mathfrak{D} = (Y, \{p_k | k \in K\})$ can be extended to a homomorphism from $\mathfrak{A}$ into $\mathfrak{D}$. Then, the elements of Z are said to be *free generators* of $\mathfrak{A}$ and $\mathfrak{A}$ is said to be *freely generated* by Z.

3. Logical meaning

Then, h is an assignment of truth values, or a valuation, into $\mathfrak{L}$.

More specifically, the algebra $\mathfrak{A}$ is similar to the set of all formulas of $\mathfrak{L}$ formed by means of the m operations of $\mathfrak{L}$ (i.e., the connectives), in other words, the algebra of $F_\mathfrak{L}$. From the semantical viewpoint, the fundamental importance of the homomorphism $hom\,(\mathfrak{L},\mathfrak{A})$ is that it is an embedding of $F_\mathfrak{L}$ in $\mathfrak{A}$ that is in fact an *interpretation* for the formulas of $\mathfrak{L}$ (in $\mathfrak{A}$).[9]

Definition 3.17. A set of functions $X \subseteq Z_n$ is said to be *functionally complete* iff every function $f \in Z_n$ can be defined by means of functions in the set X.

In other words, the finite mapping $f : G_n^m \longrightarrow G_n$ can be represented as a composition of the operations $f_1, ..., f_m$. Then, we say that an algebra $\mathfrak{A}_n = (G_n, f_1, ..., f_m)$ is functionally complete. This reduces the problem of functional completeness of logical systems to that of the definability of all unary and binary connectives:

Fact 3.18. *Given $n \in \mathbb{N}$, in a n-valued logic any given place in the truth table can be occupied by n truth values. For a k-place connective, the truth table has room for entries*

$$\underbrace{n \times n \times ... \times n}_{k \text{ times}} = n^k$$

and for each there will be any of n possibilities, so that we can have

$$\underbrace{n \times n \times ... \times n}_{n^k \text{ times}} = n^{n^k}$$

possible k-place truth tables for a given n-valued logic.

We thus have it that the number of unary and binary connectives equals n^n and n^{n^2}, respectively (see Example 3.13). This explains why truth tables, though they are decision algorithms, find only limited application in logic, classical or non-classical. In effect, given the truth-values set W, we have 2^k for k atomic formulas, which means that a truth-table for classical logic has exponential growth. (Below, in Section 3.4, it should be clear why truth tables are essentially restricted to propositional languages.)

---

[9] Let $\mathfrak{A} = (U, O)$ and $\mathfrak{B} = (A, O)$ be two similar algebras. An *embedding* (also: *monomorphism* or *injection*) of $\mathfrak{A}$ into $\mathfrak{B}$ is a one-to-one homomorphism $hom\,(\mathfrak{A}, \mathfrak{B})$.

3.3. Compositionality and truth-functionality

$\to_{Ł3}$	t	i	f
t	t	i	f
i	t	t	i
f	t	t	t

$\to_{KW3}$	t	i	f
t	t	i	f
i	i	i	i
f	t	i	t

$\to_{Rn3}$	t	i	f
t	t	f	f
i	t	t	t
f	t	t	t

Figure 3.3.2.: Truth tables for the connective $\to$ in the 3-valued logics Ł$_3$, K$_3^W$, and Rn$_3$.

Example 3.19. Let $A, B \in F_{L0}$. The following are the truth tables for the classical $\heartsuit_i^k$ connectives with $k = 1, 2$:[10]

A	$\neg A$
t	f
f	t

A	B	$A \wedge B$	$A \vee B$	$A \to B$	$A \leftrightarrow B$
t	t	t	t	t	t
t	f	f	t	f	f
f	t	f	t	t	f
f	f	f	f	t	t

Definition 3.20. *Logical equivalence 1* – Two formulas $A, B \in$ L0 are said to be (logically) equivalent, denoted by $A \equiv_{(L0)} B$, iff both are valuated to the same truth value under every assignment of truth values, i.e., if their truth tables for the main logical connective are identical.

Example 3.21. We show the classical equivalence

$$(P \to Q) \equiv (\neg P \vee Q)$$

by comparing the truth table for the right side of this equivalence (cf. Def. 3.4) with the truth table of Example 3.3. As can be seen, the valuations for the main connectives are exactly the same in the four rows of both truth tables.

---

[10] By "the classical $\heartsuit_i^k$ connectives" we mean the connectives of the classical logical system CL whose foundations we are elaborating on. (See Part II for a fuller treatment of these connectives.) Figure 3.3.2 shows the truth table for the connective $\to$ in three different 3-valued logics, i.e. logics with $W_3 = \{\mathtt{f}, \mathtt{i}, \mathtt{t}\}$, where i denotes indeterminacy or even nonsense. These logics are Łukasiewicz's 3-valued logic (denoted by Ł$_3$), Kleene's 3-valued logic known as "weak" (K$_3^W$), and Piróg-Rzepecka's 3-valued logic (Rn$_3$). (Augusto (2020c) provides many further examples.) Compare with the table above for the classical connective $\to$.

3. Logical meaning

P	Q	¬P	∨	Q	≡	P	→	Q
1	1	0	1	1		1	1	1
1	0	0	0	0		1	0	0
0	1	1	1	1		0	1	1
0	0	1	1	0		0	1	0

In terms of functional completeness, it is obvious from Example 3.21 that the connective $\to$ can be defined by means of the subset $O' = \{\neg, \vee\}$, i.e. $P \to Q \equiv \neg P \vee Q$. Figure 2.2.1 provides a clue for the fact that $\leftrightarrow$ can be defined in terms of $\to$ and $\wedge$; in fact, we have $P \leftrightarrow Q \equiv (P \to Q) \wedge (Q \to P)$.

In Section 4.1 we elaborate further on this topic, namely from the viewpoint of classical logic.

▶ *Do Exercise 3.3.*

3.4. Classical interpretations and valuations

We next define some important semantical notions for a logical language L1 that follow from the exposition above. We begin by stating that every row of a truth table is an *interpretation*.

Example 3.22. In the truth table of Example 3.13, there are two interpretations that evaluate the propositional Boolean formula $\psi(x_1, x_2, x_3)$ to 1: when $val(x_1) = 0$, $val(x_2) = 0$, $val(x_3) = 1$ (7th row), and when $val(x_1) = 0$, $val(x_2) = 0$, $val(x_3) = 0$ (8th row). All the remaining six interpretations evaluate $\psi(x_1, x_2, x_3)$ to 0.

Importantly, the valuations $val_\mathcal{I}$ of propositional formulas are fixed for every interpretation $\mathcal{I}$ in the sense that they depend only on the immediate sub-formulas of a formula ϕ.

Proposition 3.23. *The following are the classical valuations for the formulas of* L0:

1. $val_\mathcal{I}(\top) = 1$ and $val_\mathcal{I}(\bot) = 0$

2. $val_\mathcal{I}(\neg \phi) = \begin{cases} 1 & \text{if } val_\mathcal{I}(\phi) = 0 \\ 0 & \text{otherwise} \end{cases}$

3. $val_\mathcal{I}(\phi \wedge \psi) = \begin{cases} 1 & \text{if } val_\mathcal{I}(\phi) = 1 \text{ and } val_\mathcal{I}(\psi) = 1 \\ 0 & \text{otherwise} \end{cases}$

3.4. Classical interpretations and valuations

4. $val_\mathcal{I}(\phi \vee \psi) = \begin{cases} 1 & \text{if } val_\mathcal{I}(\phi) = 1 \text{ or } val_\mathcal{I}(\psi) = 1 \\ 0 & \text{otherwise} \end{cases}$

5. $val_\mathcal{I}(\phi \rightarrow \psi) = \begin{cases} 1 & \text{if } val_\mathcal{I}(\phi) = 0 \text{ or } val_\mathcal{I}(\psi) = 1 \\ 0 & \text{if } val_\mathcal{I}(\phi) = 1 \text{ and } val_\mathcal{I}(\psi) = 0 \end{cases}$

6. $val_\mathcal{I}(\phi \leftrightarrow \psi) = \begin{cases} 1 & \text{if } val_\mathcal{I}(\phi) = val_\mathcal{I}(\psi) \\ 0 & \text{otherwise} \end{cases}$

Proof: $\top$ and $\bot$ are valuated as 1 and 0, respectively, by definition. For the other connectives, we build their truth tables. **QED**

We now need to generalize the notion of interpretation, as well as the concept of valuation, beyond truth tables, as these are not appropriate for FO formulas, but these, too, require interpretations and valuations in the formal sense. A series of definitions will help us to pave the way for this generalization.

Definition 3.24. A *frame* for an object language L1 with an alphabet Σ and truth-value set W is a pair $(\mathscr{D}, \Theta)$ where $\mathscr{D}$ is a non-empty *domain of discourse* and Θ is a *signature interpretation*, i.e., a mapping assigning the functions $\mathscr{D}^n \longrightarrow \mathscr{D}$ and $\mathscr{D}^n \longrightarrow W$ to each n-place function symbol and to each n-place predicate symbol of Σ_{L1}, respectively.

Definition 3.25. An *interpretation* $\mathcal{I}$ for L1 is a triple $(\mathscr{D}, \Theta, \varpi)$ where $(\mathscr{D}, \Theta)$ is a frame and ϖ is a *variable assignment* $\varpi : Vi \longrightarrow \mathscr{D}$. We say that $\mathcal{I}$ is based on the frame $(\mathscr{D}, \Theta)$.

Definition 3.26. For $\mathcal{I} = (\mathscr{D}, \Theta, \varpi)$, $\mathcal{I}$ is an interpretation for L1, and W, there is a corresponding *(e)valuation function* $val_\mathcal{I} : F_{L1}, \Sigma_{L1} \longrightarrow W$ defined inductively as follows:

1. $val_\mathcal{I}(x) = \varpi(x)$ for all x in Σ.

2. $val_\mathcal{I}(f(t_1, ..., t_n)) = \Theta(f)(val_\mathcal{I}(t_1), ..., val_\mathcal{I}(t_n))$ for all n-place function symbols f, $n \geq 0$, in Σ.

3. $val_\mathcal{I}(P(t_1, ..., t_n)) = \Theta(P)(val_\mathcal{I}(t_1), ..., val_\mathcal{I}(t_n))$ for all n-place predicate symbols P, $n \geq 0$, in Σ.

4. $val_\mathcal{I}(\heartsuit_i(\phi_1, ..., \phi_n)) = \widetilde{\heartsuit}_i(val_\mathcal{I}(\phi_1), ..., val_\mathcal{I}(\phi_n))$ for all logical connectives $\heartsuit_i$, $n \geq 0$, in Σ.

5. $val_\mathcal{I}(\blacklozenge_i x(\phi)) = \widetilde{\blacklozenge}_i(\Delta_{\mathcal{I},x}(\phi))$ for all quantifiers $\blacklozenge_i$ in Σ, where $\Delta_{\mathcal{I},x}(\phi)$ is the distribution of ϕ in $\mathcal{I}$ with respect to x, denoted by the set $\Delta_{\mathcal{I},x}(\phi) = \{val_{\mathcal{I}_a^x}(\phi) \,|\, a \in \mathscr{D}\}$, and $\mathcal{I}_a^x$ is the interpretation identical to $\mathcal{I}$ when setting $\varpi(x) = a$.

Despite the complexity above, there are in fact some fixed valuations for FO *atomic* formulas. The following proposition extends Proposition 3.23 in the sense that if ϕ, ψ are FO atomic formulas, i.e. formulas of the type $\blacklozenge_1 x_1, ..., \blacklozenge_n x_n(\phi)$, then their valuation follows the fixed interpretations of Proposition 3.23 once the valuations for the quantifiers have been carried out. These are as follows:

Proposition 3.27. *The following classical valuations for FO atomic formulas are fixed for any interpretation* $\mathcal{I} = (\mathscr{D}, \Theta, \varpi)$:

1. $val_\mathcal{I}(\forall x \phi) = \begin{cases} 1 & \text{if } val_{\mathcal{I}_a^x}(\phi) = 1 \text{ for all } a \in \mathscr{D} \\ 0 & \text{otherwise} \end{cases}$

2. $val_\mathcal{I}(\exists x \phi) = \begin{cases} 1 & \text{if } val_{\mathcal{I}_a^x}(\phi) = 1 \text{ for at least one } a \in \mathscr{D} \\ 0 & \text{otherwise} \end{cases}$

Proof: (Sketch) Recall Definition 3.2.2. Points 1 and 2 above agree with the distribution function for the two quantifiers $\forall, \exists$ given the truth-values set $W = \{0, 1\}$:

$$\widetilde{\forall}(\{1\}) = 1$$
$$\widetilde{\forall}(\{0, 1\}) = \widetilde{\forall}(\{0\}) = 0$$
$$\widetilde{\exists}(\{0, 1\}) = \widetilde{\exists}(\{1\}) = 1$$
$$\widetilde{\exists}(\{0\}) = 0$$

This, in turn, has as an algebraic basis points 2 and 3 of Definition 3.5, as we have the classical definition of the quantifiers given by

$$\widetilde{\forall} = f(\forall x \phi(x)) = min\,\{f(\phi(a)) \,|\, a \in \mathscr{D}\}$$

and

$$\widetilde{\exists} = f(\exists x \phi(x)) = max\,\{f(\phi(a)) \,|\, a \in \mathscr{D}\}.$$

QED

It should be remarked that by introducing the distribution function for classical valuations with the quantifiers of L1 we are actually making

3.4. Classical interpretations and valuations

of classical logic a special–the simplest non-trivial–case of many-valued logics, namely when $|W| = 2$.[11]

Example 3.28. Let there be given the FO formula

$$\chi = \forall x \forall y \exists z \, (L \, (f \, (x, y), z) \to L \, (x, y)).$$

We consider the interpretation $\mathcal{I} = (\mathcal{D}, \Theta, \varpi)$ such that $\mathcal{D} = \mathbb{Z}$, $\varpi(z) = 0 \in \mathbb{Z}$, $f(x, y) = x - y$, and $L(x, y) = \{(x, y) \in \mathbb{Z} \times \mathbb{Z} | x < y\}$. Then, we have the formula

$$\chi^\mu = \forall x \forall y \, ((x - y < 0) \to x < y)$$

and

$$val_\mathcal{I}(\chi) = 1$$

as we have

$$val_\mathcal{I}(\forall x, y \, (\chi)) = \widetilde{\forall} (\Delta_{\mathcal{I}, x, y}(\chi)) = 1 \text{ for } \Delta_{\mathcal{I}, x, y}(\chi) = \{val_{\mathcal{I}_a^{x,y}}(\chi) | a \in \mathbb{Z}\}$$

and

$$val_\mathcal{I}(\exists z \, (\chi)) = \widetilde{\exists} (\Delta_{\mathcal{I}, z}(\chi)) = 1 \text{ for } \Delta_{\mathcal{I}, z}(\chi) = \{val_{\mathcal{I}_0^z}(\chi) | 0 \in \mathbb{Z}\}.$$

In particular, we can rewrite χ as

$$\chi' = \underbrace{\forall x \forall y \exists z \, (L \, (f \, (x, y), z))}_{\phi} \to \underbrace{\forall x \forall y \, (L \, (x, y))}_{\psi}$$

and apply Propositions 3.27 and 3.23.5: if $val_{\mathcal{I}_a^{x,y}} (val_{\mathcal{I}_0^z}(\phi)) = 1$ and $val_{\mathcal{I}_a^{x,y}}(\psi) = 1$, then $val_\mathcal{I}(\chi') = 1$. In other words, given any two integers x and y, if $x - y < 0$, then $x < y$.

Note that it is impossible to build a truth table for χ in Example 3.28, as the domain $\mathcal{D} = \mathbb{Z}$ is infinite. Given a finite, small, domain, it may be possible to build a truth table for a FO formula, but in general other means are preferred to "decide" on its truth value.[12]

Remark 3.29. With respect to Definition 3.26, we have the following observations: (i) Free variables must be assigned a value in $\mathcal{D}$ and all free occurrences of a variable x must be assigned the same value in $\mathcal{D}$,

---

[11] See Augusto (2020c) for this important topic that we must skip here.
[12] A FO language is essentially undecidable, being at best semi-decidable. This holds for the standard classical FO language. (See Chapter 10.)

3. Logical meaning

and (ii) every constant must be assigned a value in $\mathscr{D}$ and all occurrences of the same constant a must be assigned the same value in $\mathscr{D}$.

We can now reformulate the notion of logical equivalence as follows:

Definition 3.30. *Logical equivalence 2* – Two formulas $A, B \in \text{L1}$ are said to be (logically) equivalent, denoted by $A \equiv_{(L1)} B$, iff both are valuated to the same truth value under every assignment of truth values, i.e., if their truth values are identical with respect to every interpretation $\mathcal{I}$.

In other words, two formulas are equivalent iff they have the same meaning. Propositions 3.31-2 below give important FO equivalences:

Proposition 3.31. *Quantifier duality* – *The following formulas are equivalent:*

1.
$$(QN_\exists) \qquad \neg \exists x\,(A) \equiv \forall x\,(\neg A)$$

2.
$$(QN_\forall) \qquad \neg \forall x\,(A) \equiv \exists x\,(\neg A)$$

Proof: Left as an exercise.

Proposition 3.32. *The following formulas are equivalent* when x does not appear as a free variable in B:

1.
$$\forall x\,(A\,(x)) \wedge B \equiv \forall x\,(A\,(x) \wedge B)$$

2.
$$\forall x\,(A\,(x)) \vee B \equiv \forall x\,(A\,(x) \vee B)$$

3.
$$\exists x\,(A\,(x)) \wedge B \equiv \exists x\,(A\,(x) \wedge B)$$

4.
$$\exists x\,(A\,(x)) \vee B \equiv \exists x\,(A\,(x) \vee B)$$

5.
$$\forall x\,(A\,(x)) \to B \equiv \exists x\,(A\,(x) \to B)$$

6.
$$\exists x \, (A(x)) \to B \equiv \forall x \, (A(x) \to B)$$

7.
$$B \to \forall x \, (A(x)) \equiv \forall x \, (B \to A(x))$$

8.
$$B \to \exists x \, (A(x)) \equiv \exists x \, (B \to A(x))$$

Proof: Left as an exercise.

Remark 3.33. With respect to Proposition 3.32, if x appears as a free variable in B, then rename the bound variable x in $\blacklozenge x \, (A(x))$ as y, obtaining the equivalent $\blacklozenge y \, (A\,[x/y]) = \blacklozenge y \, (A(y))$.

▶ *Do Exercises 3.4-13.*

3.5. Meaning and form

The principle of compositionality of meaning (cf. Prop. 3.12) shows that there is an intimate relation between the meaning of a formula $\phi = \phi(x_1, ..., x_n)$ and its form. Besides allowing us to evaluate ϕ, its truth table also provides us with a method to obtain its DNF and CNF.

As seen above, Theorem 2.44 guarantees the existence of a formula in DNF that is equivalent to a formula A. In order to find this formula, it suffices to compute a truth table for A. By considering the rows in which A is true, we obtain a formula in DNF equivalent to A.

Example 3.34. Let $A = (B \vee C) \wedge ((\neg B \wedge C) \vee D)$. The truth table for A is as follows:

B	C	D	A	$\neg A$
t	t	t	t	f
t	t	f	f	t
t	f	t	t	f
t	f	f	f	t
f	t	t	t	f
f	t	f	t	f
f	f	t	f	t
f	f	f	f	t

3. Logical meaning

Then, the DNF of A is

$$(B \wedge C \wedge D) \vee (B \wedge \neg C \wedge D) \vee (\neg B \wedge C \wedge D) \vee (\neg B \wedge C \wedge \neg D)$$

Likewise, by negatively considering the rows in which A is false, i.e., by considering the DNF for $\neg A$ and subsequently negating it, we obtain a formula in CNF equivalent to A. Thus, the DNF of $\neg A$ is

$$(B \wedge C \wedge \neg D) \vee (B \wedge \neg C \wedge \neg D) \vee (\neg B \wedge \neg C \wedge D) \vee (\neg B \wedge \neg C \wedge \neg D)$$

By Proposition 2.42, the negation of this DNF produces the CNF

$$(\neg B \vee \neg C \vee D) \wedge (\neg B \vee C \vee D) \wedge (B \vee C \vee \neg D) \wedge (B \vee C \vee D)$$

However, this method, too, may lead to an exponential growth of the atoms, reason why the Tseitin algorithm may be preferable.

▶ Do Exercise 3.14.

Exercises

Exercise 3.1. Show all the properties of a Boolean algebra $\mathfrak{B} = (A, ', \wedge, \vee, 0, 1)$ for all $x, y, z \in A$ and for the distinguished elements 0 and 1 (cf. Fig. 3.5.1).

Exercise 3.2. Evaluate the following Boolean formulas:

1. $((x_1 \vee x_3) \wedge \neg x_4) \wedge \neg x_2 \wedge (\neg x_1 \vee x_2)$ for $x_1 = 1$, $x_2 = 1$, $x_3 = 0$, $x_4 = 1$.

2. $(p_2 \vee \neg p_2 \vee p_3) \wedge p_1 \wedge (\neg p_1 \vee \neg p_3)$ for $p_1, p_2, p_3 = 1$.

3. $p \vee (q \wedge \neg r) \vee (\neg p \wedge \neg q \wedge r)$ for $p = 0$, $q = 1$, $r = 0$.

4. $(x_2 \wedge \neg x_1) \vee (x_1 \vee (\neg x_2 \wedge x_3))$ for $x_1, x_2, x_3 = 0$.

5. $x_2 \wedge x_1 \wedge (x_3 \vee \neg x_4) \wedge \neg x_2$ for $x_1 = 0$, $x_2 = 1$, $x_3 = 0$, $x_4 = 1$.

6. $(p \vee \neg q) \wedge (\neg p \vee \neg q) \wedge (p \vee q) \wedge (\neg p \vee q)$ for $p, q = 1$.

Exercise 3.3. Construct the truth tables for the formulas in Exercise 3.2.

3.5. Meaning and form

a.	$x \wedge y = y \wedge x$	Commutativity of $\wedge$
	$x \vee y = y \vee x$	Commutativity of $\vee$
b.	$x \wedge (y \wedge z) = (x \wedge y) \wedge z$	Associativity of $\wedge$
	$x \vee (y \vee z) = (x \vee y) \vee z$	Associativity of $\vee$
c.	$x \wedge (y \vee z) = (x \wedge y) \vee (x \wedge z)$	Distributivity of $\wedge$
	$x \vee (y \wedge z) = (x \vee y) \wedge (x \vee z)$	Distributivity of $\vee$
d.	$x \wedge (x \vee y) = x$	Absorption for $\wedge$
	$x \vee (x \wedge y) = x$	Absorption for $\vee$
e.	$0' = 1$	Complementation
	$1' = 0$	Complementation
f.	$(x')' = x$	Complementation
g.	$x \wedge x' = 0$	Complementation
	$x \vee x' = 1$	Complementation
h.	$x \wedge 1 = x$	Identity element of $\wedge$
	$x \vee 0 = x$	Identity element of $\vee$

Figure 3.5.1.: The properties of a Boolean algebra.

Exercise 3.4. Let there be given the formula $\phi = \exists x \, (S(g(x), a))$ and the interpretation $\mathcal{I} = (\mathcal{D}, \Theta, \varpi)$ such that $\mathcal{D} = \mathbb{Z}$, $a \in \mathcal{D} = 5$, $g(x) = x^2$, and $S(x, y) = \{(x, y) \in \mathbb{Z} \times \mathbb{Z} | x = y\}$.

1. Rewrite ϕ as ϕ^μ.

2. Evaluate ϕ in the given interpretation.

Exercise 3.5. Let there be given the formula $\phi = \exists x \forall y \, (Q(f(x), y))$ and the interpretation $\mathcal{I} = (\mathcal{D}, \Theta, \varpi)$ such that $\mathcal{D} = \mathbb{N}^+$, $\varpi(x) = 2$, $f(x) = x - 1$, and $Q(x, y) = \{(x, y) \in \mathbb{Z} \times \mathbb{Z} | x \leq y\}$.

1. Rewrite ϕ as ϕ^μ.

2. Evaluate ϕ in the given interpretation.

3. Let now ϕ be the formula $\phi = \forall y \exists x \, (Q(f(x), y))$. Evaluate ϕ in the given interpretation.

Exercise 3.6. Let there be given the interpretation $\mathcal{I} = (\mathcal{D}, \Theta, \varpi)$ such that $\mathcal{D} = \mathbb{R}$ and $P(x, y) = \{(x, y) | f(x) = y\}$ for $f(x) = x^2 - 1$. Evaluate the following formulas:

3. Logical meaning

1. $\forall x \forall y \, (P(f(x), y))$

2. $\forall x \exists y \, (P(f(x), y))$

3. $\forall y \exists x \, (P(f(x), y))$

4. $\exists x \forall y \, (P(f(x), y))$

5. $\exists x \exists y \, (P(f(x), y))$

Exercise 3.7. Let there be given the interpretation $\mathcal{I} = (\mathcal{D}, \Theta, \varpi)$ such that $\mathcal{D}$ = all the people, and $F(x, y) = \{(x, y) \,|\, x \text{ is the father of } y\}$. Evaluate the following formulas:

1. $\forall x \exists y \, (F(x, y))$

2. $\forall y \exists x \, (F(x, y))$

3. $\exists x \forall y \, (F(x, y))$

4. $\exists y \forall x \, (F(x, y))$

5. $\exists x \exists y \, (F(x, y))$

6. $\exists y \exists x \, (F(x, y))$

7. $\forall x \forall y \, (F(x, y))$

8. $\forall y \forall x \, (F(x, y))$

Exercise 3.8. With respect to the interpretation above for Exercise 3.7, let us assume that John is the father of Sara. We write the individual constants for John and Sara as *john* and *sara*, respectively. Evaluate the following formulas:

1. $F(john, sara)$

2. $\forall x \, (F(x, sara))$

3. $\exists x \, (F(x, sara))$

4. $\exists y \, (F(john, y))$

5. $\forall y \, (F(john, y))$

3.5. Meaning and form

Exercise 3.9. For each of the following formulas find an interpretation that makes it evaluate to **true**:

1. $\forall x\, (P(x) \to \neg Q(x))$
2. $\forall x \forall y\, (R(x,y) \lor R(y,x))$
3. $\forall x \forall y\, (\neg R(x,y) \lor R(y,x))$
4. $\exists x \exists y\, (R(x,y) \land R(y,x))$
5. $\forall x \exists y\, (P(x) \to R(x,y))$
6. $\forall x \forall y\, ((Q(x) \land \neg Q(y)) \to R(x,y))$

Exercise 3.10. Let there be given an interpretation $\mathcal{I}^\emptyset = (\emptyset, \Theta, \varpi)$.

1. Give examples of $\mathscr{D} = \emptyset$. (Hint: empty denotations.)
2. Given $\mathcal{I}^\emptyset$, then for any formula ϕ and any variable x, we have $val_{\mathcal{I}^\emptyset}(\exists x\,(\phi)) = \mathtt{f}$ and $val_{\mathcal{I}^\emptyset}(\forall x\,(\phi)) = \mathtt{t}$. Give the rationale for this.

Exercise 3.11. Complete the proof of Proposition 3.27.

Exercise 3.12. Prove Propositions 3.31-2.

Exercise 3.13. Recall from Definition 2.16.3 the notions of universal and existential closure. Let now $\phi = \phi(x_1, ..., x_n)$ be a formula with all $x_1, ..., x_n$ free variables. Then, the universal closure of ϕ preserves its validity, denoted by $\equiv_\top$, and the existential closure of ϕ preserves solely its (un)satisfiability, denoted by $\equiv_{sat}$, i.e.

$$\forall x_1, ..., \forall x_n\,(\phi) \equiv_\top \phi$$

and

$$\exists x_1, ..., \exists x_n\,(\phi) \equiv_{sat} \phi$$

Give the rationale for this.

Exercise 3.14. Obtain the DNFs and CNFs of the formulas in Exercise 3.2 by means of their truth tables.

4. Logical consequences

Above, we saw how formulas in a logical language L are well formed by means of formation rules (Chapter 2) and how they are given meaning (Chapter 3). This allows us to formalize assertions in natural language with a view to interpreting them from a logical viewpoint. This aims more often than not at the verification whether some reasoning instance is correct or valid, i.e. whether in an argument the conclusion does indeed follow from, or is entailed by, the premises. This, *derivation* or *entailment*, is a property that is specific to a *logical system*, which in turn requires a notion of *logical consequence*. This comes in two versions, syntactical and semantical, and when a logical system has both such they coincide–a common desideratum–it is characterized as adequate.

This is a Chapter wholly on the metatheory of the classical formalization of logic; in other words, on the metalogic of formal classical logic. Readers with more practical concerns (e.g., making proofs) can skip it, but not entirely, being advised to read at least the definitions of *proof, derivability, soundness, validity,* and *satisfiability*. Additionally, the concepts of *soundness* and *completeness* are also relevant. (Figure 4.1.2 in Section 4.1.4 gives a schematic summary of this Chapter.)

In this Chapter, we touch upon general aspects of logical consequence and its connection to the central notions of *inference* and *deduction*, as well as to *logics* and *logical theories*. In this context, it is interesting to consider the mathematical basis of logical consequence as a complete lattice, and we give this topic a very brief treatment; for an extensive elaboration on the central topic of logical consequence largely from an order-relation perspective, see Augusto (2017). Below, in Chapter 6, we elaborate on *classical consequence*, but the elaboration in it does not explicitly require this mathematical basis.

4.1. Logical consequence: A central notion

4.1.1. Consequences and systems logical, inferential, and deductive

Recall Definition 2.23; α is a construct of the object language L, but there are several other ways to represent α in the metalanguage μ^{L}, of

4. Logical consequences

which we introduce the following two to express formally the fact that $B \in F_L$ is a *logical consequence* of the set of premises $\{A_1, ..., A_n\} \subseteq F_L$:

$$(Cn) \qquad B \in Cn(\{A_1, ..., A_n\})$$

$$(\Vdash) \qquad \{A_1, ..., A_n\} \Vdash B$$

From the viewpoint of the notion of logical consequence, denoted above by Cn or $\Vdash$, a logical system is the theory of what formulas (more generally: assertions or sentences) *follow from* or *are entailed by* (i.e. *are consequences of*) which other formulas.

Definition 4.1. A *logical system* is a pair $L = (L, Cn)$, where L is a logical language and Cn is a *consequence operation*. Equivalently, $L = (L, \Vdash)$, where $\Vdash$ is a (syntactical or semantical) *consequence relation*.

Definition 4.2. Given a logical system $L = (L, Cn)$, for every $X \subseteq F_L$ the set $Cn_L(X)$ is the set of all the consequences of X in L. For a formula $\phi \in F_L$, $\phi \in Cn_L(X)$ denotes the fact that ϕ is a consequence of X in L, or, which is the same, that ϕ can be inferred from X in L.

We shall often omit the subscript in Cn, in particular when the logical system is left unspecified, or we know the logical system being considered.

Definition 4.3. For a logical system $L = (L, Cn)$, we define a *(logical) consequence operation* Cn as a mapping $Cn : 2^F \longrightarrow 2^F$ satisfying the following conditions for any $X, Y \subseteq F_L$:

(C1)	$X \subseteq Cn(X)$	*Inclusion*
(C2)	$Cn(Cn(X)) = Cn(X)$	*Idempotency*
(C3)	$Cn(X) \subseteq Cn(Y)$ whenever $X \subseteq Y$	*Monotonicity*

The following are basic properties of a consequence operation Cn:

Definition 4.4. Let Cn be a consequence operation on a set of formulas F_L. Let $X \subseteq F_L$ be given.

1. If for all substitutions σ it is the case that

$$\sigma Cn(X) \subseteq Cn(\sigma X)$$

   then Cn is said to be *structural*.

4.1. Logical consequence: A central notion

2. Cn is *finitary* if it is the case that

$$Cn(X) = \bigcup \{Cn(X') \mid X' \text{ is a finite subset of } X\}.$$

Otherwise, Cn is *infinitary*.

3. Cn is said to be *standard* if it is both finitary and structural.

4. The strongest consequence operation on L is the operation Cn such that $Cn(X) = F = \mathsf{L}$. This is called the *inconsistent* or *trivial consequence operation* on F.

5. The weakest consequence operation on L is the operation Cn defined by $Cn(X) = X$. This is called the *idle consequence operation* on F.

6. We say that a set $X \subseteq F_\mathsf{L}$ is *closed* under a consequence operation Cn if $X = Cn(X)$. If $X = Cn(X)$ or, equivalently, $X = Cn(Y)$ for some Y, then X is a *theory of Cn*, denoted by Θ_{Cn}.

7. Let $\mathscr{F}$ be a family of sets. For each consequence operation Cn, Θ_{Cn} is a *closure system*, i.e. for each $\mathscr{F} \subseteq \Theta_{Cn}$, $\bigcap \mathscr{F} \in \Theta_{Cn}$, with $Cn(\emptyset)$ and F as the least and greatest elements thereof. The former is called the *base theory* and the latter the *trivial theory* of Cn.

8. Let $\mathscr{F} = 2^F$. Then $\mathscr{F}$ is a *closure base* for a consequence operation Cn iff for each $X \subseteq F_\mathsf{L}$,

$$Cn(X) = \bigcap \{Y \in \mathscr{F} \mid X \subseteq Y\}.$$

Note that Θ_{Cn} is the greatest closure base for Cn.

9. If Θ_{Cn} is a closure system, then the pair $(\mathscr{F}, \subseteq)$ is a *complete lattice* such that for every $\mathscr{F} \subseteq \Theta_{Cn}$,

   a) $inf(\mathscr{F}) = \bigcap \mathscr{F}$,
   b) $sup(\mathscr{F}) = inf \{X \in \Theta_{Cn} \mid \bigcup \mathscr{F} \subseteq X\}$.

10. Two consequence operations can be ordered according to their *strength* (denoted by $\preccurlyeq$): we say that $Cn_1 \preccurlyeq Cn_2$ (Cn_2 is stronger than Cn_1) iff, for all $X \subseteq F_\mathsf{L}$, $Cn_1(X) \subseteq Cn_2(X)$. In fact, the following conditions are equivalent

    a) $Cn_1 \preccurlyeq Cn_2$

4. Logical consequences

b) $\Theta_{Cn_2} \subseteq \Theta_{Cn_1}$

c) $Cn_1(X) \subseteq Cn_2(X)$

The following theorem summarizes the above:

Theorem 4.5. *All consequences in a given logical language* L *form a complete lattice under* $\leq$.

Proof: (Idea) A *partially ordered set* (or *poset*) is a pair $\mathcal{R} = (A, \leq)$ where A is a non-empty set and $\leq$ is a binary relation on A such that for every $x, y, z \in A$ we have

$$x \leq x$$

$$(x \leq y) \wedge (y \leq x) \Rightarrow (x = y)$$
$$(x \leq y) \wedge (y \leq z) \Rightarrow (x \leq z)$$

properties known as *reflexivity*, *anti-symmetry*, and *transitivity*, respectively. $\mathcal{R}$ constitutes a *complete lattice* if every subset $B \subseteq A$ has a supremum and an infimum, denoted respectively by $sup(B)$ and $inf(B)$, where

$$sup(B) = x$$

denotes that x is an upper bound of B (i.e. $z \leq x$ for every $z \in B$) and $x \leq x'$ for any other upper bound x' of B,

$$inf(B) = y$$

denotes that y is a lower bound of B (i.e. $y \leq z$ for every $z \in B$) and $y' \leq y$ for any other lower bound y' of B. Let now for any $x, y \in A$, $sup(x, y) := x \cup y$ and $inf(x, y) := x \cap y$, called the *union* and *intersection* of x and y, respectively. Let further the pair $\mathcal{S} = (2^A, \subseteq)$ be given, where 2^A denotes the power set of A, and $\subseteq$ denotes *set inclusion*. Then, $\subseteq$ is a partial ordering, and the structure $\mathcal{L} = (\mathcal{S}, \cup, \cap)$ is a complete (distributive) lattice with $\bigcap A = \emptyset$ and $\bigcup A = \{a, b, c\}$, where $\bigcap A$ and $\bigcup A$ denote the intersection and the union, respectively, of the elements of A (see Fig. 4.1.1). Note now that $\mathcal{S} = (2^A, \subseteq)$ is a Boolean lattice and recall that every Boolean lattice is a Boolean algebra (and vice-versa). Consider then the operations $\wedge$ and $\vee$ in a Boolean algebra $\mathfrak{B} = (A,', \wedge, \vee, 0, 1)$ such that $\bigwedge A = 0$ and $\bigvee A = 1$, where $\bigwedge A$ and $\bigvee A$ denote the meet and the join, respectively, of the elements of A. Consider now $Cn(\emptyset)$ and $Cn(A) = A \subseteq F_L$, as well as $Cn(A')$ for all $A' \subset A$, from the viewpoints of both logical consequence and a Boolean

4.1. Logical consequence: A central notion

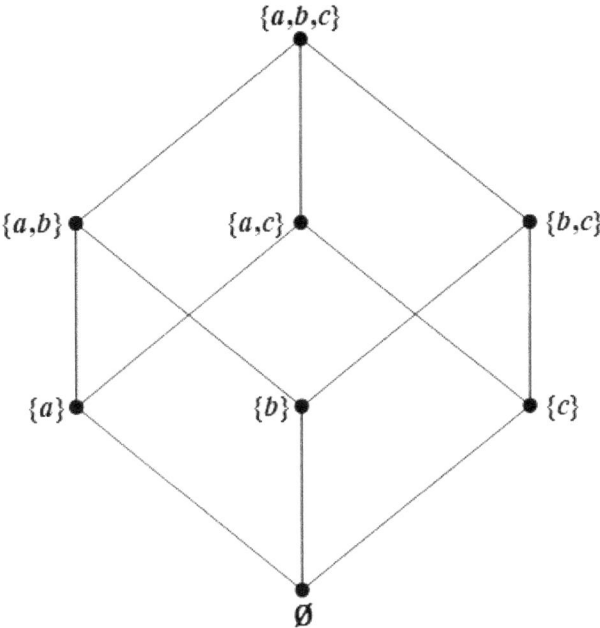

Figure 4.1.1.: The complete lattice $\mathcal{S} = (2^A, \subseteq)$ for $A = \{a, b, c\}$.

lattice. (Hint: Definitions 4.4.9-10. In particular, let the binary relation $\preceq$ satisfy exactly the same properties as $\leq$.) **QED**

With the consequence operation Cn in a logical system L there is naturally associated a consequence relation $\Vdash$:

Proposition 4.6. *The consequence operation Cn induces a* consequence relation

$$(\Vdash_{Cn}) \quad (X, \phi) \in \Vdash \quad \text{iff} \quad \phi \in Cn(X)$$

and in turn $\Vdash$ induces the consequence operation

$$(Cn_{\Vdash}) \quad Cn(X) = \{\phi \in F_{\mathsf{L}} | (X, \phi) \in \Vdash\}.$$

Proof: Left as an exercise.

Definition 4.7. For a logical system $\mathsf{L} = (\mathsf{L}, \Vdash)$, we define a *(logical) consequence relation* $\Vdash$ as a relation $\Vdash \subseteq 2^F \times F$ satisfying the following conditions for any sets $X, Y \subseteq F_{\mathsf{L}}$ and for arbitrary formulas $\phi, \chi \in F_{\mathsf{L}}$:

4. Logical consequences

(R1) If $\phi \in X$, then $(X, \phi) \in \Vdash$
(R2) If $(X, \phi) \in \Vdash$ and $X \subseteq Y$, then $(Y, \phi) \in \Vdash$
(R3) If $(X, \phi) \in \Vdash$ and $(Y, \chi) \in \Vdash$ for every $\chi \in X$, then $(Y, \phi) \in \Vdash$

R1-3 can be reformulated so that they more clearly convey three important properties of a logical system, to wit, reflexivity (R), monotonicity (M), and transitivity (T).

Proposition 4.8. *We rewrite R1-R3 as follows:[1]*

(R) $\phi \Vdash \phi$
(M) If $X \Vdash \phi$, then $X, Y \Vdash \phi$
(T) If $X \Vdash \phi$ and $X, \phi \Vdash \psi$, then $X \Vdash \psi$

Proof: Left as an exercise.

Reflexivity is the case when $X = \{\phi\}$, and it expresses the fact that every formula is a logical consequence of itself. *Monotonicity* is an important property of some logical systems, prominently so of classical logic, and it conveys the fact that the addition of formulas to a theory does not change its set of consequences. *Transitivity* expresses the fact that if ϕ is a lemma in the proof of a theorem, then we are allowed to "cut" ϕ, i.e. substitute it by its proof, reason why transitivity is also spoken of as *cut*.[2]

Given a logical consequence operation or relation, we can also, or alternatively, speak of *logical inference* and of an *inference system*.

Definition 4.9. Let L be a logical language. An *inference* is a couple (X, ψ) such that $X \subseteq F \subseteq L$ and $\psi \in F$. An alternative notation is $\psi \in Cn(X)$ or $X \Vdash \psi$. In effect,

---

[1] Note the following abbreviation: we write $X, \phi \Vdash \psi$ for $X \cup \{\phi\} \Vdash \{\psi\}$.
[2] In particular, R3 expresses the fact that if $X \Vdash \phi$ and $Y \Vdash \chi$ for every $\chi \in X$, then χ can be used (as a lemma) to obtain ϕ from Y. Other, not necessarily equivalent, formalizations of T/R3 are

(T')   If $X \Vdash \phi$ and $\phi \Vdash \psi$, then $X \Vdash \psi$
(T'')   If $X \Vdash \phi$ and $Y, \phi \Vdash \psi$, then $X \cup Y \Vdash \psi$
(T''') If $X \cup \{\phi_1, ..., \phi_n\} \Vdash \psi$ and $X \Vdash \phi_i$ for $i = 1, ..., n$, then $X \Vdash \psi$

1. a consequence operation Cn (relation $\Vdash$) on F_L is called an *inference operation* (*inference relation*) on L if it satisfies C1 and C2 (R1 and R2, respectively) for every $X \subseteq F \subseteq L$ and $\psi \in F$.

2. an *inference system* on L is a pair (L, Cn) (a pair $(L, \Vdash)$) where Cn ($\Vdash$, respectively) is an inference operation (relation) on L.

Clearly, a logical system is an inference system.[3]

Definition 4.10. A *deductive system* is a system of Cn. In other words, a deductive system of $Cn(X)$ is the least theory of Cn containing X.

As a matter of fact, any set of sentences X of a logical language (a theory) that contains all its consequences (i.e. $X \supseteq Cn(X)$) can be seen as a deductive system (Tarski, 1930). In particular:

Definition 4.11. A logical system whose consequence relation (operation) satisfies R1-3 (C1-3, respectively) and in addition satisfies the following conditions for finite $X' \subseteq X \subseteq F_L$ and $\psi \in F_L$,

$$\left(\Vdash'\right) \qquad \text{If } X \Vdash \psi, \text{ then } X' \Vdash \psi$$

and

$$(Cn') \qquad Cn(X) \subseteq \bigcup \{Cn(X') \mid X' \subseteq X\}$$

is a deductive system.

Conditions $\Vdash'$ and Cn' define the property of *compactness*.

▶ *Do Exercises 4.1-7.*

4.1.2. Syntactical consequence and proof theory

In order to fully understand the notion of *syntactical consequence* we require such fundamental notions as *proof* and *proof system*. These are objects of the subfield of logic known as *proof theory*. The most central notion of proof theory is that of *inference rule*. We provide the most important aspects of these notions in the definitions that follow. We focus on the syntactical consequence relation, but all the statements below can be reformulated equivalently for the syntactical consequence operation.

---

[3] We are here taking *inference* in a broad sense that allows for either a syntactical or a semantical characterization. Nevertheless, the expression *rule of inference* (or *inference rule*) should be taken strictly in the syntactical sense (cf. Section 4.1.2 below).

4. Logical consequences

Definition 4.12. Given an inference system L= (L, Cn) that is also a logical system, an *inference rule* **r** on a set of formulas $F \subseteq L$ is a mapping assigning to some finite sequence $\chi_1, ..., \chi_n \in X$, $n \geq 0$, of formulas (the *premises*) a formula ψ (the *conclusion*), i.e. $\mathbf{r} : X \longrightarrow F$ where $X \subseteq F^m$ for some $m = 1, 2, ...$ We write $\mathbf{r}(\chi_1, ..., \chi_n) = \psi$, $\mathbf{r}(\{\chi_1, ..., \chi_n\}, \psi)$, or more commonly,

$$(\mathbf{r}) \quad \frac{\chi_1, ..., \chi_n}{\psi}$$

or

$$(\mathbf{r}) \quad \chi_1, ..., \chi_n / \psi$$

We denote by RI the set of inference rules $\{\mathbf{r}_1, ..., \mathbf{r}_n\}$.

1. Given a substitution σ, **r** is called a *structural inference rule* of L if it has the form

$$(\mathbf{r}) \quad \sigma\chi_1, ..., \sigma\chi_n / \sigma\psi$$

2. A rule **r** is said to *preserve* a set of formulas X and X is said to be *closed under* **r** iff for all $X' \subseteq X$ and for all formulas ψ, if $\mathbf{r}(X', \psi)$, then $\psi \in X$.

3. A subset $F_0 \subseteq F$ is said to be *closed under a rule of inference* **r** provided that $(\chi_1, ..., \chi_n) \in F_0^m \cap X$ implies that $\mathbf{r}(\chi_1, ..., \chi_n) \in F_0$.

Definition 4.13. Given an inference of the form X/ψ,

1. we say that ψ is an *axiom* when $X = \emptyset$ in (the application of) an inference rule $\mathbf{r}(X, \psi)$.

2. X can comprise a set AX of *axiom schemata*, formulas in the set $F^\mu \subseteq \mu^L$ that can be replaced by formulas of L.

Definition 4.14. A *proof system* (or *proof calculus*) $\mathcal{P}$ for a logical language L is a pair (RI, AX) where either RI or AX–but not both–can be empty, i.e. a system of rules of inference and/or axiom schemata. The triple (L, RI, AX) is an instance of a *formal system*, namely an *inference system*.[4]

---
[4] Insofar as the pair (RI, AX) gives rise to a consequence relation (see Def. 4.20 below).

4.1. Logical consequence: A central notion

Definition 4.15. Let $\mathcal{P} = (\mathsf{L}, RI, AX)$ be a proof system. A *proof* of ψ in $\mathcal{P}$ is a finite collection of rules of inference and/or axioms of $\mathcal{P}$ that leads to concluding that ψ is a member of F_L.

We make Definition 4.15 formally more precise by means of the notion of *provability* or *derivability*.

Definition 4.16. A formula ψ is *provable, or derivable, from* a (possibly empty) set of formulas X by means of axioms in the set AX and/or rules in the set RI iff there is a finite sequence of formulas $\psi_1, ..., \psi_n$ that is a *proof* or a *derivation of* ψ *from* X, i.e. there is a finite sequence $\psi_1, ..., \psi_n$ such that

1. $\psi_1 \in (X \cup AX \cup RI)$;

2. for every $1 < i \leq n$, either $\psi_i \in (X \cup AX \cup RI)$ or ψ_i is the conclusion of one of the rules of inference $\mathbf{r}_j$, $j = 1, ..., k$ of which the premises are some of the $\psi_1, ..., \psi_{i-1}$;

3. $\psi_n = \psi$.

Definition 4.17. Let ψ be a conclusion in a proof (a derivation). When proven (derived) by means of a rule of inference $\mathbf{r}$ from $X = \emptyset$, then ψ is a *theorem*. If proven (derived) by means of a rule of inference $\mathbf{r}$ from the axioms of a theory, then ψ is called a *theorem* of the theory. In any case, *theorems* are always provable (derivable) formulas.

Recall now Definition 4.13. Obviously, every axiom is a theorem, i.e. is derivable or provable (from $\emptyset$).

Definition 4.18. Two inference rules are particularly important:

1. *Modus ponens*:
$$(\text{MP}) \quad \frac{\phi, \phi \rightarrow \psi}{\psi}$$

2. The *substitution rule*
$$(\text{SUB}) \quad \frac{\phi}{\sigma\phi}$$
which states that if a formula ϕ is a theorem, then any of its substitution instances (i.e., extensionally equivalent formulas) is also a theorem.

4. Logical consequences

Given the definitions above, it is now an easy matter to define a syntactical consequence relation. We first provide a very general definition (cf. Def. 4.19) and then specify it in terms of a binary and a ternary relation.

Definition 4.19. For a logical system L, a *syntactical consequence relation*, denoted by $\vdash_L$, specifies what conclusions are *derivable* or *provable* in L. Let us denote a proof by the symbol ■. Let X be a (possibly empty) set of formulas and ψ a formula. We say that ψ is *derivable* from (i.e. is a conclusion of) the set of premises (or assumptions) X iff there is a *proof* ■ such that we have the relation $X \vdash_{\blacksquare} \psi$. Otherwise, we write $X \nvdash_{\blacksquare} \psi$ and call ■ a *counter-proof* or *refutation*.

We often abbreviate $X \vdash_{\blacksquare} \psi$ as $X \vdash \psi$, especially when ϕ is an axiom, i.e. when we have $\vdash \phi$. We likewise abbreviate $X \nvdash_{\blacksquare} \psi$ as $X \nvdash \psi$.

Definition 4.20. Let a proof system $\mathcal{P}$ be given for L, and let $F \subseteq$ L. Then, we can define the *syntactical consequence relation* as the binary relation $\vdash \subseteq \mathcal{P} \times$ L such that for an arbitrary formula $\psi \in F$ and a (possibly empty) set of formulas $X \subseteq F$ we have

$$(\vdash) \qquad X \vdash \psi \quad \text{iff} \quad X \vdash_{\mathcal{P}} \psi.$$

This is more precisely the *syntactical consequence relation induced by* $\mathcal{P}$ as, given $\mathcal{P}$, there is some proof ■ and a ternary relation $\vdash \subseteq \mathcal{P} \times \blacksquare \times$ L such that we have

$$X \vdash_{\mathcal{P}} \psi \quad \text{iff} \quad X \vdash_{\mathcal{P}, \blacksquare} \psi.$$

Unless otherwise stated, we shall take the syntactical consequence relation as coinciding with derivability or provability. Clearly, a rule of inference is an instance of the syntactical consequence relation (cf. Definition 4.12), and we can now write

$$(\mathbf{r}) \qquad \{\chi_1, ..., \chi_n\} \vdash \psi.$$

actually abbreviating $\chi_1 \wedge ... \wedge \chi_n \vdash \psi$. We can also write MP and SUB as

$$(\text{MP}) \qquad \{\phi, \phi \to \psi\} \vdash \psi$$

and

$$(\text{SUB}) \qquad \phi \vdash \sigma\phi.$$

Obviously, α in Definition 2.23 can be rewritten as

$$(\alpha) \qquad \{A_1, ..., A_n\} \vdash B$$

4.1. Logical consequence: A central notion

This reformulation should give the reader a clearer notion of argument form and arguments whose conclusions follow logically from the premises by virtue of form alone (cf. Figure 2.3.1).

Consistency is another property of relevance in proof theory. We first define it and then show its importance for the notion of proof-theoretical deduction.

Definition 4.21. We say that a set $X = \{\chi_1, ..., \chi_n\}$ (a theory Θ) is *inconsistent* iff we can derive from it both a formula ψ and its negation, i.e.

$$X \vdash (\psi \wedge \neg \psi).$$

Otherwise, the set X (the theory Θ) is said to be *consistent*.[5]

Important results concerning (in)consistency are expressed in the following two theorems, whose proofs are left as exercises.

Theorem 4.22. *If X is inconsistent, then $X \vdash \psi$ for any formula ψ.*

From Theorem 4.22, a theorem of *reductio ad absurdum* (RA) follows:

Theorem 4.23. *If $X \cup \{\neg \psi\}$ is inconsistent, then $X \vdash \psi$.*

The following theorems, dependent on Theorems 4.22-3, are central in proof theory, in particular in the proof theory of classical logic.

Theorem 4.24. *($\vdash$-Deduction theorem 1) $X \vdash \psi$ iff $X \cup \{\neg \psi\}$ is inconsistent.*

Theorem 4.25. *($\vdash$-Deduction theorem 2) Given a set of formulas $X = \{\chi_1, ..., \chi_n\}$ and a formula ψ, ψ is a syntactical consequence of X iff the formula $((\chi_1 \wedge ... \wedge \chi_n) \to \psi)$ is a theorem, i.e. iff*

$$\text{if } (\chi_1 \wedge ... \wedge \chi_n) \vdash \psi, \text{ then } \vdash (\chi_1 \wedge ... \wedge \chi_n) \to \psi.$$

Example 4.26. Recall Definition 2.23. Theorem 4.25 allows us to rewrite any argument α in the form

$$(\alpha') \quad (A_1 \wedge ... \wedge A_n) \to B$$

so that we can test whether an argument α is a theorem by applying some proof method (cf. Parts IV and V) to

---

[5] For simplicity, we make $\Theta = X$; below, in Section 4.2, we specify formally the notion of a (logical) theory Θ. Note that X can be a singleton, in which case we speak of a formula χ being (in)consistent.

$$\vdash (A_1 \wedge ... \wedge A_n) \to B.$$

Proposition 4.27. *In fact, Theorem 4.25 generalizes in the following way:*

$$\text{if } (\chi_1 \wedge ... \wedge \chi_n) \vdash \psi, \text{ then } \vdash (\chi_1 \wedge ... \wedge \chi_n) \to \psi,$$

$$\text{then } (\chi_1 \wedge ... \wedge \chi_{n-1}) \vdash (\chi_n \to \psi),$$

$$\text{then } (\chi_1 \wedge ... \wedge \chi_{n-2}) \vdash (\chi_{n-1} \to (\chi_n \to \psi)),$$

$$\vdots$$

$$\text{then } \vdash \chi_1 \to (... \to (\chi_{n-2} \to (\chi_{n-1} \to (\chi_n \to \psi)))).$$

Theorem 4.28. *A formula ψ is a logical consequence of a set of formulas $X = \{\chi_1, ..., \chi_n\}$ iff the formula $(\chi_1 \wedge ... \wedge \chi_n \wedge \neg\psi)$ is inconsistent.*

The proofs of the above theorems are left as exercises.

Theorems are of fundamental importance for the definition, given a logical system L, of the logic **L**:

Definition 4.29. *Given a logical system $\mathsf{L} = (\mathsf{L}, \vdash)$, where L is a logical language and $\vdash$ is a syntactical consequence relation, the* logic *of L, denoted by* **L**, *is defined as*

$$\mathbf{L} := \{\phi \in \mathsf{L} | \vdash_\mathsf{L} \phi\}$$

where $\vdash \phi$ abbreviates $\emptyset \vdash \phi$.

In other words, the logic **L** is the set of theorems generated by the logical system L. However, we shall often relax this definition and consider a logic **L** to be a pair $(F, \vdash_\mathsf{L})$ where $F \subseteq \mathsf{L}$ is a set of formulas of a logical language L and $\vdash_\mathsf{L}$ denotes the syntactical consequence relation specified by the logical system L.[6]

▶ *Do Exercises 4.8-12.*

---

[6]This relaxation is not without grounds; see Augusto (2017), p. 64-5, for some critical remarks.

4.1. Logical consequence: A central notion

4.1.3. Semantical consequence and model theory

For a logical system L, the *semantical consequence relation*, denoted by $\models_L$, specifies the class of *valid inferences* in L, or, in other terms, which inferences in L *preserve truth*, or are *deductively valid*. We next provide definitions and theorems that specify the consequence relation from the semantical viewpoint. As above, we concentrate on the semantical consequence relation; equivalent formulations can be given for the semantical consequence operation.

The fundamental concepts of semantics *validity* and *satisfiability* depend on the notions of *valuation* and *interpretation* (cf. Def.s 3.1 and 3.25). These, in turn, are naturally associated to the most central concept of *model theory*, to wit, *model*. In very general terms, *a model for a logical language* L is any system that gives meaning to (i.e. valuates or interprets) the statements of that language; in a stricter sense, *a model for a sentence* $\phi \in F_L$ (*a theory* $\Theta \subseteq F_L$) is an interpretation $\mathcal{I}$ such that $val_\mathcal{I}(\phi) = \mathbf{t}$ ($val_\mathcal{I}(\Theta) = \mathbf{t}$, respectively). To avoid ambiguity, we shall reserve the term *model* for this latter usage.

In what follows, let $X \subseteq F_L$ and $\psi \in F_L$, L is a logical language.

Definition 4.30. *Satisfiability* – An interpretation $\mathcal{I}$ is said to *satisfy*

1. a formula ψ, and ψ is said to be *satisfiable*, iff there is a valuation $val_\mathcal{I} : F_L \longrightarrow W$ such that $val_\mathcal{I}(\psi) = \mathbf{t}$ and we denote this satisfiability relation by $\models_\mathcal{I} \psi$. Otherwise, we write $\not\models_\mathcal{I} \psi$. If there is no interpretation $\mathcal{I}$ that satisfies ψ, then ψ is *unsatisfiable*, and we write $\not\models \psi$.

2. a set of formulas (a theory) $X = \{\chi_1, ..., \chi_n\}$, and X is said to be *satisfiable*, iff there is a valuation $val_\mathcal{I} : F_L \longrightarrow W$ such that $val_\mathcal{I}(\chi_i) = \mathbf{t}$ for all $\chi_i \in X$, and we denote this satisfiability relation by $\models_\mathcal{I} X$.[7] Otherwise, we write $\not\models_\mathcal{I} X$. If there is no interpretation $\mathcal{I}$ that satisfies X, then X is *unsatisfiable*, and we write $\not\models X$.

Definition 4.31. We say that an interpretation $\mathcal{I}$ is a *model* of

1. ψ, and write $\models_\mathcal{M} \psi$, iff $\models_\mathcal{I} \psi$; otherwise, we write $\not\models_\mathcal{I} \psi$ to denote that there is a *counter-model*.

---

[7] This implies that X can be viewed as a conjunction of the formulas χ_i, i.e., $X = \bigwedge_{i=1}^{n} \chi_i$.

4. Logical consequences

2. X, and write $\models_\mathcal{M} X$, iff $\models_\mathcal{I} X$; otherwise, we write $\not\models_\mathcal{I} X$ to denote that there is a *counter-model*.

Thus, a formula ψ is said to be satisfiable iff there is an interpretation that is a model of ψ, and a set of formulas (a theory) X is satisfiable iff all its elements (formulas, respectively) have a common model. A counter-model is the case when $val_\mathcal{I}(\psi) = \mathtt{f}$, or $val_\mathcal{I}(\chi_i) = \mathtt{f}$ for at least one $\chi_i \in X$.

Example 4.32. In Example 3.21, the interpretation in which we have $val_\mathcal{I}(P) = \mathtt{f}$ and $val_\mathcal{I}(Q) = \mathtt{t}$ (i.e. the third row of the truth table) is a model for the formula $\neg P \vee Q$; the interpretation in which $val_\mathcal{I}(P) = \mathtt{t}$ and $val_\mathcal{I}(Q) = \mathtt{f}$ (i.e. the second row of the truth table) is a counter-model for the formula $\neg P \vee Q$.

Definition 4.33. A *semantics* $\mathfrak{S}$ for a logical language L is an infinite set of (classes of) models.

Definition 4.34. *Validity* – Let X be a (possibly empty) set of formulas and ψ a formula entailed from X. We say that ψ is *valid* iff there is no interpretation assigning the value $\mathtt{t}$ to all the members of X and $\mathtt{f}$ to ψ, and we write $X \models \psi$ (abbreviating $X \models_\mathfrak{S} \psi$, or $X \models_{\mathcal{M}_i} \psi$ for all $\mathcal{M}_i \subseteq \mathfrak{S}$). A formula is said to be *invalid*, written $X \not\models \psi$, iff it is not valid.

Given these definitions, it is now an easy matter to define a semantical consequence relation as a binary or ternary relation.

Definition 4.35. Let $\mathfrak{S} = \{\mathcal{M}_1, \mathcal{M}_2, ...\}$ where the $\mathcal{M}_i$ are models. Then, we can define the *semantical consequence relation* as the binary relation $\models \subseteq \mathfrak{S} \times \mathsf{L}$ such that for an arbitrary formula $\psi \in F_\mathsf{L}$ and a (possibly empty) set of formulas $X \subseteq F_\mathsf{L}$ we have, for $\mathfrak{S}$,

$$(\models) \qquad X \models \psi \quad \text{iff} \quad X \models_\mathfrak{S} \psi.$$

This is more precisely the *semantical consequence relation induced by the semantics* $\mathfrak{S}$ as, given a model $\mathcal{M}_i \in \mathfrak{S}$, we have a ternary relation $\models \subseteq \mathfrak{S} \times \mathcal{M}_i \times \mathsf{L}$ such that

$$\models_\mathfrak{S} X \quad \text{iff} \quad \models_{\mathfrak{S}, \mathcal{M}_i} \psi.$$

Obviously, Definition 4.35 means that ψ is a semantical consequence of X iff every semantics (model) that satisfies X also satisfies ψ. Unless otherwise stated, we shall consider the semantical consequence relation as coinciding with the satisfiability relation.

4.1. Logical consequence: A central notion

Definition 4.36. A formula ϕ is said to be

1. a *tautology* iff every interpretation $\mathcal{I}$ of ϕ is also a model of ϕ. In other words, ϕ is a tautology if it uniformly takes the truth value $\mathbf{t}$ for any and every assignment of truth values to its atoms. We consequently have the set $Taut\,(\mathsf{L})$ of all the tautologies in a logical system L with a language L:

$$Taut\,(\mathsf{L}) = \{\phi \in F_\mathsf{L} | val\,(\phi) = \mathbf{t} \text{ for every } val : F_\mathsf{L} \longrightarrow W_n\}$$

2. a *contradiction* iff there is no $\mathcal{I}$ of ϕ that is a model of ϕ. In other words, ϕ is said to be a contradiction if it uniformly takes the truth value $\mathbf{f}$ for any and every assignment of truth values to its atoms, i.e. for the same system we have

$$Cont\,(\mathsf{L}) = \{\phi \in F_\mathsf{L} | val\,(\phi) = \mathbf{f} \text{ for every } val : F_\mathsf{L} \longrightarrow W_n\}$$

3. *contingent* iff it is neither a tautology nor a contradiction.

We shall often denote an arbitrary tautology by $\top$ and an arbitrary contradiction by $\bot$.

It is now obvious:

Proposition 4.37. *It is now obvious that a formula ϕ is valid iff*

1. *every interpretation of ϕ is also a model of ϕ, and*

2. *ϕ is a tautology.*

Proof: Left as an exercise.

The valid formulas of a logical system L are fundamental to the definition of the logic **L**:

Definition 4.38. Given a logical system $L = (\mathsf{L}, \models)$, where L is a logical language and $\models$ is a semantical consequence relation, the *logic of* L, denoted by **L**, is defined as

$$\mathbf{L} := \{\phi \in \mathsf{L} | \models_L \phi\}$$

where $\models \phi$ abbreviates $\emptyset \models \phi$.

4. Logical consequences

In other words, **L** is the set of tautologies generated by L. However, just as in the case of the syntactical consequence relation, we shall often relax this definition and consider a logic **L** to be a pair $(F, \models_L)$ where $F \subseteq \mathsf{L}$ is a set of formulas of a logical language L and $\models_L$ denotes the semantical consequence relation specified in the logical system L.[8]

It is also obvious:

Proposition 4.39. *A formula ϕ is*

1. *valid iff its negation ($\neg\phi$) is unsatisfiable.*

2. *unsatisfiable iff its negation is valid.*

3. *invalid iff there is at least one interpretation that falsifies it.*

4. *satisfiable iff there is at least one interpretation that makes it true.*

Proof: Left as an exercise.

Proposition 4.40. *If a formula is valid, then it is satisfiable (but not vice-versa). If a formula is unsatisfiable, then it is invalid (but not vice-versa).*

Proof: Left as an exercise.

Theorem 4.41. ($\models$-*Deduction theorem 1*) $X \models \psi$ *iff* $X \cup \{\neg\psi\}$ *is unsatisfiable.*

Proof: ($\Rightarrow$) For any interpretation $\mathcal{I}$, either $\mathcal{I}(\chi_i) = \mathsf{t}$ for all $\chi_i \in X$ and $\mathcal{I}(\psi) = \mathsf{t}$ (hence, $\mathcal{I}(\neg\psi) = \mathsf{f}$), or $\mathcal{I}(\chi_i) = \mathsf{f}$ for some $\chi_i \in X$. Either way, $\mathcal{I}(X \cup \{\neg\psi\}) = \mathsf{f}$.[9]

($\Leftarrow$) For any interpretation $\mathcal{I}$, either $\mathcal{I}(\chi_i) = \mathsf{t}$ for all $\chi_i \in X$ and $\mathcal{I}(\neg\psi) = \mathsf{f}$ (hence, $\mathcal{I}(\psi) = \mathsf{t}$), or $\mathcal{I}(\chi_i) = \mathsf{f}$ for some $\chi_i \in X$. Therefore, $\mathcal{I}(\psi) = \mathsf{t}$ whenever $\mathcal{I}(X) = \mathsf{t}$, and thus $X \models \psi$. **QED**

It will be useful to provide an equivalent formulation of the deduction theorem:

---

[8] See Augusto (2017), p. 64-5, for some critical remarks on this definition.
[9] $\mathcal{I}(\phi)$ abbreviates $val_\mathcal{I}(\phi)$.

4.1. Logical consequence: A central notion

Theorem 4.42. ($\models$-*Deduction theorem 2*) *Given a set of formulas* $X = \{\chi_1, ..., \chi_n\}$ *and a formula* ψ, ψ *is a logical consequence of* X *iff the formula* $((\chi_1 \wedge ... \wedge \chi_n) \to \psi)$ *is valid, i.e. iff*

$$\text{if } (\chi_1 \wedge ... \wedge \chi_n) \models \psi, \text{ then } \models (\chi_1 \wedge ... \wedge \chi_n) \to \psi.$$

Proof: Left as an exercise.

Example 4.43. Recall Definition 2.23. Theorem 4.42 allows us to rewrite any argument α in the form

$$(\alpha') \qquad (A_1 \wedge ... \wedge A_n) \to B$$

so that we can test whether an argument α is valid by applying some validity testing method (cf. Part IV) to

$$\models (A_1 \wedge ... \wedge A_n) \to B.$$

The above allows a reformulation of the definition of *logical equivalence*:

Definition 4.44. We say that two formulas ϕ and ψ are *equivalent*, and we write $\phi \equiv \psi$, iff

1. they have exactly the same truth value in every interpretation.
2. $\phi \models \psi$ and $\psi \models \phi$.
3. $\phi \leftrightarrow \psi$ is a tautology, i.e. $\models (\phi \leftrightarrow \psi)$.

Theorem 4.45. *A formula ψ is a logical consequence of a set of formulas* $X = \{\chi_1, ..., \chi_n\}$ *iff the formula* $(\chi_1 \wedge ... \wedge \chi_n \wedge \neg\psi)$ *is unsatisfiable.*

Proof: It follows from Theorem 4.28 that ψ is a logical consequence of $X = \{\chi_1, ..., \chi_n\}$ iff the negation of $((\chi_1 \wedge ... \wedge \chi_n) \to \psi)$ is unsatisfiable.[10] In effect,

$$\neg((\chi_1 \wedge ... \wedge \chi_n) \to \psi) \equiv \chi_1 \wedge ... \wedge \chi_n \wedge \neg\psi$$

---
[10] This is so because *satisfiability* is the semantical counterpart of *consistency*.

4. Logical consequences

by the equivalences $\phi \to \psi \equiv \neg\phi \vee \psi$, $\neg(\phi \vee \psi) \equiv \neg\phi \wedge \neg\psi$ and $\neg\neg\phi \equiv \phi$, and by the property of associativity of $\wedge$. **QED**

Theorem 4.46. *A formula ϕ is unsatisfiable iff it entails a contradiction, i.e. iff we have*
$$\phi \models (\psi \wedge \neg\psi).$$

Proof: By the $\models$-deduction theorem, we have (i) $\phi \models (\psi \wedge \neg\psi)$ iff we have (ii) $\models \phi \to (\psi \wedge \neg\psi)$. In turn, we have (ii) iff, for every interpretation $\mathcal{I}$, (a) $val_{\mathcal{I}}(\phi) = \mathtt{f}$ or (b) $val_{\mathcal{I}}(\phi) = \mathtt{t}$ and $val_{\mathcal{I}}(\psi \wedge \neg\psi) = \mathtt{t}$. But the truth table of $\psi \wedge \neg\psi$ shows that it is a contradiction, and thus for every interpretation $\mathcal{I}$ we have $val_{\mathcal{I}}(\psi \wedge \neg\psi) = \mathtt{f}$. Therefore, we must have (a). Thus, $\models \phi \to (\psi \wedge \neg\psi)$ iff ϕ is unsatisfiable, and $\phi \models (\psi \wedge \neg\psi)$ iff ϕ is unsatisfiable. **QED**

▶ *Do Exercises 4.13-15.*

4.1.4. Adequateness of a deductive system

Now suppose that our logical system L at hand is equipped with both a syntactical and a semantical consequence relation such that we have the deductive system $\mathsf{L} = (\mathsf{L}, \Vdash)$ satisfying the following theorem:

Theorem 4.47. *(Deduction theorem) For a (possibly empty) set $X \subseteq F_\mathsf{L}$ and for any formulas $\phi, \psi \in F_\mathsf{L}$,*

(DT)   If $X, \phi \Vdash \psi$, then $X \Vdash \phi \to \psi$

Proof: Proofs for $\vdash$ and $\models$ were given or sketched above in this Chapter. **QED**

Any logical system $\mathsf{L} = (\mathsf{L}, \Vdash)$ in which DT holds is indeed a deductive system. Consider now that the symbol $\Vdash$, otherwise employed to denote indifferently a syntactical consequence relation $\vdash$ or a semantical consequence relation $\models$, denotes the *coincidence* of both these relations into a single consequence relation. Then, we say that the logical system L is both sound and complete.

We next give the fundamental concepts of *soundness* and *completeness* of a logical system formulated as theorems. We consider DT in the two versions for $\vdash$ and $\models$, denoted by $DT_\vdash$ and $DT_\models$, respectively.

Theorem 4.48. *L is sound if, if $X \vdash_\mathsf{L} \phi$, then $X \models_\mathsf{L} \phi$.*

4.1. Logical consequence: A central notion

Proof: (Sketch) We first sketch a proof when $X = \emptyset$. Assume that $\vdash_L \phi$. Then, there is a sequence $\phi_1, ..., \phi_n$ that is a proof of ϕ in L in which every step is either an axiom or derived from previous steps by means of rules of inference (cf. Def. 4.16). The idea is to show, by induction, that every step of the proof is a tautology. Assume that all steps up to ϕ_i are tautologies; it is now necessary to show that ϕ_i itself is a tautology. But ϕ_i is either an axiom, in which case it is a tautology (a truth-table verification of the axioms of the deductive system at hand may elucidate this), or is derived from a previous result by means of a rule of inference, which assures us that it is also a tautology. Therefore, $\models_L \phi$ by proof induction up to $\phi_n = \phi$. For $X = \{\chi_1, ..., \chi_n\}$, we have $\vdash (\chi_1 \to (\chi_2 \to ... (\chi_n \to \phi)))$ by multiple applications of $DT_\vdash$ from $\{\chi_1, ..., \chi_n\} \vdash \phi$ (cf. Prop. 4.27); by soundness, we conclude $\models (\chi_1 \to (\chi_2 \to ... (\chi_n \to \phi)))$, and hence $\{\chi_1, ..., \chi_n\} \models \phi$. **QED**

As can be seen in its proof, Theorem 4.48 is actually a corollary of the case when $X = \emptyset$. In order to segregate this from the case when $X \neq \emptyset$ we may, for disambiguation, speak of *strong soundness* in the latter case.

Theorem 4.49. L *is strongly complete if, if* $X \models_L \phi$, *then* $X \vdash_L \phi$.

Proof: (Sketch) Let us assume that $X \models_L \phi$. Then, $X \cup \{\neg\phi\}$ does not have a model in L. This means that we can have $X \cup \{\neg\phi\} \vdash \bot$ for some contradiction $\bot$, and $X \cup \{\neg\phi\}$ is inconsistent. Hence, by Theorems 4.22-3, we have $X \vdash_L \phi$. **QED**

Just as in the case of Theorem 4.48, this is actually the corollary to the *completeness theorem*, which states that

$$\text{if } \models_L \phi, \text{ then } \vdash_L \phi.$$

Theorem 4.48 expresses the fact that everything that is provable in L (i.e. every theorem of L) is also logically true: L does not prove falsities. As for Theorem 4.49, it expresses the fact that every logical truth in L is provable in L (i.e. is a theorem of L); that is to say that L requires no additional inference rules to prove every logical truth in L, being thus complete in this sense. Hence, with Gödel (1930) we say that a logical system or a theory is complete (incomplete) if every logical truth of the system/theory is (not, respectively) a theorem thereof. Together, soundness and completeness express the fact that in a deductive system L one can derive everything one should (i.e., the system is complete), and

4. Logical consequences

nothing one should not (the system is sound). However, it is obvious that if one has to choose, then one should choose soundness; completeness is a nice property of a logical system, but one has to make sure first and foremost that one does not prove falsities.

Given Theorems 4.48-9, the following proposition is obvious.

Proposition 4.50. *L is sound if* $\vdash_L \subseteq \models_L$ *and complete if* $\models_L \subseteq \vdash_L$.

Proof: Left as an exercise.

Definition 4.51. Given a deductive system $L = (L, \Vdash)$, we say that the axiomatization $\mathcal{P}$ of L is *adequate* whenever the set of theorems of L coincides with the set of tautologies of L, i.e.

$$\vdash_{L,\mathcal{P}} \psi \quad \text{iff} \quad \models_L \psi.$$

We say that a semantics $\mathfrak{S}$ is *adequate* for a deductive system L whenever the set of tautologies of L coincides with the set of theorems of L, i.e.

$$\models_{L,\mathfrak{S}} \psi \quad \text{iff} \quad \vdash_L \psi.$$

Proposition 4.52. *A deductive system* $L = (L, \Vdash)$ *is adequate iff, given a (possibly empty) set of formulas X and a formula ψ, $X \subseteq F_L$ and $\psi \in F_L$, we have*

$$X \vdash_{L,\mathcal{P}} \psi \quad \text{iff} \quad X \models_{L,\mathfrak{S}} \psi.$$

Proof: Trivial from the above. **QED**

The property of *adequateness* is especially desirable when talking about theories, i.e. deductively closed sets of formulas (see Section 4.2.1 below). We thus say that a deductive system L is adequate for, say, group theory. It is also a central property for (automated) theorem proving, as we can answer questions related to validity by ultimately checking for derivability (cf. Fig. 4.1.2). In particular, Herbrand's theorem (see Section 13.3) relates validity with satisfiability, and relates the latter with consistency by appealing to computable functions.

Theorem 4.53. $\chi_1, ..., \chi_n \Vdash \psi$ *iff* $\{\chi_1, ..., \chi_n, \neg\psi\}$ *is unsatisfiable / inconsistent.*

Proof: Left as an exercise.

4.1. Logical consequence: A central notion

MODEL THEORY	PROOF THEORY
Validity Given a semantics $\mathfrak{S}$, we have $\chi_1, ..., \chi_n \models \psi$ iff, if $\models_\mathfrak{S} \chi_1, ..., \chi_n$, then $\models_\mathfrak{S} \psi$	*Derivability* Given a proof ■, we have $\chi_1, ..., \chi_n \vdash \psi$ iff $\chi_1, ..., \chi_n \vdash_■ \psi$
Satisfiability Given a model $\mathcal{M}$, $\{\chi_1, ..., \chi_n\}$ is satisfiable iff $\models_\mathcal{M} \chi_1$ ... and ... and $\models_\mathcal{M} \chi_n$	*Consistency* $\{\chi_1, ..., \chi_n\}$ is consistent iff $\{\chi_1, ..., \chi_n\} \nvdash (\psi \wedge \neg\psi)$
$\Downarrow$ $\models$-deduction theorems DT$_\models$	$\Downarrow$ $\vdash$-deduction theorems DT$_\vdash$
$\xrightarrow{Completeness}$	$\xleftarrow{Soundness}$
THEOREMS	**4.47** and **4.53**

Figure 4.1.2.: Adequateness of a deductive system $\mathsf{L} = (\mathsf{L}, \Vdash)$.

4. Logical consequences

Example 4.54. Recall from Examples 4.26 and 4.43 that we can rewrite any argument α in the form

$$(\alpha') \qquad (A_1 \wedge ... \wedge A_n) \to B.$$

Adequateness allows us to test for theoremhood or validity by applying either a proof method or a validity testing method to

$$\Vdash (A_1 \wedge ... \wedge A_n) \to B.$$

In particular, Theorem 4.53 allows us to test an argument α for theoremhood or validity by testing whether with respect to $\neg \alpha'$ we have

$$(\neg \alpha_{ref}) \qquad \not\Vdash A_1 \wedge ... \wedge A_n \wedge \neg B.$$

If $\neg \alpha_{ref}$ holds, then α' is a theorem or a valid formula, and hence α is a correct or valid argument.

Importantly, if a deductive system L for a specific logic is adequate, then it is equivalent to any other adequate deductive system S for the same logic, i.e. any proof in L can be converted to a proof in S.

We now present an important result with respect to the adequateness of CFOL. Recall Theorem 4.47. We have the following important lemma:

Lemma 4.55. *If* $\Vdash \phi(x) \to \psi(x)$, *then* $\Vdash \forall x \phi(x) \to \forall x \psi(x)$.

Proof: We give the semantical version of the proof. Assume that we have $\models \phi(x) \to \psi(x)$. Let $\mathcal{I}$ be an interpretation and let $\varpi \in \mathcal{I}$ be a variable assignment with $val_\varpi(\forall x \phi(x)) = \mathbf{t}$. Then, for all x-variants ϖ' of ϖ we have $val_{\varpi'}(\phi(x)) = \mathbf{t}$. By assumption, we have $val_{\varpi'}(\psi(x)) = \mathbf{t}$, and consequently $val_\varpi(\forall x \psi(x)) = \mathbf{t}$. **QED**

▶ *Do Exercises 4.16-18.*

4.2. Logical theories and decidability

As seen above, the adequateness of a logical system is so with respect to the theories that are formalizable in it. We begin in Section 4.2.1 by providing a few precisions with respect to logical theories, so far treated largely as unspecified sets of formulas. This done, in Section 4.2.2 we elaborate on the central problem of the undecidability of FOL theories.

4.2. Logical theories and decidability

4.2.1. Theories, subtheories, and extensions

The following definitions make the notion of a *logical theory* precise.

Definition 4.56. A *(logical) theory* Θ is a deductively closed (sub)set $F_{\mathsf{L}}^{(\prime)}$ of formulas of some logical language L, i.e.

$$\Theta = F_{\mathsf{L}}^{(\prime)} \cup AX \cup RI.$$

1. A theory Θ is said to be *consistent* if it is *not* the case that we have *both* $\Theta \vdash \phi$ *and* $\Theta \vdash \neg\phi$ for some formula ϕ. Otherwise, Θ is *inconsistent*.

2. A theory Θ is *complete* if it is the case that *either* $\Theta \vdash \phi$ *or* $\Theta \vdash \neg\phi$ for some formula ϕ. Otherwise, Θ is *incomplete*.

Recall now Definition 4.13.

Definition 4.57. Let $\Theta = F_{\mathsf{L}}^{(\prime)} \cup AX \cup RI$ be a theory.

1. An axiom $\psi \in AX$ is said to be a *logical axiom* if it is an axiom schema of the utmost generality, i.e. with no domain specified.

2. An axiom $\psi \in AX$ is a *non-logical* or *proper axiom* if it is formulated for a specific domain.

Definition 4.58. Let $\Theta = F_{\mathsf{L}}^{(\prime)} \cup AX \cup RI$ be a theory. If all the $\psi \in AX$ are logical axioms, then Θ is a *calculus*.

Clearly, a theory Θ is a calculus Θ' together with the proper axioms $AX(\Theta) - AX(\Theta')$.

Example 4.59. We give an example of a theory and a calculus:

- In Exercise 2.10.4, (1)-(3) are the proper axioms of *group theory*, denoted by $\Theta_{\mathcal{G}}$. If the conclusion does indeed follow from the premises (it does), then it too is part of the theory $\Theta_{\mathcal{G}}$, i.e. is a theorem thereof, as are the rules of inference–if any–that were employed to prove it. (However, only exceptionally do these rules feature explicitly in $\Theta_{\mathcal{G}}$.)

4. Logical consequences

- Below, in Chapter 11, $\mathcal{L}$1-3 (cf. Prop. 11.1) and $\mathcal{Q}$1-2 (Prop. 11.11) are logical axioms; these five axioms constitute–together with the rules MP and GEN–the *Frege-Lukasiewicz calculus*.

We expand now on the group theory $\Theta_\mathcal{G}$ from the viewpoint of logical theories, their extensions, and their subtheories.

Definition 4.60. Let Θ be a (proper) theory. Θ^* is said to be an *extension* of Θ if $\Theta \subset \Theta^*$, and Θ' is said to be a *subtheory* of Θ if $\Theta' \subset \Theta$.

Example 4.61. Consider the group theory $\Theta_\mathcal{G}$. The model for $\Theta_\mathcal{G}$ is (called) a *group*. Obtain the extension $\Theta_\mathcal{G}^*$ by adding to $\Theta_\mathcal{G}$ the following proper axiom:[11]

$$\forall x \forall y \, (x \star y = y \star x)$$

Then, $\Theta_\mathcal{G}^*$ is the theory of Abelian groups, and a model for $\Theta_\mathcal{G}^*$ is (called) an *Abelian group*. Restrict now $\Theta_\mathcal{G}$ to the axiom of associativity; then, the subtheory $\Theta_\mathcal{G}'$ obtained is the theory of semigroups and a model for $\Theta_\mathcal{G}'$ is (called) a *semigroup*. Extend $\Theta_\mathcal{G}'$ with the axiom of the identity element; then, $\Theta_\mathcal{G}'^*$ is the theory of monoids and a model for $\Theta_\mathcal{G}'^*$ is (called) a *monoid*. Let $\Theta_\mathcal{G}'^* = \Theta_\mathcal{G}''$; then,

$$\Theta_\mathcal{G}' \subset \Theta_\mathcal{G}'' \subset \Theta_\mathcal{G} \subset \Theta_\mathcal{G}^*$$

which entails that $\Theta_\mathcal{G}^*$ has at least one model. This, in turn, by the compactness property entails that $\Theta_\mathcal{G}^*$ has a model.

The following theorem on consistent theories will play an important role in the proof of completeness of the classical FO predicate calculus (see Chapter 8).

Theorem 4.62. *(Lindenbaum's Theorem)* If Θ is a consistent theory, then there is a consistent, complete extension Θ^* of Θ.

Proof: Left as an exercise.

▶ *Do Exercises 4.19-21.*

---

[11] For convenience, we introduce the symbol for equality (=) in L1, *but see Section 7.2 below*.

4.2.2. FOL theories and decidability

Given a formula ϕ and a theory $\Theta = \{\theta_1, ..., \theta_n\}$, we often wish to know $\phi \stackrel{?}{\in} \Theta$, i.e. whether $\phi \in \Theta$. Because Θ is a logical theory, we are actually asking, given a logical system L,

$$\phi \stackrel{?}{\in} Cn_L(\Theta)$$

or, equivalently,

$$\Theta \stackrel{?}{\Vdash_L} \phi.$$

The reply to this question may be given in syntactical terms, in semantical terms, or in either of these indifferently if L is an adequate logical system. We assume that L is indeed an adequate logical system, and focus on the semantical aspects of this which is known as a *decision problem* (see Section 1.4).

Recall now the discussion on the semantical logical consequence, in particular the notions of validity and satisfiability (cf. Section 4.1.3).

Definition 4.63. A *(logical) decision problem* is defined as the set

$$DP(\Theta, \phi) = \{(\Theta, \phi) \mid \Theta \models \phi\}$$

where Θ is a logical theory (or, more simply, a–possibly empty–set of logical formulas) and ϕ is a logical formula.

1. If we wish to know whether $\Theta \models \phi$ holds in every interpretation, then we speak of the *validity problem*, and abbreviate it as *VAL*.

2. If we wish to know whether $\Theta \models \phi$ holds in at least one interpretation, then we speak of the *satisfiability problem*, and abbreviate it as *SAT*.

Definition 4.64. Given a finite theory $\Theta = \{\theta_1, ..., \theta_n\}$ and a formula ϕ, we say that $DP(\Theta, \phi)$ is *decidable* if there is a decision procedure for it, i.e., an algorithm that terminates with a "Yes/No" answer. Otherwise, $DP(\Theta, \phi)$ is said to be *undecidable*.

1. In a stricter sense, we say that *theory* Θ *is decidable* if there is a decision procedure for Θ. Otherwise, we say that Θ is undecidable.

Definition 4.65. A procedure Ψ is a *decision procedure* for a theory Θ if it is sound *and* complete with respect to Θ.

4. Logical consequences

These two terms, *sound* and *complete*, are already well known to us, but we now specify them with respect to theories and decision procedures therefor:

Definition 4.66. Let Ψ be a procedure for some theory Θ. Given an input (Θ, ϕ),

1. Ψ is said to be *sound* if when it returns $\Theta \vdash \phi$, it is the case that $\Theta \models \phi$.

2. Ψ is said to be *complete* if

   a) it always terminates, and
   b) it returns $\Theta \models \phi$ when it is the case that $\Theta \vdash \phi$.

Example 4.67. A *Turing machine* is a decision procedure.[12] Let there be given the decision problem

$$DP(S, x) = \{(S, x) \,|\, x \in S, S \subset \mathbb{N}, S \text{ is finite}\}.$$

Then, given that S is a finite subset of the natural numbers, a Turing machine is expected to terminate after a finite amount of time with a "Yes/No" answer to the question, for a given $x \in \mathbb{N}$, $x \stackrel{?}{\in} S$. More formally, a Turing machine performs the *characteristic function* $\chi : \mathbb{N} \longrightarrow \{0, 1\}$ such that

$$\chi_S(x) = \begin{cases} 1 & \text{if } x \in S \\ 0 & \text{otherwise} \end{cases}.$$

There is, however, yet another possibility: the algorithm (i.e. the Turing machine) may not stop. If the theory Θ is actually or virtually infinite–for instance, it has an infinite domain of discourse–, then it should not be surprising that we have no final answer; that is, the theory is *undecidable*. Otherwise, we should expect an answer, but it may be the case that, if in fact $\Theta \nvdash \phi$, we end up with no answer. This is known as *semi-decidability*.

Example 4.68. Consider the set

$$S = \{p \,|\, p \text{ is a polynomial over } x \text{ with an integral root}\}.$$

---
[12]See Section 1.4 above.

4.2. Logical theories and decidability

We ask whether S is decidable. Given a polynomial p over x such as $2x^3 + 6x^2 - x + 5$, we ask $p \stackrel{?}{\in} S$. In other words, we want to know if $x = n, n \in \mathbb{Z}$. There is indeed an algorithm to test for this decision problem: evaluate p for the values $0, 1, -1, 2, -2, 3, ...$ one at a time successively; if for some $n \in \mathbb{Z}$ it is the case that $p = 0$, then $\chi_S(p) = 1$. Nevertheless, we have no guarantee that this algorithm will ever terminate: it will only terminate if indeed $p \in S$; otherwise, it is obvious that it will run forever, because $\mathbb{Z}$ is an infinite set. In computational jargon, we say that S is *recursively enumerable*, i.e. semi-decidable, but $DP(S, p)$ is in fact not computable in case $p \notin S$.

Although there are undecidable propositional theories, propositional logic is generally decidable. That is to say that there is a decision procedure for it. Given a formula $\phi \in$ L0 and a theory / empty set $\Theta \in$ L0, we can always say whether $\Theta \Vdash \phi$ or $\Theta \nVdash \phi$ by means of a truth table, for instance. FOL is generally undecidable, being at best semi-decidable. This feature of FOL is directly connected to its *expressiveness*, a property it owes to quantification associated with infinite domains of discourse. As a matter of fact, the undecidability of FOL constitutes a field of its own, with its specific decision problem being known as the *Entscheidungsproblem*, owing to its original formulation in German by the mathematician D. Hilbert (Hilbert & Ackermann, 1928).

In Chapter 10, we give a brief overview of the *Entscheindungsproblem* and then elaborate on Turing's negative result at some length. We next discuss two important aspects of FOL decidability. We refer the reader to, for instance, Börger, Grädel, & Gurevich (2001), for a comprehensive discussion of this topic.

▶ *Do Exercises 4.22-26.*

4.2.2.1. Finite satisfiability and ground extensions

In fact, semi-decidability is a significant improvement with relation to undecidability, because some procedure Ψ *always* terminates on input (Θ, ϕ) if indeed we have $\Theta \models \phi$. This said, the fact that Ψ may never stop if $\Theta \nvDash \phi$ significantly decreases this improvement. Thus, what we require is a complete deductive system, i.e. a system in which there is a complete procedure for the decision problem $DP(\Theta, \phi)$ (cf. Def. 4.66.2). In Chapter 8, we prove the completeness of the classical FO deductive system, but we begin now by proving the completeness of an arbitrary FO deductive system with the logical language L1.

But we want to do more than prove FO completeness; we want to do this while at the same time reducing this property to propositional com-

4. Logical consequences

pleteness, as this alone guarantees decidability, rather than only semi-decidability. With this objective in mind, if we present completeness as a consequence of compactness,[13] then we want to reduce FO compactness to propositional compactness, too. In order to do this some further elaboration on contents already discussed is required, namely on satisfiability; some anticipations are also needed, namely with respect to material from Sections 9.2 and, especially, 13.3. Internal references aid the reader in these recallings and anticipations.

Recall from Definition 4.30 that a set of formulas X is satisfiable iff there is an interpretation $\mathcal{I}$ that satisfies all the elements of X; in other words, X is satisfiable iff every element thereof has a (common) model.

Definition 4.69. *Finite and infinite satisfiability* – Let $X \subseteq F_{L1}$ be a (possibly infinite) satisfiable set of FO formulas. We say that X is *finitely satisfiable* iff every finite subset $X'_i \subseteq X$ has a model; otherwise, X is said to be *infinitely satisfiable*.

Obviously, a satisfiable finite set of formulas is trivially finitely satisfiable, and a finite set of propositional formulas is always finitely satisfiable if it is satisfiable. The importance of finite satisfiability resides in the well-known property of the infinity of models for FO formulas. This infinity is, in turn, accounted for by the infinity of domains for a FO language.[14] We know that a valuation $val_{\mathcal{I}}(\phi)$ for any FO formula $\phi(x_1, ..., x_n)$ given any interpretation $\mathcal{I} = (\mathcal{D}, \Theta, \varpi)$ depends on the variable assignment ϖ that replaces all bound or free variables $\vec{x} = x_1, ..., x_n$ of ϕ by some constants $\vec{a} = a_1, ..., a_n$ in the domain $\mathcal{D}$; thus, this valuation is more precisely a valuation $val_{\mathcal{I}^{\vec{x}}_{\vec{a}}}(\phi)$, so that a FO formula $\phi(\vec{x})$ is said to be *(un)satisfiable in a domain* $\mathcal{D}$. If the domain is infinite, the interpretations $\mathcal{I}^{\vec{x}}_{\vec{a}}$ are so, too, and so are the models of ϕ.

So, we have it that a satisfiable FO formula ϕ can be either finitely or infinitely satisfiable. If we can determine that ϕ is finitely satisfiable, this is an improvement over tackling infinite satisfiability, which is of little to no practical use.[15] Is there a way, given some signature Υ_ϕ, to fix the interpretations of ϕ such that ϕ has a *fixed* interpretation that is independent of $\mathcal{D}$ and thus of ϖ, too? In other words, is there a way to make a FO formula ϕ behave like a propositional formula?

---

[13] For convenience, we choose to support our proof of FO completeness on an important property already mentioned: *compactness* (cf. Def. 4.11).

[14] This infinity with respect to the domains is double: for any FO language, there may be arbitrarily finitely or infinitely many domains, and these domains may themselves be arbitrarily finitely or infinitely large.

[15] Say we consider models as states of some database; clearly, infinite states simply cannot be stored, let alone manipulated.

Indeed there is. All we need to do is, given a FO formula, firstly remove all the quantifiers and substitute all individual variables by ground terms, and secondly, substitute all obtained ground atoms by propositional variables. That is, we first need a ground extension, and then a transfer function mapping ground atoms to propositional variables. We next explain this jargon.[16]

Definition 4.70. Let us consider the FO formula ϕ in the SNF

$$\forall x_1...\forall x_n (\phi_M)$$

where ϕ_M is quantifier-free, only atoms are negated in ϕ_M, and the only connectives in ϕ_M are $\wedge$ and $\vee$.[17] Let us consider also the substitution set $\sigma = \{x_1 \mapsto t_1, ..., x_n \mapsto t_n\}$, where every $t_i \in T_g(\phi)$ and $T_g(\phi)$ is the set of ground terms in ϕ. Then, the *ground extension* of ϕ is defined as

$$GE(\phi) = \{\phi_M [x_1/t_1, ..., x_n/t_n] \mid t_i \in T_g(\phi)\}.$$

Example 4.71. Let there be given the formula B in SNF

$$B = \forall x \forall y \forall u \forall v (\neg T(x, y, u, v) \vee P(x, y, u, v)).$$

Given $T_g(B) = \{a, b, c, d\}$ and $\sigma = \{x \mapsto a, y \mapsto b, u \mapsto c, v \to d\}$, we

---

[16] We apply here somehow loosely what is known as the *transfer principle*, which allows FO assertions in a (algebraic) structure $\mathcal{A}$ to be transferred or translated to another (algebraic) structure $\mathcal{B}$, called its *(elementary) extension*, by means of a transfer or translation function $\tau : \mathcal{A} \longrightarrow \mathcal{B}$ in such a way that

$$\mathcal{B} = \tau(\mathcal{A}).$$

This is a principle with important applications, namely in non-standard analysis, where it is applied to guarantee that there is a non-standard model $\mathcal{Q} = \tau(\mathcal{R})$ of the standard model $\mathcal{R}$ of the reals. An elaboration on this topic is outside the scope of this book, but two aspects need specification: firstly, because we often identify the structure $\mathcal{A}$ with its universe A, the transfer principle can be applied to (infinite) sets; secondly, the function τ is more rigorously applied to *superstructures*, i.e. $\tau : S_\infty(A) \longrightarrow S_\infty(B)$, given $S_0(A) = A$, $S_n(A) = \mathscr{P}\left(\bigcup_{k=0}^{n-1} S_k(A)\right)$ for $\mathscr{P}(A) = 2^A$, and $S_\infty(A) = \bigcup_{k=0}^{\infty} S_k(A)$, $S_\infty(A)$ is the superstructure over A. In more logical terms, the identity $\mathcal{B} = \tau(\mathcal{A})$ entails that $\mathcal{B}$ inherits all the FO properties of $\mathcal{A}$ in such a way that, given a FO model $\mathcal{A}$, we have $X \models_\mathcal{A} \phi$ iff $X \models_\mathcal{B} \phi$. The principle can then be stated as: for any FO formula $\phi(x_1, ..., x_n)$ with bounded variables and any sequence $a_1, ..., a_n \in S_\infty(A)$, the formula $\phi(a_1, ..., a_n)$ is *true* in $\mathcal{A}$ iff $\phi(\tau(a_1), ..., \tau(a_n))$ is *true* in $\mathcal{B}$.

[17] Cf. Definition 2.37. Note that only the universal quantifier is allowed to occur in the SNF of ϕ. The reason for this restriction should be clear to the reader by the end of this Section.

4. Logical consequences

have the ground formula

$$B^g = \neg T(a, b, c, d) \vee P(a, b, c, d)$$

and

$$GE(B) = \{\neg T(a, b, c, d) \vee P(a, b, c, d)\}.$$

By this means, we have now actually a set of quantifier-free formulas $\Phi^g = GE(\phi)$ to which the following lemma can be applied:

Lemma 4.72. *Let ϕ be a FO formula. Then, ϕ is satisfiable iff $GE(\phi)$ is satisfiable.*

Proof: (Idea) We know that satisfiability of a universally-quantified formula $\phi(x_1, ..., x_n) = \phi(\vec{x})$ depends on there being at least one model $\mathcal{M}$ in an interpretation $\mathcal{I}_{\vec{t}}^{\vec{x}}$ in which every variable in $\vec{x}$ is replaced in $\vec{t}$ by a ground term in $T_g(\phi) = \{t_1, ..., t_n\}$. In assigning to every variable in $\vec{x}$ a term in $\vec{t}$, the quantifiers are eliminated and we obtain a quantifier-free formula ϕ^g only with ground terms–i.e. a ground formula–that is equi-satisfiable to ϕ, a relation we denote by $\phi \equiv_{sat} \phi^g$ and which means that ϕ is satisfiable iff ϕ^g is. The proof of this equisatisfiability relation is by induction on the single-variable FO formula $\phi(x)$, for which we have $\models_\mathcal{M} \forall x \phi(x)$ iff $\models_\mathcal{M} \phi_t^x$ (cf. Prop. 3.27.1).[18] **QED**

But if $T_g(\phi) = T_g(\phi_M)$ is infinite, then $GE(\phi)$ is also infinite, and we have ended up with infinite satisfiability.[19] Can we do better than this?

Definition 4.73. *Let Φ^g be the set of quantifier-free formulas in $GE(\phi)$; then, all the (sub-)formulas in Φ^g are atoms $A_i \in At_g(\Phi^g)$, where $At_g(\Phi^g)$ is the set of the ground atoms of Φ^g. For arbitrary (possibly*

---

[18] We actually leave the proof of this lemma as an exercise in Section 9.2, but the idea provided should allow the reader a working grasp of this lemma for the reading of this Section. Importantly, in order to invoke Proposition 3.27.1 we need to consider the implicit identity $T_g(\phi) = \{t_1, ..., t_n\} = \mathscr{D}(\phi)$ and implicitly take it that there is some interpretation $\mathcal{I}$ such that $\varpi \in \mathcal{I}$ corresponds to some substitution σ. In Section 9.2, we shall see how the finite set $T_g(\phi) = \{t_1, ..., t_n\}$ corresponds explicitly to the Herbrand universe H_ϕ of a function-free formula $\phi = \forall x_1...\forall x_n(\phi_M)$ and $At_g(\phi^g)$ (see below) does so with respect to the Herbrand base $H(\phi)$. (For a set Φ of function-free FO formulas $\phi = \forall x_1...\forall x_n(\phi_M)$, we have $T_g(\Phi)$ and $At_g(\Phi^g)$, and correspondingly in Herbrand semantics H_Φ and $H(\Phi)$.)

[19] Note that in Example 4.71 we obtained a singleton set $\Phi^g = GE(\phi)$, because there was only one substitution considered. Generally, the cardinality of Φ^g will depend on the number of substitutions; Φ^g is infinite if the substitutions to be considered are so. Besides infinite domains, the occurrence of function symbols in a FO formula also implicates infinite sets $\Phi^g = GE(\phi)$. For instance, given the formula $A = \forall x (P(f(x)) \vee \neg Q(a))$, we have $T_A = \{a, f(a), f(f(a)), f(f(f(a))), ...\}$

4.2. Logical theories and decidability

infinite) sets At_g and At_p of ground atoms and propositional variables, respectively, we define a *transfer function* $\tau : At_g \longrightarrow At_p$ by means of which each ground atom A_i is replaced by some propositional atom $P_i \in At_p = V$ such that for the finite sets $\{\phi\}_g = \phi^g$ and Φ^g we have

$$\tau(\Phi^g) = \{\phi[A_1/P_{A_1}, ..., A_n/P_{A_n}] \mid \phi \in \Phi, \{A_1, ..., A_n\} \subseteq At_g(\phi^g)\}$$

and

$$\tau(At_g(\phi^g)) = \{P_{A_i} \mid A_i \in At_g(\Phi^g)\}$$

where we write P_{A_i} to denote the propositional atom obtained from the ground atom A_i.

Example 4.74. Let us consider the argument of Example 2.25. One of the premises is already a ground formula, and so is the conclusion. The first premise is the ground formula B^g of Example 4.71 above, and we obtain the ground formula of the second premises by the same method. We thus have the sets

$$C^g = \left\{ \begin{array}{c} \neg T(a,b,c,d) \vee P(a,b,c,d), \\ \neg P(a,b,c,d) \vee E(a,b,d,c,d,b), \\ T(a,b,c,d), \\ E(a,b,d,c,d,b) \end{array} \right\}$$

and

$$At_g(C^g) = \{T(a,b,c,d), P(a,b,c,d), E(a,b,d,c,d,b)\}.$$

We apply the transfer function τ in a very simple way by assigning to each atom in $At_g(C^g)$ the corresponding predicate name, and we obtain the set of propositional atoms

$$\tau(At_g(C^g)) = At_p(C^g) = \{T, P, E\}.$$

---

and $At_g(A^g) = \{P(a), Q(a), P(f(a)), Q(f(a)), ...\}$; this entails

$$A^g = \left\{ \begin{array}{c} P(a) \vee \neg Q(a), \\ P(f(a)) \vee \neg Q(a), \\ P(f(f(a))) \vee \neg Q(a), \\ P(f(f(f(a)))) \vee \neg Q(a), \\ \vdots \end{array} \right\}$$

and $|A^g| = \infty$.

4. Logical consequences

We terminate with the (satisfiable) set of clauses[20]

$$\tau(C^g) = \{\|\neg T, P\|, \|\neg P, E\|, T, E\}.$$

Proposition 4.75. *All the elements of $\tau(\Phi^g)$ are (well-formed) propositional formulas (over $\tau(At_g(\phi^g)))$.*

Proof: (Sketch) Apply structural induction on $\tau(a) = p$ for some $a \in At_g(\phi^g)$ and $p \in V$. **QED**

Lemma 4.76. *Let ϕ be a FO formula. Then ϕ is (FO-)satisfiable iff $\tau(\Phi^g)$ is (propositional-)satisfiable.*

Proof: (Sketch) By Lemma 4.72, ϕ is satisfiable iff Φ^g is satisfiable. So we need only show now that Φ^g is satisfiable iff $\tau(\Phi^g)$ is. This we can do by showing that

$$\models_\mathcal{I} \Phi^g \quad \text{iff} \quad \models_{val} \tau(\Phi^g)$$

for some interpretation $\mathcal{I}$ and some propositional valuation val, i.e. a valuation independent of an interpretation $\mathcal{I} = (\mathcal{D}, \Theta, \varpi)$. **QED**

Definition 4.77. *Let us now define the ground extension of Φ as*

$$GE(\Phi) = \bigcup \{GE(\phi) \mid \phi \in \Phi\}.$$

We now have the following fundamental result:

Theorem 4.78. *Let Φ be a (possibly infinite) set of FO formulas and let $\Phi^g = GE(\Phi)$ be the corresponding set of quantifier-free formulas. Then the following holds:*

1. *Φ is (FO-)satisfiable iff Φ^g is (FO-)satisfiable.*

2. *If Φ is finitely (FO-)satisfiable, then Φ^g is finitely (FO-)satisfiable.*

3. *Φ^g is (FO-)satisfiable iff $\tau(\Phi^g)$ is (propositional-)satisfiable.*

4. *If Φ^g is finitely (FO-)satisfiable, then $\tau(\Phi^g)$ is (propositional-) satisfiable.*

---

[20] We obtain here a set $\tau(C^g)$ of propositional clauses because we started with a set of ground clauses C^g. However, Φ need not be a set of clauses.

Proof: Easy from the above by induction or generalization on the proofs for Φ where $|\Phi| = 1$. **QED**

We are now ready to give the FO compactness result that will allow us to prove the completeness of FO languages.

Theorem 4.79. *(Compactness of FOL) Let $X \subseteq F_{L1}$ be a (possibly infinite) set of formulas. Then, X is satisfiable iff X is finitely satisfiable.*

Proof: ($\Rightarrow$) Trivial.
($\Leftarrow$) (Hint: Theorem 4.78.) **QED**

Lemma 4.80. *For X a set of FO formulas and some $\phi \in X$, we have $X \models \phi$ iff $X \cup \{\neg\phi\}$ is unsatisfiable.*

Proof: (Idea) This is Theorem 4.41 for FO formulas. Think now in terms of the models for X, ϕ, and $\neg\phi$, i.e. $\mathcal{M}_X$, $\mathcal{M}_\phi$, and $\mathcal{M}_{\neg\phi}$. **QED**

Corollary 4.81. *For X a set of FO formulas and some $\phi \in X$, we have $X \models \phi$ iff there is a finite set $X' \subseteq X$ such that $X' \models \phi$.*

Proof: Easy, by Lemma 4.80 and Theorem 4.79. **QED**

Finally:

Theorem 4.82. *(Completeness of FOL) For X a (possibly infinite) set of FO formulas and some $\phi \in X$, if $X \models \phi$, then $X \vdash \phi$.*

Proof: (Idea) Firstly, prove that FOL is complete if,

$$\text{if } \chi_1, ..., \chi_n \models \phi, \text{ then } \chi_1, ..., \chi_n \vdash \phi.$$

Make it a lemma to Theorem 4.49. Invoke then Corollary 4.81. **QED**

We have thus proved the *completeness* of FOL. Although we have tamed the *undecidability* of FO languages, we have not eliminated their intrinsic expressiveness. This would be possible only if we would be ready and willing to remove all function symbols from a FO language. In other words, the above results are essentially only for some signature $\Upsilon - Fun$.

This removal is in fact possible, but, as the reader may deduce from (especially) Section 2.3 above, this is not always feasible or even desirable.[21] In Section 13.3, we show how *refutation-completeness* considered in a particular semantics–Herbrand semantics (Section 9.2)–does effectively eliminate this problem.

▶ *Do Exercises 4.27-8.*

---
[21] Note that the formulas of Examples 2.25 and 4.74 have no function symbols; this is

4. Logical consequences

4.2.2.2. Finite models and prefix classes

However, function symbols are not the only factor that impacts on FO decidability. In fact, the form of the quantifier prefix in a prenex normal form, the arity of the predicates, and the inclusion of the symbol "=" for equality in a FO alphabet all determine the decidable or undecidable character of a logical theory. And indeed, though FOL theories are generally undecidable, some have been proven to be decidable. We give here the standard complete characterization.

Above, we were concerned with *finite satisfiability*; now, *finite models* are our main concern.[22]

Definition 4.83. Given an interpretation $\mathcal{I} = (\mathscr{D}, \Theta, \varpi)$, a model $\mathcal{M}$ is said to be *finite* if the cardinality of $\mathscr{D}$ is finite; otherwise, it is said to be *infinite*.

Definition 4.84. Let $\mathscr{F}$ be a class of FO formulas. We say that $\mathscr{F}$ has the *finite-model property (FMP)* if, for every formula ϕ in $\mathscr{F}$, either ϕ is unsatisfiable or it has a finite model.

Proposition 4.85. *Let a* L1*-formula be in the form*

$$(\dagger) \qquad \blacklozenge_1 x_1 ... \blacklozenge_n x_n (M)$$

where every $\blacklozenge_i x_i$, $i = 1, ..., n$ *is either* $\forall x_i$ *or* $\exists x_i$, *and* M *is a formula containing no quantifiers.* $\blacklozenge_1 x_1 ... \blacklozenge_n x_n$ *is the* prefix *and* (M) *is the* matrix *of A. There are no function symbols in* M. *Let now* $\blacklozenge, \blacklozenge^n, \blacklozenge^*$ *denote the single occurrence of a quantifier* $\blacklozenge$, *the sequence of n occurrences of a quantifier* $\blacklozenge$, *a sequence of zero or more occurrences of a quantifier* $\blacklozenge$, *respectively. Then, the following prefix forms determine* prefix classes *of FO formulas for which there are decision procedures:*

1. $\exists^* \forall^*$

---

so, to begin with, because there were no function symbols in the FO formalization of the theory (cf. Example 2.25). But, as stated in this Example, this is a very simple mathematical argument that can be adequately formulated in $\Upsilon_{L1} - Fun$. This is not always the case, with some mathematical theories actually requiring a 2nd order language. Additionally, the original FO formalization employed solely the universal quantifier, and thus the ground formulation thereof dispenses with function symbols. (Cf. Algorithm 2.2.)

[22] Don't let the word "finite" trick you into thinking that these are equivalent; in effect, compactness, for instance, fails over finite models, thus making finite model theory a field distinct from model theory.

4.2. Logical theories and decidability

2. $\exists^*\forall\exists^*$

3. $\exists^*\forall^2\exists^*$ (if the symbol $=$ does not occur in M)

Proof: (Idea) The proof is based on the fact that these classes have the FMP. See also Exercise 4.31. **QED**

Proposition 4.86. *The following prefix patterns determine prefix classes of undecidable FO formulas:*

1. $\forall^3\exists$

2. $\forall\exists\forall$

Proof: Left as a research exercise.

Let now $\bar{\blacklozenge}$ denote the dual of the quantifier $\blacklozenge$, i.e. $\bar{\forall} = \exists$ and $\bar{\exists} = \forall$. Then we have the following important result:

Proposition 4.87. *Let $\mathscr{P}$ be a prefix class and $\overline{\mathscr{P}}$ be its dual class obtained by replacing each quantifier $\blacklozenge$ by its dual $\bar{\blacklozenge}$. Then, the following statements are equivalent:*

1. *The VAL for $\mathscr{P}$ is decidable.*

2. *The SAT for $\overline{\mathscr{P}}$ is decidable.*

Proof: Left as an exercise. (Hint: Proposition 4.39.1.) **QED**

▶ *Do Exercises 4.29-33.*

Exercises

Exercise 4.1. Show that conditions C1-3 together are equivalent to the following condition for all $X, Y \subseteq F_\mathsf{L}$:

$$(\text{C0}) \qquad X \subseteq Cn\left(Cn\left(X\right)\right) \subseteq Cn\left(X\right) \subseteq Cn\left(X \cup Y\right)$$

Exercise 4.2. Let us define a *closure operation* on a set A as a function $\mathsf{C} : 2^A \longrightarrow 2^A$ satisfying the following conditions for all $X, Y \subseteq A$:

133

4. Logical consequences

C1	$X \subseteq C(X)$	Inclusion
C2	$C(C(X)) = C(X)$	Idempotency
C3	If $X \subseteq Y$, then $C(X) \subseteq C(Y)$	Monotonicity

Show that if C0 (cf. Exercise 4.1) is satisfied by all $X, Y \subseteq F_L$, then Cn is a closure operation on F_L.

Exercise 4.3. The following are the properties of a *(Hausdorff) topological closure* for some sets $X, Y \subseteq A$:

HC1	$X \subseteq C(X)$
HC2	$C(C(X)) = C(X)$
HC3	$C(X \cup Y) = C(X) \cup C(Y)$
HC4	$C(\emptyset) = \emptyset$

A mapping $C : 2^A \longrightarrow 2^A$ on a set A verifying conditions HC1-4 is called a *hull operator*. Is Cn a hull operator? Why (not)?

Exercise 4.4. Reflect on the significance of compactness (cf. Def. 4.11) for a logical system.

Exercise 4.5. Complete the proof of Theorem 4.5.

Exercise 4.6. Prove Propositions 4.6 and 4.8.

Exercise 4.7. The topic of logical consequence, though clearly formulable in a mathematical language, remains for many a problematic topic in philosophical logic and also in applied logic. Give some thought to the following statements. (You might want to read Sections 4.1.2-4 before.)

1. The [consequence] problem is that consequence relations are specifiable by truth conditions, or by proof-theoretic constraints, independently of anything that might be true of any actual agent. (Gabbay & Woods, 2003)

2. [O]ver a large body of logic, the closure structure is basically its essence and its syntactic and linguistic features are secondary, useful for purposes of understanding though they may be. (Martin & Pollard, 1996)

3. A deductive system provides only a way to study a language's consequence relation, to prove results about it, perhaps even mechanize it. But it does not determine or give rise to that relation. This is why the question of whether a particular deductive system for a particular language is sound and complete is always a sensible, and indeed important, one to ask. (Etchemendy, 1999)

4. The crux of the matter is ... the definition of the term "logical consequence." Until this term has been explained, one does not have an opinion as to the nature of mathematics at all. (Curry, 1963)

5. I do not mean to suggest that "thinking" can proceed very far without something like "reasoning." We certainly need (and use) something like syllogistic deduction, but I expect mechanisms for doing such things to emerge in any case from processes for "matching" and "instantiation" required for other functions. Traditional formal logic is a technical tool for discussing either *everything that can be deduced from some data* or *whether a certain consequence can be so deduced*; it cannot discuss at all what ought to be deduced under ordinary circumstances. (Minsky, 1974)

6. [W]hat distinguishes a logic from an algebra is a concept of logical consequence. (Cleave, 1991).

Exercise 4.8. Rewrite $X \nvDash \psi$ in terms of

1. consistency.

2. inconsistency.

Exercise 4.9. Prove the following statement: Any $\nvDash$-relation rests on a refutation.

Exercise 4.10. Give the rationale for Proposition 4.27.

Exercise 4.11. Consider Definition 4.29. In which way(s) may this be too limiting a definition?

Exercise 4.12. Prove Theorems 4.22-5 and 4.28.

Exercise 4.13. Prove Propositions 4.37, 4.39, and 4.40.

Exercise 4.14. Generalize Theorem 4.42 and give the rationale for this generalization.

Exercise 4.15. Give all the steps of the proof of Theorem 4.45.

Exercise 4.16. A stronger theorem than DT (Theorem 4.47) is the *deduction-detachment theorem*, formalized as

$$\text{(DDT)} \quad X, \phi \Vdash \psi \quad \text{iff} \quad X \Vdash \phi \to \psi$$

1. Explain informally why DDT is stronger than DT.

4. Logical consequences

2. Prove DDT.

Exercise 4.17. The sketchy proof of Theorem 4.49 above is a summary of the proof first given by L. Henkin (1949) for CFOL. We give the complete proof in Part II. Research into other proofs of completeness of FOL and give a summary of each.

Exercise 4.18. Prove Proposition 4.50 and Theorem 4.53.

Exercise 4.19. Prove the following statement: *Every FO predicate calculus is consistent.*

Exercise 4.20. Let $\Theta \nvdash \neg\phi$, ϕ is a ground formula. Prove that if we extend Θ by adding ϕ as a new axiom, then Θ^* is consistent.

Exercise 4.21. Prove Lindenbaum's Theorem.

Exercise 4.22. Let Θ be an axiomatizable theory. Show that if Θ is complete, then Θ is decidable.

Exercise 4.23. Show that the theory of groups $\Theta_\mathcal{G}$ is decidable by outlining a decision procedure Ψ for it.

Exercise 4.24. Give examples of undecidable theories.

Exercise 4.25. Give examples of theories not expressible in FOL.

Exercise 4.26. The *Löwenheim-Skolem Theorem* states that if a set X of formulas has a model, then X has an enumerable model.

1. This theorem can also be referred to as the *Löwenheim-Skolem Transfer Theorem*. Explain why.

2. This theorem is a corollary of compactness. Explain.

3. What happens if X is a set of formulas of a non-enumerable language?

4. In spite of item 2 above, this theorem can be given an independent proof. Sketch such a proof.

Exercise 4.27. Consider formula A and the set $T_g(A) = \{a, b, c, d\}$ such that there are substitutions $\sigma = \{x \mapsto a, y \mapsto c\}$ and $\theta = \{x \mapsto b, y \mapsto d\}$. Obtain the corresponding set C^g by means of the ground expansion $GE(A)$ of A and the application of the transfer function τ.

$$A = \forall x \forall y \left(\left(P(x) \land \neg Q(y) \right) \to R(x, y) \right)$$

4.2. Logical theories and decidability

Exercise 4.28. Complete the proofs of the statements in Section 4.2.2.1.

Exercise 4.29. Besides the formulas in Proposition 4.85, it has also been proven that *monadic FO* formulas, i.e. formulas whose predicates are unary and have no function symbols, are decidable. Research into the proof for this result.

Exercise 4.30. Propositions 4.85-6 give us a complete classification of the prefix classes in terms of decidability for formulas without function symbols. Why is this classification complete?

Exercise 4.31. Prove that there is a decision procedure for *VAL* for all FO formulas with the FMP.

Exercise 4.32. Prove (or complete the proof of) the statements in Section 4.2.2.2.

Exercise 4.33. Compare the use of the symbol $\vdash$ to denote the syntactical logical consequence with its use in computation theory (cf. Def. 1.10.1 and Example 1.13). In particular, comment on possible equivalences.

Part II.
The System CL and the Logic CL

5. The language of classical logic

The logical system known as *classical logic* (abbreviated: CL) has a well-known and well-studied characterization from diverse perspectives. From the algebraic viewpoint, CL is *Boolean*, i.e. a Boolean algebra provides it with a mathematical foundation for both its syntax and its semantics. To say that CL is Boolean in terms of syntax means that all formulas of CL can be rewritten using only the subset of Boolean operators (cf. Section 3.2). With respect to its semantics, to say the same means that CL has a bivalent semantics (see Section 9.3). This, in turn, has an impact on the set-theoretical foundations: the sets of CL are *crisp*, with crispiness–a term arising in the context of fuzzy and/or rough sets–being the property of a set membership to which is decidable by a "Yes/No" answer (but recall that FOL is essentially undecidable; see Section 4.2). These properties, more immediately connected to the classical notion of logical consequence, are also intimately connected to the *language* of CL. In this Section, we discuss this latter topic.

5.1. Some preliminary remarks

It is not an easy matter to segregate a logical language with respect to the logical systems that employ it. For instance, both classical logic and a plethora of many-valued logics employ L1, but while a formula such as $\phi \wedge \neg \phi$ is a contradiction in classical terms, it need not be so in a many-valued logic. Yet another example: L1 is a functionally complete language if employed in CL and in some many-valued logics as well, but it is not so in Łukasiewicz's 3-valued logic, as in this we have, for instance, $\phi \vee_{\text{Ł3}} \psi :\neq \neg_{\text{Ł3}} \phi \rightarrow_{\text{Ł3}} \psi$ (see below). To be sure, we are considering meaning in these examples, but this is hardly a negligible aspect, as a logical language is frequently employed with expression in view. If we want to express the fact that a proposition P is not only true, but necessarily so, then L1, or any of its classical extensions, is not an adequate language, as we need an extra, non-classical, unary operator to this end.[1]

---

[1] We speak here of the modal unary operator denoted by $\square$ (e.g., $\square P$ is read "necessarily P"). We say that this particular connective is non-classical, because its

5. The language of classical logic

This is so because, in mathematical terms, the operators of a logical language actually express functions and relations among propositions and formulas, and while this entails a syntax that rules the well-formedness of these, it alone does not suffice to provide an adequate means of expression. For instance, the connective $\wedge$ is employed to express the binary function $\wedge(\phi, \psi)$ where the formulas $\phi, \psi \in \mathsf{L1}$ are the arguments, but this is actually a *truth function* (cf. Def. 3.2.1), namely the truth function

$$\widetilde{\wedge}(\phi, \psi) = \begin{cases} 1 & \text{if } val(\phi) = 1 \text{ and } val(\psi) = 1 \\ 0 & \text{otherwise} \end{cases}.$$

This is an aspect that fundamentally distinguishes a logical language from all other formal languages. In effect, in these we can derive well-formed, or legal, strings of symbols from equally well-formed, or legal, strings of symbols, and we do so in a purely syntactical way (see Examples 1.2-3). In a logical language, on the contrary, derivability entails, if not compositionality of meaning (cf. Section 3.3), then provability. Thus, the metalanguage symbol $\vdash$, here denoting the classical syntactical consequence relation, is in principle not to be interpreted as the same symbol employed in the context of non-logical formal languages. Let there be given some non-logical formal language L and a computing machine M such that, given the derivation rule $A ::= bB$ in the grammar $G(L)$ corresponding to L, we have the following computation where q, p are any states and x, y are substrings:

$$q, xAy \vdash_M p, xbBy$$

This means that in a formal grammar we can derive the legal string $xbBy$ from the legal string xAy in one step (see Sections 1.1 and 1.4). When, given some logical system $\mathsf{L} = (L, \vdash)$, we write $\phi \vdash_\mathsf{L} \psi$, we are actually stating that $\vdash_\mathsf{L} \phi \to \psi$, and this in turn means that there is a proof ■ such that $\vdash_\blacksquare \phi \to \psi$. As seen, this in turn entails that there is a proof system $\mathcal{P} = (L, AX, RI)$ for L such that we have $\vdash_\blacksquare \phi \to \psi$ iff we have $\vdash_{\mathcal{P},\blacksquare} \phi \to \psi$. Regardless of whether or not we see this as *in turn* entailing some meaning, this makes it that a formula of the form

if A_1 and A_2 and ... and A_n, then B

---

use extends the set of tautologies of classical logic; a connective is equally non-classical if its use reduces this set. This said, the introduction of the operator for equality (=) in CL does indeed extend the set of classical tautologies (e.g., $s = t \vee s \neq t$) without for that changing the classicality status of **CL** (see Section 7.2). Therefore, we say that $O_{\mathsf{L1}} \cup \{=\}$ is a *classical extension* of $\mathsf{L1}$.

5.1. Some preliminary remarks

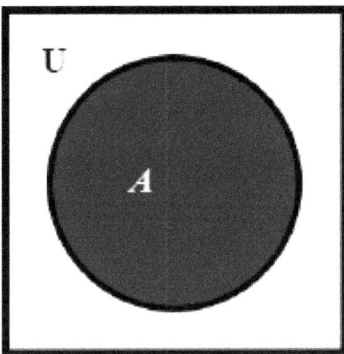

Figure 5.1.1.: Venn diagram of the set A.

where the **bold** font expresses the main connective, in a logical system or in a logic-based programming language or computation system (e.g., Prolog) is interpreted in a wholly different way from its occurrence in, say, a not logic-based programming language or computation system (for instance, in a production system; see Augusto, 2020a, p. 13).[2]

This said, we think that a diagrammatic presentation of the L1 connectives and quantifiers, while not being properly semantical or syntactical, can serve both meaning and form in classical logic. We eschew any metatheoretical discussion on this topic, relying solely on its intuitive appeal. We shall use both Euler and Venn diagrams, and shall further employ these in informally proving the validity of syllogisms (Section 5.3.1). Besides this application of L1, we also elaborate on a central application of a subset of L1, to wit, in logic programming (Section 5.3.2). Given its intrinsic interest, we also elaborate on an extension of O_L that has practical applications–logic design–while remaining classical through and through (Section 5.3.3).[3]

▶ *Do Exercise 5.1.*

---

[2] See Section 5.3.2 below for the interpretation of a conditional proposition of the form $(A_1 \wedge ... \wedge A_n) \to B$ in logic programming.

[3] We give details of the algebra of sets behind these when appropriate. Recall that a *Venn diagram* of a set A is a circle (or some other closed curve) inside a quadrilateral closed area; this closed area depicts the universe **U** (see Fig. 5.1.1). If there are two or more sets, then their corresponding circles overlap in such a way that their intersection and/or union can be considered. Typically, the presence of members is depicted by filling in areas, or by some other means (e.g., the symbol ×). An *Euler diagram* of a set A just is a circle, without consideration of the universe (see Fig. 1.2.1). Below, we elaborate on both kinds of diagrams.

143

5. The language of classical logic

5.2. L1 and classical subsets/extensions thereof

We shall consider the language of classical logic to be L1, as firstly introduced in Section 2.1.1, with the specifications that follow.

Definition 5.1. Let there be given the sets $O_L = \{\neg^1, \wedge^2, \vee^2, \rightarrow^2, \leftrightarrow^2\}$ and $Q_{L1} = \{\forall, \exists\}$ of connectives and quantifiers, respectively. If we define terms, atoms and formulas (whose corresponding sets are T, At, and F, respectively) over the signature $\Upsilon_{L1} = Pred_{L1} \cup Fun_{L1} \cup Cons_{L1}$, with $Pred_{L1} \cap Fun_{L1} = \emptyset$, inductively in the Backus-Naur notation as

Terms	t	::=	$x \mid a \mid f(t_1, ..., t_n)$
Atoms	$P\ (Q, ...)$	::=	$p \mid P(t_1, ..., t_n)$
Formulas	$A\ (B, ...)$	::=	$P \mid \neg A \mid A \wedge B \mid A \vee B \mid A \rightarrow B \mid A \leftrightarrow B \mid$ $\forall x\,(A) \mid \exists x\,(A)$

then we have the syntax of the logical connectives $\neg$ (*negation*), $\wedge$ (*conjunction*), $\vee$ (*disjunction*), $\rightarrow$ (*material implication*, or *conditional*), and $\leftrightarrow$ (*material equivalence*, or *biconditional*), and of the quantifiers $\forall$ (*universal quantifier*) and $\exists$ (*existential quantifier*) that we shall refer to as L1.

5.2.1. The classical connectives

Often, we actually require the set $O_L = \{\top^0, \bot^0, \neg^1, \wedge^2, \vee^2, \rightarrow^2, \leftrightarrow^2\}$, where the additional 0-ary connectives stand for an arbitrary tautology and an arbitrary contradiction. This is a very frequent addition to the connectives of L, so much so that we shall not consider it to be an extension thereof.

Definition 5.2. Let $A, B \in L0$ be two formulas and let the set O_L be given. Then, we can define the classical connectives of O_L by means of Venn diagrams as shown in Figure 5.2.1.

We henceforth omit the superscripts indicating arity for the connectives of O_L.

As seen above, L1 is an extension of L0. In terms of orders, L0 provides the language for the *classical propositional logic* (CPL), and L1 does so for the *classical first-order logic* (CFOL). We denote by L the language

5.2. L1 and classical subsets/extensions thereof

Operator	Formula	Truth Table	Diagram	Set Algebra
$\top$	$\top$	1		$A \cup \overline{A} = \mathbf{U}$ $\overline{\emptyset} = \mathbf{U}$
$\bot$	$\bot$	0		$A \cap \overline{A} = \emptyset$ $\overline{\mathbf{U}} = \emptyset$
	A	10		$A \cap \mathbf{U} = A$
$\neg$	$\neg A$	01		$\overline{A}$
$\wedge$	$A \wedge B$	1000		$A \cap B$
$\vee$	$A \vee B$	1110		$A \cup B$
$\rightarrow$	$A \rightarrow B$	1011		$\overline{A} \cup B$
$\leftrightarrow$	$A \leftrightarrow B$	1001		$(\overline{A} \cup B)$ $\cap$ $(\overline{B} \cup A)$

Figure 5.2.1.: Diagrammatic representations of the connectives of O_L.

5. The language of classical logic

of classical logic when it is not necessary to specify the order; by this, we usually mean L1, though higher orders might also be meant.

Often, formulas of L are simplified in the sense that parentheses are omitted. This, however, is only possible because there are agreed-on precedence rules for the connectives in O_L.

Fact 5.3. *The following is the precedence of the logical connectives in O_L:*

1. $\neg$
2. $\wedge$
3. $\vee$
4. $\rightarrow$
5. $\leftrightarrow$

Example 5.4. The formula $((((P \rightarrow Q) \rightarrow (R \rightarrow S)) \wedge (\neg P)) \vee R)$ can be simplified as

$$((P \rightarrow Q) \rightarrow (R \rightarrow S)) \wedge \neg P \vee R.$$

Definition 5.5. The language L is functionally complete (cf. Def. 3.17), as it has at least one set of connectives that is functionally complete.

Example 5.6. The subset $O'_{\neg,\rightarrow} = \{\neg, \rightarrow\}$ of L is functionally complete. In effect, we have the following inductive definitions of the remaining connectives of O_L:

1.
$$(\vee_{df}) \qquad A \vee B \quad := \quad \neg A \rightarrow B$$

2.
$$(\wedge_{df}) \qquad A \wedge B \quad := \quad \neg(A \rightarrow \neg B)$$

3.
$$(\leftrightarrow_{df}) \quad A \leftrightarrow B \quad := \quad (A \rightarrow B) \wedge (B \rightarrow A) \, ; \, (A \wedge B) \vee (\neg A \wedge \neg B)$$

As a matter of fact:

Proposition 5.7. *The subsets $O'_{\neg,\vee}$ and $O'_{\neg,\wedge}$ of L are also functionally complete. $O' = \{\neg, \wedge, \vee\}$ is also functionally complete.*

146

5.2. L1 and classical subsets/extensions thereof

Proof: Left as an exercise.

Although $O'_{\neg,\rightarrow}$, $O'_{\neg,\vee}$, and $O'_{\neg,\wedge}$ are already very small sets indeed, if we extend O_L as shown next we can actually obtain one-element functionally complete connective sets.

Definition 5.8. Let $O_G = \{\uparrow^2, \downarrow^2, \leftrightarrow\!\!\!/\,^2\}$ be a set of binary logical connectives whose definitions are given in Figure 5.2.2.

Operator	Formula	Truth Table	Diagram	Set Algebra
$\uparrow$	$A \uparrow B$	0111		$\overline{A \cap B}$
$\downarrow$	$A \downarrow B$	0001		$\overline{A \cup B}$
$\leftrightarrow\!\!\!/$	$A \leftrightarrow\!\!\!/ B$	0110		$(A \cap \overline{B}) \cup (\overline{A} \cap B)$

Figure 5.2.2.: Diagrammatic representations of the logical connectives $O_G = \{\uparrow^2, \downarrow^2, \leftrightarrow\!\!\!/\,^2\}$.

Proposition 5.9. *The sets $\{\uparrow\}$ and $\{\downarrow\}$ are functionally complete with respect to O_L.*

Proof: With respect to $\uparrow$ (also: $|$), called *Sheffer stroke*, we have the following equivalences:
$$\neg A \equiv A \uparrow A$$
$$A \rightarrow B \equiv A \uparrow (B \uparrow B)$$
$$A \wedge B \equiv (A \uparrow B) \uparrow (A \uparrow B)$$
$$A \vee B \equiv (A \uparrow A) \uparrow (B \uparrow B)$$
$$A \leftrightarrow B \equiv ((A \uparrow (B \uparrow B)) \uparrow (B \uparrow (A \uparrow A))) \uparrow ((A \uparrow (B \uparrow B)) \uparrow (B \uparrow (A \uparrow A)))$$

We leave $\downarrow$, called *Peirce arrow*, as an exercise. **QED**

5. *The language of classical logic*

Proposition 5.10. *The set $O_L^* = O_L \cup O_G$ is a classical extension of O_L.*

Proof: Left as an exercise.

▶ *Do Exercises 5.2-6.*

5.2.2. The quantifiers of CFOL

In CFOL, there are only two quantifiers: $\forall$ and $\exists$. Recall from Figure 2.2.2 the main applications of these quantifiers.

Definition 5.11. Let there be given the set of quantifiers $Q_{L1} = \{\forall, \exists\}$. Then, for x an arbitrary member, the classical definition of these two quantifiers is given in Figure 5.2.3 by means of Euler diagrams. This is complemented with the corresponding set-theoretical interpretations.[4]

▶ *Do Exercise 5.7.*

5.3. Applications of L1

5.3.1. Logical arguments: Categorical syllogisms

The diagrammatic representations of the quantifiers of Q_{L1} not only provide intuitive definitions of these quantifiers, but also constitute an equally intuitive means to check the validity of arguments. In Parts IV and V, we shall do the latter task by carrying out fully formal proofs in proof calculi, but it will be interesting–and useful–to familiarize the reader with this simpler, more intuitive, method of "proof" that does not require much metatheory and is essentially a direct application of L1.

Both Euler and Venn diagrams can be used for evaluating arguments. Although this latter diagrammatic representation is more frequent today, some readers find Euler diagrams more intuitive, as these only show the relevant relationships among sets. We thus give examples of both.

Our interest in this Chapter is not so much argumentation per se, but the expressiveness of L1, and these two kinds of diagrams are useful to getting familiarized with quantification over individual variables. Therefore, we shall restrict our brief discussion to *categorical syllogisms*, a very restricted form of argument in which the premises and the conclusion

---

[4]With respect to a set, we denote all members thereof by "x" and one/some member(s) thereof by "$\bullet x$". Furthermore, assume that no set is empty (though an intersection may be so), so that if there is no "x", then there may be some other variable.

5.3. Applications of L1

	Formula	Diagram	Set Theory
∀	$\forall x\,(R(x))$	(R)	$R = \{x_i \mid i \in I\}_{i=1}^{n}$
	$\forall x\,(R(x) \to S(x))$	(R inside S)	$R \subseteq S$
	$\forall x\,(R(x) \to \neg S(x))$	(R, S disjoint)	$R \cap S = \emptyset$
∃	$\exists x\,(R(x))$	(x in R)	$x \in R$
	$\exists x\,(R(x) \land S(x))$	(x in R∩S)	$R \cap S = x$
	$\exists x\,(R(x) \land \neg S(x))$	(x in R, not S)	$x \in R,\, x \notin S$

Figure 5.2.3.: Euler diagrams for the classical quantifiers.

5. The language of classical logic

are *categorical propositions*, i.e. statements beginning with *all*, *no*, or *some*. These are the four statements that constitute the famous *square of opposition*, which has its roots in Aristotle. The four statements are of the forms:[5]

| A | All S are P. | No S is P. | E |

| I | Some S is P. | Some S is not P. | O |

A and **E** are called *universal*, **I** and **O** *particular* categorical statements or propositions. Statements **A** and **O**, as well as **E** and **I**, are said to be *contradictory*. The relations **A** – **E** and **I** – **O** are said to be *contrary* and *subcontrary*, respectively, and the relations **A** – **I** and **E** – **O** are called *subaltern*.

The so-called *modern* square of opposition, based on G. Boole's algebraic approach, only considers the relations of contradiction, all the others being logically undetermined. Importantly, this approach is formal through and through in the sense that form alone matters, and the (philosophical) question of whether or not S has actually any member, known as *existential implication*, is of no importance whatsoever. We here consider the Boolean approach.

The method of evaluating an argument in L1 consists in representing the premises by means of an Euler or Venn diagram: If the conclusion is already represented in the diagram, then the argument is considered valid; otherwise, it is invalid. If there is a counter-example, i.e. a diagrammatic representation that actually denies the conclusion, the argument is equally invalid. Thus, we can say that an argument is valid if the premises state at least as much as the conclusion.

5.3.1.1. Evaluating arguments with Euler diagrams

Euler diagrams are compositions of–typically two to four–closed curves showing only the relevant relations among them. The diagrams in Figures 1.2.1 and 5.2.3 are Euler diagrams.

Example 5.12. Let there be given the following argument A1:

1. All cats are finicky animals.
2. Some finicky animals have a tail.
3. Therefore, some cats have a tail.

---
[5]See Figure 5.2.3.

5.3. Applications of L1

We need two sets for premise 1, say, C the set of cats and F the set of finicky animals. For premise 2, we need an additional set, say, T of animals with a tail (i.e. tailed animals). Formalizing the above premises and conclusion in L1, we have

1. $\forall x\, (C(x) \to F(x))$
2. $\exists x\, (F(x) \land T(x))$
3. $\exists x\, (C(x) \land T(x))$

We evaluate the argument by drawing the diagram for the premises and then checking if it does *necessarily* represent the conclusion, too. We build our diagram accordingly to Figure 5.2.3. Actually, we can draw two diagrams, one (Fig. 5.3.1.2) in which $(C \cap T) \cap F \neq \emptyset$, representing the fact that indeed some cats do have a tail, and another (Fig. 5.3.1.1) in which $T \cap F \neq \emptyset$ but $C \cap T = \emptyset$, representing the fact that no cat has a tail (or that no tailed animal is a cat), which is clearly the denial of the conclusion of the argument. Thus, Figure 5.3.1.1 constitutes a *counter-example* to our conclusion, and the argument is *invalid*.

We now reformulate our argument as A2:

1. All cats are finicky animals. $\forall x\, (C(x) \to F(x))$
2. Some tailed animals are cats. $\exists x\, (T(x) \land C(x))$
3. Hence, some tailed animals are finicky. $\exists x\, (T(x) \land F(x))$

This time, there is only one way to represent diagrammatically the two premises, and that corresponds to Figure 5.3.1.2. Not only does this representation agree with the conclusion of the argument, but also there is no other diagrammatic representation possible, so that the conclusion follows necessarily from the premises. This argument is thus *valid*.

5.3.1.2. Evaluating arguments with Venn diagrams

More intuitive as they might be, Euler diagrams are not as adequate for L1 as they are for natural language. In effect, given three sets X, Y, and Z all whose members are xs, ys, and zs, respectively, it is not a matter of fact to visualize their inclusion and their exclusion in a Euler diagram. As shown in Figure 5.3.2, this is not the case in a Venn diagram. In this Figure, inclusion is represented by x (or y, etc.), and exclusion is represented by x' (or y', etc.). Given n curves, a Venn diagram then must represent the 2^n zones of inclusion and exclusion, called *minterms*.[6]

---

[6] A minterm is actually a conjunction of disjunctions. In any case, xyz, say, abbreviates $x \land y \land z$.

5. The language of classical logic

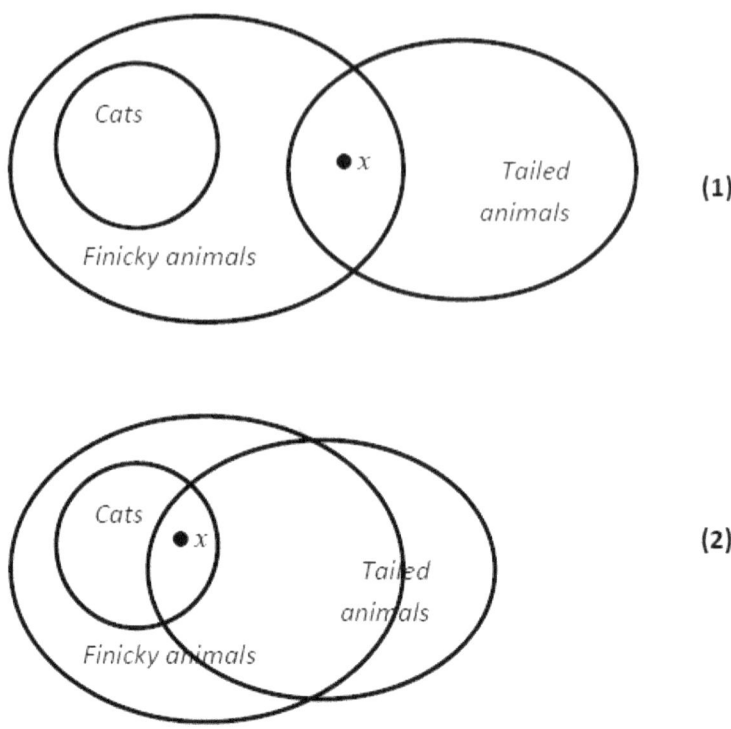

Figure 5.3.1.: Euler diagrams of an invalid (1) and a valid (2) argument.

We adopt this technique: We shade an area with no members.[7] For example, to represent "All X are Y" we shade the minterms $xy'z'$ and $xy'z$, denoting with this that it is not possible to have the case xy', i.e. no members of X are outside Y. To represent the inclusion of some member(s), we draw some shape. For example, to represent "Some Y is Z" we draw a cross in the area for yz. Because there are in fact two minterms for this case, to wit, xyz and $x'yz$, we draw the cross on the line dividing both minterms.

Example 5.13. We construct the Venn diagram for argument A1. We make X be the set of cats, Y be the set of finicky animals, and Z be the set of tailed animals. We shade the area corresponding to "All cats are finicky animals" and draw a cross for the premise "Some tailed animals are cats," as shown in Figure 5.3.3. As it can be seen in this diagram, the cross is on the division between xyz and $x'yz$, which means that

---

[7] Note that this is exactly the opposite of what we did above when drawing Venn diagrams for the connectives of L1. However, we are now concerned with quantifiers alone. See Exercise 5.8.

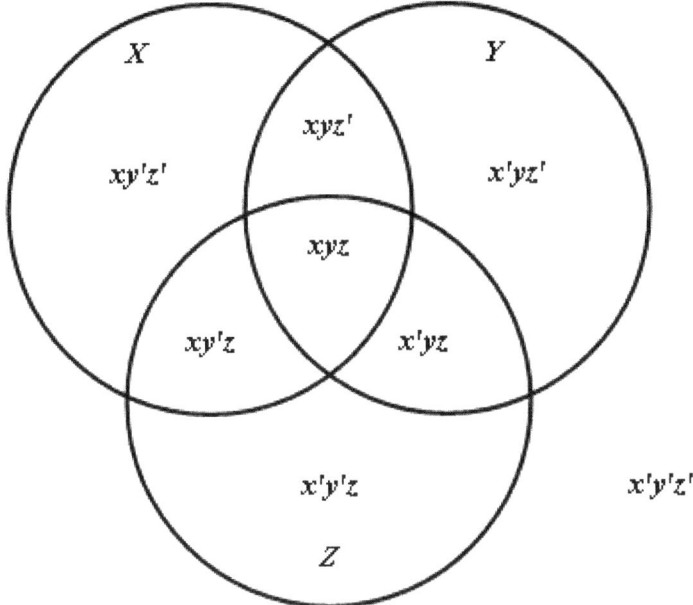

Figure 5.3.2.: Venn diagram with eight minterms.

though it is indeed possible that some cats have a tail (i.e. xyz), it is also possible that no cat does have a tail (i.e. $x'yz$), and the argument is thus invalid.

Example 5.14. We do now the same for argument A2. When drawing the cross, we actually have to do it in the minterm xyz, as the other possibility, in the minterm $xy'z$, has already been ruled out as having no members.[8] In Figure 5.3.4, looking at the Venn diagram representing the premises, we see that it also represents the conclusion. A2 is thus a valid argument.

▶ *Do Exercises 5.8-10.*

5.3.2. Logic programming (I): Prolog

Logic programming, a programming paradigm whose main (family of) language(s) is **Prolog**, is based on the non-functionally complete subset $O'_L = \{\rightarrow, \wedge\}$. Our interest falls on precisely the fact that a FO logical language of such frugality and essentially non-functionally complete can

---

[8]Thus, if only one of the premises is a universal proposition, the corresponding shading must be completed first of all.

5. The language of classical logic

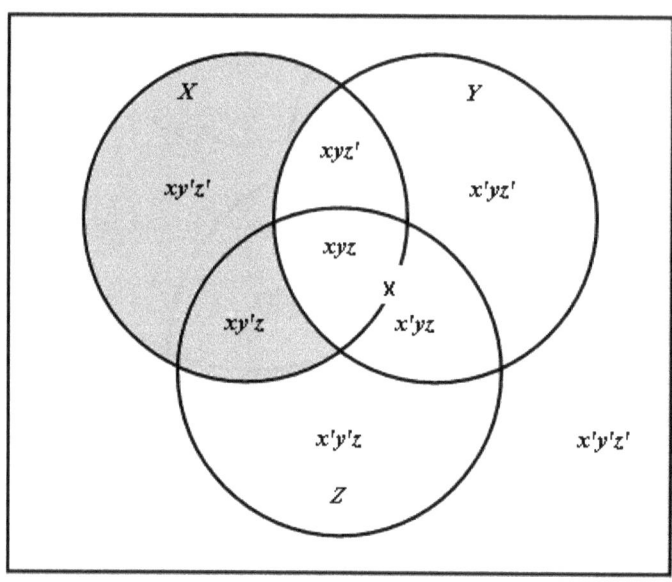

Figure 5.3.3.: A Venn-diagram representation of argument A1.

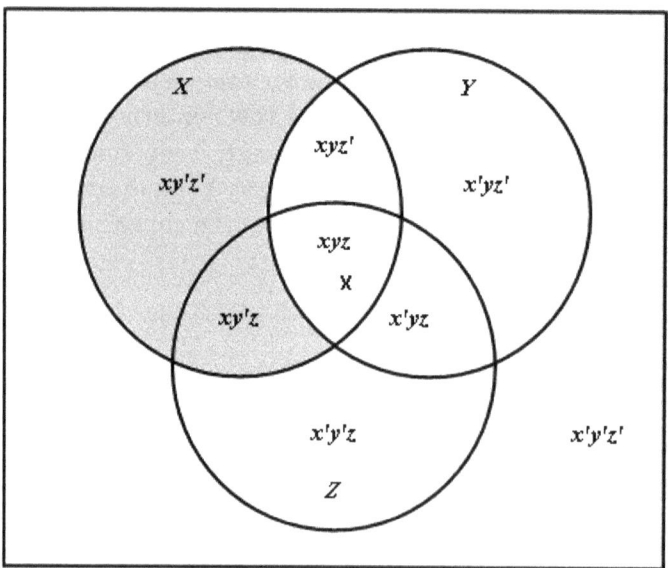

Figure 5.3.4.: A Venn-diagram representation of argument A2.

154

5.3. Applications of L1

be so expressive as **Prolog**, as well as on the fact that Prolog is an adequate deductive system.[9]

Prolog can be used at a purely propositional level, but then it is a programming system with very limited applications. On the other hand, it can also be used at a second-order level, but this is outside our scope. Thus, we concentrate here on the above reduct from the viewpoint of a FO language, namely of a FO predicate language that is classical through and through.[10] In the present Section we focus on the aspect of expressiveness, and in Section 14.2.2 we elaborate on Prolog as a sound and complete system. However, there is much more to Prolog than we discuss here. For more contents on Prolog, see Augusto (2020a, Chapter 9), where further, more specialized, literature is also cited.

5.3.2.1. The language (of) Prolog

A Prolog formula is a clause, more specifically so a Horn clause.

Definition 5.15. A *Prolog clause* is a formula of the form

$$(\mathcal{C}_P) \qquad \forall x_1...\forall x_l \, (A \leftarrow B_1, ..., B_n) \, .$$

A Prolog clause is typically simplified as

$$(\mathcal{C}_P) \qquad A \leftarrow B_1, ..., B_n$$

where A is called the *head*, and $B_1, ..., B_n$ is the *body*. This is in fact a Horn clause, as inverting the symbol $\leftarrow$ we have the formula

$$B_1 \wedge ... \wedge B_n \rightarrow A.$$

Applying now $\rightarrow_{df}$ and DeM$_\wedge$, we have

$$B_1 \wedge ... \wedge B_n \rightarrow A \equiv \neg (B_1 \wedge ... \wedge B_n) \vee A \equiv \neg B_1 \vee ... \vee \neg B_n \vee A.$$

Obviously, $A \leftarrow B_1, ..., B_n$ is a *definite clause*.

---

[9] We henceforth write "Prolog" indifferently for the language and the programming system. This common practice in the field of programming avoids a fastidious distinction between both a programming language and the corresponding system(s). Moreover, we shall consider Prolog as a synonym for logic programming.

[10] If we consider Prolog as an algebra of formulas $\mathfrak{P} = (F, \rightarrow, \wedge)$, then it is a reduct of the algebra of formulas known as classical logic $\mathfrak{K} = (F, \neg, \wedge, \vee, \rightarrow, \leftrightarrow)$. Besides being classical, this reduct allows for *Turing-completeness* of Prolog, where by this we mean that Prolog has a computational power equivalent to that of a Turing machine. Note that Prolog actually includes the connective for negation, but only implicitly–and problematically–so (see below).

155

5. The language of classical logic

Definition 5.16. The basic constructs of Prolog are *terms* and *statements*.

1. A Prolog *term* can be simple or compound. A *simple term* is a variable or a constant. A *compound term* comprises a *functor* and a sequence of one or more terms called *arguments*. A compound term of arity n has the form $p(t_1, ..., t_n)$, where p is the *name* of the functor and $t_1, ..., t_n$ are the arguments of p. A functor p with arity n is denoted by p/n. A functor can be a *relation symbol* or a *function symbol*. The name of a functor is an *atom*. A constant just is a functor of arity 0; hence, it is also an atom.

2. Prolog *statements* can be *facts*, *goals*, *rules*, and *queries*. Let $\mathcal{C}_P$ be given; then, the unit clause A is a fact, the B_i are goals, and $\mathcal{C}_P$, for $n \geq 0$, is a rule. Thus, a fact is the special case of a rule when $n = 0$. When $n = 1$, we have an *iterative clause*. A *query* is a clause with a question mark.

It should be obvious that *relation symbol* just is another name for *predicate (symbol)*, and we shall favor the latter over the former for consistency reasons.

Definition 5.17. Let there be given the two-element set

$$O_P = \left\{ \begin{array}{c} :- \\ , \end{array} \right\} = \left\{ \begin{array}{c} \leftarrow \\ \wedge \end{array} \right\}$$

of operators and the punctuation marks "," (between arguments) and ".", and left and right parentheses. If we define inductively terms, atoms, and statements over a signature $\Upsilon = (Pred, Fun, ar)$ with $ar \geq 0$ denoted by $\cdot/n$ for $t_1, ..., t_n$ terms, as

Terms (t)	Variables	X	::=	e.g., X | John | _john
	Constants	c	::=	e.g., a | john
	Functors	p; f	::=	p/n; f/n
Atoms		$A\,(B, ...)$	::=	$p \mid f \mid X$
Statements	Facts	A	::=	A.
	Rules	r	::=	$A : -B_1, ..., B_n.$
	Goals	$G\,(A, ...)$	::=	$B_1, ..., B_n$

5.3. Applications of L1

then we have the language (basic FO) Prolog and its syntax.

In the set O_P, the connective : − just is the connective ←, which in turn just is the conditional connective of L1 reverted; the comma is the connective for conjunction. As seen above, Prolog's conditional connective behaves exactly like the conditional connective of L1, but this is not the case for Prolog's conjunction: this is a non-commutative connective. In effect, in L1 we have $A \wedge B = B \wedge A$ (cf. Fig. 3.5.1), but this is generally not the case in Prolog, reason why the order of both the goals in the body of rules and the rules in a program is crucial.

Example 5.18. X, Y, John, john, sara, father, father (X, Y), and male(john) are terms of Prolog.

- X, Y and John are individual variables, john and sara are atoms (constants, or names of individuals), and father (X, Y), as well as male(john), are functors whose arity is denoted by father/2 and male/1, respectively. It is easy to see that individual variables are written with initial uppercase letters and atoms are written with initial lowercase letters; variables can also start with an underscore "_".[11]

- father (X, Y) is a non-ground atom and male(john) is a ground atom.

- father (X, sara). and male(john). are facts built from the atoms father (X, sara) and male (john).

- father(john, sara)? is a query asking whether the relation "X is the father of Y" holds between John and Sara, i.e. whether John is Sara's father.

- daughter (X, Y) : −father (Y, X), female (X). is a rule for the relation "X is the daughter of Y." In this rule, father (Y, X) and female (X) are the goals, but the head, daughter (X, Y), can also be a goal in a query, and as such it corresponds to the right side of : −, being thus (implicitly) a negative literal.

---
[11] Variables with "_" are called *anonymous variables* and each such occurrence in a clause or query denotes a different variable.

5.3.2.2. Increased expressiveness and ambivalent syntax

Contrary to the language L1, in Prolog the same functor name can be used for functions or predicates of different arity, a feature that is responsible for what is spoken of as *ambivalent syntax*. This is a useful feature when there is a natural relation between predicate and function symbols (see Example 5.19), but it can be problematic in some cases (see Example 5.20). In particular, it violates the condition in L1 (cf. Def. 5.1) that the intersection of the sets *Pred* and *Fun* must be empty.

Example 5.19. The atom father(john, father(rita)) is a legal atom of Prolog. With it, we express the fact that John is the father of the father of Rita, i.e. John is Rita's grandfather. In this atom, thus, the functor father/n is a predicate and a function name at the same time.

In this Example, we have the predicate father/2 and the function father/1, but in fact we can have the same functor with the same arity both as a predicate symbol and as a function symbol.

Example 5.20. The following is a legal rule of Prolog:

$$p(t_1, ..., t_n) : -not(p(t_1, ..., t_n)).$$

In this rule, p/n is a predicate symbol in the head, but it is also a function symbol in the body.

In Definition 5.17, note that an individual variable can also be an atom. We call such a variable a *metavariable*.

Example 5.21. In Prolog, given a predicate p/1, it is possible to have the legal rule

$$p(X) : -X.$$

In the head of the rule, X is the individual variable of the predicate p/1; in the body, X is a metavariable. Examples of such predicates are the in-built Prolog predicates call/1 and solve/1.

Importantly, the occurrence of both cases above can be combined.

Example 5.22. For instance, for a predicate functor p/1, we can have the rules

$$p(p(X)) : -p(X).$$
$$p(X) : -X.$$

While this and the above Examples show that Prolog has an increased expressive power with respect to L1, deduction in this programming system is bound to be more error-prone if we accept the fact that Prolog has an ambivalent syntax.

5.3. Applications of L1

5.3.2.3. Programs and substitutions

Definition 5.23. A *Prolog program*, or *logic program*, in its simplest form is a finite set of facts. More typically, a logic program is a finite set of rules.

Example 5.24. The following facts constitute the program *Fatherhood*:

```
father(john, sara).      male(john).
father(john, peter).     male(rick).
father(john, rick).      male(peter).
father(rick, carl).      male(carl).
father(harry, louis).    male(harry).
father(harry, mary).     male(louis).
father(harry, jane).     female(sara).
                         female(mary).
                         female(jane).
```

The program *Fatherhood* in Example 5.24 contains facts about three fathers, to wit, John, Rick, and Harry, and their respective children. As seen above, a goal is typically an element in the body of a rule, but it can also be a fact. If we enter the query

$$\text{father}(\text{john}, \text{sara})?$$

in some Prolog implementation, we obtain the answer "Yes" or "True". However, if we enter the query

$$\text{father}(\text{rick}, \text{sara})?$$

we get "No" or "False" as a reply. As limited a program as *Fatherhood* is, it suggests the applications of fact-only programs in the case of large databases. In particular, we can obtain answers to queries respecting relations between individuals, as well as to queries concerning individuals. For instance, given the program *Fatherhood*, we can obtain replies to the questions "Who is the father of Sara?", "Is Peter a male?", "Is Mary John's child?", and "Who is a female?" by entering the corresponding queries

$$\text{father}(X, \text{sara})?$$

$$\text{male}(\text{peter})?$$

$$\text{father}(\text{john}, \text{mary})?$$

5. The language of classical logic

female (Y)?

The first and the last queries above make it evident that *substitution* is a fundamental operation in Prolog. Indeed, queried father(X, sara)?, Prolog checks the first argument of the predicate father/2 given Sara as the second argument thereof, and deduces X = john. When queried female(Y)?, Prolog gives all the instances of Y in the program, to wit, Sara, Mary, and Jane.[12]

Definition 5.25. A Prolog *substitution* is a set of pairs of the form $X_i = t_i$, $0 \leq i \leq n$, where X_i is a variable and t_i is a term, $X_i \neq X_j$ for every $i \neq j$, and X_i does not occur in t_j for any i and j.

1. Let σ be a substitution and A a term; then the result of applying substitution σ to term A, denoted by $A\sigma$, is the term obtained by substituting t for every occurrence of X in A for every pair $(X = t) \in \sigma$.

2. We say that B is an *instance* of A if there is a substitution σ such that $A\sigma = B$.

Example 5.26. Consider the program *Fatherhood*. Let there be given the substitution $\sigma = \{X = \text{john}, Y = \text{sara}\}$. Then, the result of applying σ to the term father(X, Y), denoted by (father(X, Y))σ, is the term father(john, sara). The goal father(john, sara) is an instance of the goal father(X, Y) (under substitution σ).

It is important to remark that variables in facts are implicitly universally quantified, whereas variables in queries are implicitly existentially quantified.

Definition 5.27. A fact $p(t_1, ..., t_n)$ reads "for all $X_1, ..., X_n$, where the X_i are variables in the fact, $p(t_1, ..., t_n)$ holds or is true," i.e.

$$\forall X_1, ..., \forall X_n \left(p(t_1, ..., t_n) \right).$$

This definition holds for rules: the variables appearing in the head are universally quantified and their scope is the whole rule. However, variables occurring in the body of a rule but not in its head are considered to be existentially quantified.

Example 5.28. grandfather(X, Y) ← father(X, Z), father(Z, Y). is read "for all X and Y, X is the grandfather of Y if there exists a Z such that X is the father of Z and Z is the father of Y."

---

[12]Prolog does so one at a time for each instance and provided we prompt it with ";", which is equivalent to a disjunction operator.

5.3. Applications of L1

Definition 5.29. A query $p(t_1, ..., t_n)$? reads "are there variables $X_1, ..., X_n$ such that $p(t_1, ..., t_n)$ holds or is true?", i.e.

$$\exists X_1, ..., \exists X_n \, (p(t_1, ..., t_n))?$$

For convenience reasons, the universal quantifiers are omitted in facts and the existential quantifiers are so in queries; both quantifiers are omitted in rules.

Example 5.30. The goal father(john, sara) implies that there exists an X such that father(X, sara) is *true*, in this case X = john. Then, father(john, sara) is a solution to the query father(X, sara)?, and the solution is represented by the substitution X = john.

Definition 5.31. For any substitution σ,

1. from a universal fact P deduce an instance $P\sigma$ of it. We call this *instantiation* and denote it $\vdash_{inst}$.

2. an existential query P? is a logical consequence of an instance $P\sigma$ of it. We call this *generalization* and denote it by $\vdash_{gen}$.

By combining 1 and 2 above, we have a reply to a query P? by means of a common instance, i.e. we have:

$$P. \vdash_{inst} P\sigma \vdash_{gen} (P?)$$

Definition 5.32. A *solution* to a query is a fact that is a (common) instance of the query.

Recall from Section 2.4.6 that a main employment of substitution is in *unification*. In effect, *the* fundamental operation of Prolog is unification, and this is carried out by obtaining common instances.

Definition 5.33. C is a *common instance* of A and B if it is an instance of A and an instance of B, i.e. if there are substitutions σ_1, σ_2 such that $A\sigma_1 \equiv B\sigma_2$.

Common instances are important because in fact we can reply to a query by finding a common instance of both a query and a fact.

▶ *Do Exercises 5.11-16.*

5. The language of classical logic

5.3.3. Logic design: Logic circuits

One of the simplest, yet most important and widespread, applications of L1 is in logic design, namely logic circuits. Although this is a topic that is largely outside the scope of this text, we find it interesting to give its basics, especially as an illustration of the more practical–actually, physical–applications of Boolean logic.

As said, logic design of circuits based on CL is simple, in the sense that it is based on Boolean functions (cf. Def. 3.5). An example of a Boolean function of interest in logic circuits is of the kind

$$f : D \longrightarrow E$$

where D and E are sets of electrical devices or components and for both sets we have

$$g : D, E \longrightarrow \{0, 1\}.$$

Example 5.34. Let us consider a switch S; then, we say that the switch is open if $x = 0$ and closed if $x = 1$, for x an input variable. Clearly, we have

$$S(x) = x.$$

In Figure 5.3.5.(i), we show the symbol for a switch. Consider now a lightbulb L. L is off if the switch is open (i.e. no current passes through to L) and on if the switch is closed (and the current passes through to L), so that now we have the state of the lightbulb as a function of the state of the switch, a strict Boolean function

$$f : \underbrace{\{0, 1\}}_{S} \longrightarrow \underbrace{\{0, 1\}}_{L}.$$

In any case, again we have

$$L(x) = x.$$

Figure 5.3.5.(ii) shows the simplest combination of a switch and a lightbulb (the symbol ♦ stands for power supply). Let us now consider two switches, S_1 and S_2, so that now we have two input variables x_1 and x_2. We apply now a Boolean function of degree 2, as $L(x)$ depends on $\{0,1\}^2$ for x_1 and x_2. As we also have $L(x) = S(x)$, we may restrict ourselves to the output of interest and consider only $L(x)$. For two input variables x_1 and x_2 we have

$$L(x_1, x_2) = x_1 \wedge x_2$$

if the two switches are connected in series (Fig. 5.3.5.(iii)), and

$$L(x_1, x_2) = x_1 \vee x_2$$

if the switches are connected in parallel (Fig. 5.3.5(iv)). By applying Definitions 3.4.1-2, we know that in the first case the light is on iff both switches are closed, whereas in the second case the light is on iff at least one of the switches is closed, and otherwise it is off. Let now three switches be given so that we have, say,

$$L(x_1, x_2, x_3) = (x_1 \wedge x_2) \vee (x_1 \wedge x_3).$$

Then, Figure 5.3.5.(v) shows this series-parallel connection. To check when the light is on or off we simply construct the truth table for $(x_1 \wedge x_2) \vee (x_1 \wedge x_3)$. It is easy to see that the light is on when $x_1 = 1$ and (a) either $x_2 = 1$ or $x_3 = 1$, or (b) both $x_2, x_3 = 1$; the light is off otherwise.

From this Example, it should be obvious that in logic circuits we very much restrict L1 to L0, and operate on individual variables as if they were propositional variables. In effect, we consider no domain of discourse, and thus the variables have fixed truth values such that $x = 1$ if $\neg x = 0$ and $x = 0$ if $\neg x = 1$. In order to design logic circuits we require the Boolean operators of O_G (cf. Def. 5.8), so that in fact we require the extension O_L^* of Proposition 5.10. The subscript G stands for "gate," because O_G is essentially used to construct *logic gates*, i.e. logic operations that are implemented electronically with transistors and constitute the basic elements in a logic circuit. Because circuits can be rather complex from the logical viewpoint, it is useful to represent them by means of *circuit diagrams* (or *schematics*), for which we require the graphical representations of the logical operators in Figure 5.3.6.

Complex circuits are implemented by a *network* of gates. Such circuits are thus called *logic networks*, or *logic circuits*. The process of determining the function performed by a given logic circuit is called *analysis*.

Example 5.35. Figure 5.3.7 shows the logic circuit implementing the logical function $f(x_1, x_2, x_3) = (x_1 \wedge x_2) \vee (x_1 \wedge x_3)$ from Figure 5.3.5.(v). Note in this Figure the use of the symbols · (often omitted) for ∧ and + for ∨.

The notion of logical equivalence (Def. 3.20) acquires a very practical importance in the case of logic circuits. In effect, the cost of a logic circuit is directly proportional to its complexity, i.e. the more complex

5. The language of classical logic

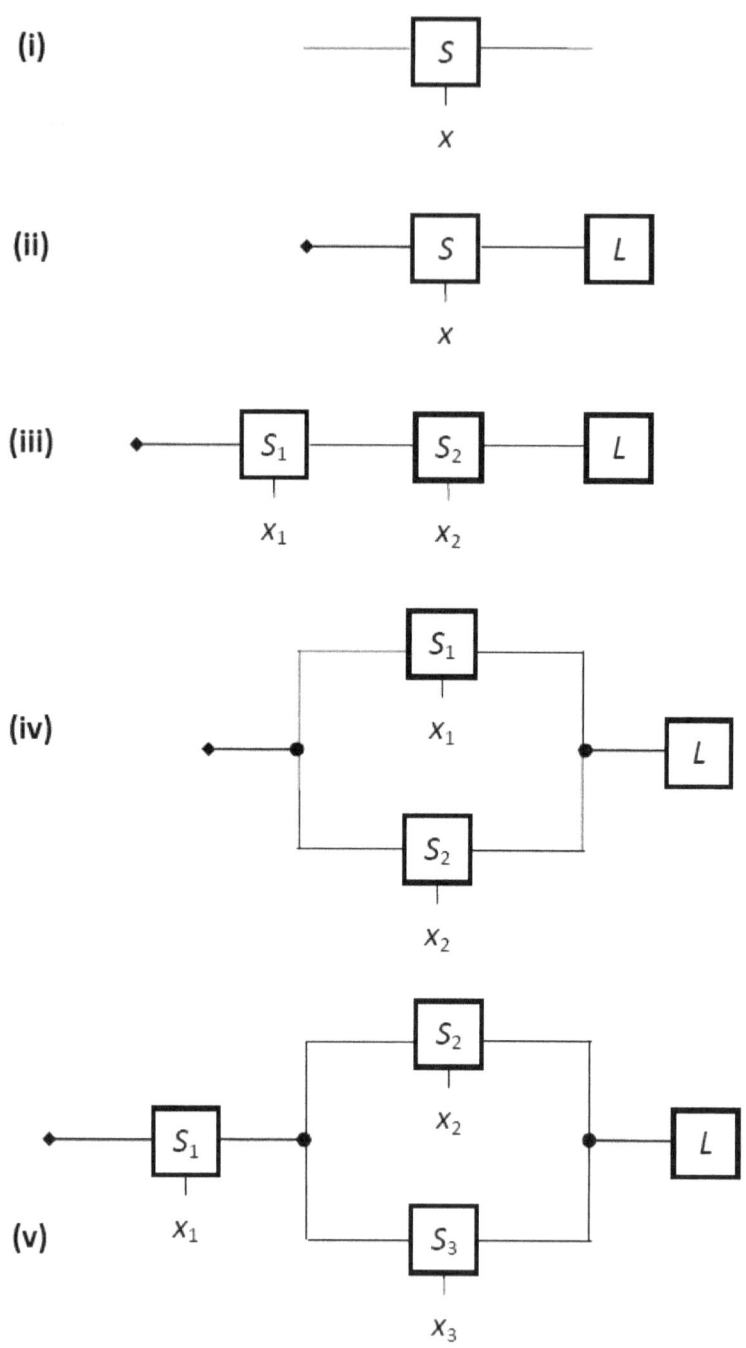

Figure 5.3.5.: From a binary switch (i) to a series-parallel connection (v).

Operator	Gate	Representation
¬	NOT	
∧	AND	
∨	OR	
↑	NAND	
↓	NOR	
↔	XOR	

Figure 5.3.6.: Logic gates and their graphical representations.

it is, the higher are the costs associated to the implementation of a logic circuit. Thus, it is desirable to reduce its complexity. By now, the reader should realize that the idea is to find less complex logic circuits that are nevertheless functionally equivalent to a given logic circuit, where by "functional equivalence" it is meant that two logic circuits output the same function. This, however, just is their truth table, so that the notions of *logical equivalence* and *functional equivalence* coincide in this particular context.

However useful truth tables are to *check* functional equivalence, in order to *find* a circuit functionally equivalent to a given one we should apply the axioms of Boolean algebra given above in Figure 3.5.1.

Example 5.36. By the property of distributivity of ∧ (Axiom *c* in Fig. 3.5.1), we have the equivalence

$$(x_1 \wedge x_2) \vee (x_1 \wedge x_3) \equiv x_1 \wedge (x_2 \vee x_3).$$

It should be easy to see that the circuit based on the right side of the equivalence is simpler than that on the left side, which is actually the

5. The language of classical logic

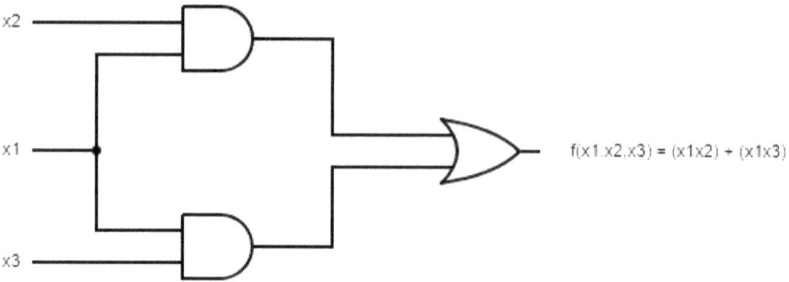

Figure 5.3.7.: A logic circuit for the function $f(x_1, x_2, x_3) = (x_1 \wedge x_2) \vee (x_1 \wedge x_3)$.

circuit of Figure 5.3.7. As a matter of fact, no simpler circuit than this, shown in Figure 5.3.8, can be found by means of the axioms of Boolean algebra, and thus this is the simplest circuit for the function $f(x_1, x_2, x_3) = x_1 \wedge (x_2 \vee x_3)$.

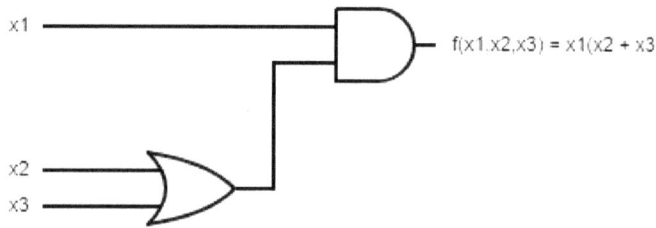

Figure 5.3.8.: A logic circuit for $f(x_1, x_2, x_3) = x_1 \wedge (x_2 \vee x_3)$.

An even more dramatic simplification is shown in the following Example:

Example 5.37. Let $f(x_1, x_2) = (\neg x_1 \wedge \neg x_2) \vee (x_1 \wedge (x_2 \vee \neg x_2))$ be given; we show that this logical function is equivalent to $f(x_1, x_2) = x_1 \vee \neg x_2$:

$(\neg x_1 \wedge \neg x_2) \vee (x_1 \wedge (x_2 \vee \neg x_2))$
$\equiv (\neg x_1 \wedge \neg x_2) \vee (x_1 \wedge 1)$ By Ax. g for $\vee$ (Fig. 3.5.1)
$\equiv (\neg x_1 \wedge \neg x_2) \vee x_1$ By Ax. h for $\wedge$ (Fig. 3.5.1)
$\equiv x_1 \vee (\neg x_1 \wedge \neg x_2)$ By Ax. a for $\vee$ (Fig. 3.5.1)
$\equiv x_1 \vee \neg x_2$ By Exercise 5.17.3

Figure 5.3.9.(i) shows the diagram for the original function and Figure 5.3.9.(ii) shows the diagram for the simplified function.

As seen above (cf. Prop. 5.9), the Sheffer stroke and the Peirce arrow, denoted by the symbols $\uparrow$ and $\downarrow$, respectively, are functionally complete in a special sense: Every connective $\heartsuit_i \in O_L$ can be defined by either one of these two connectives, i.e. both $\{\uparrow\}$ and $\{\downarrow\}$ are functionally complete sets with respect to O_L. It should be obvious by now that this functional completeness has as a practical consequence significant minimizations in logic circuits using their logic circuit corresponding gates, to wit, the NAND and NOR gates.[13] Figure 5.3.10 shows the definition of these gates by means of the De Morgan's laws.

Although the employment of the XOR gate can also lead to significant simplifications in terms of circuit complexity, our interest in it falls mainly on its algebraic properties. These we give in Figure 5.3.11, and leave their proofs as an exercise. Note in Figure 5.3.11 the use of the alternative symbol $\oplus$.

a.	$x \oplus y = y \oplus x$	*Commutativity of $\oplus$*
b.	$x \oplus (y \oplus z) = (x \oplus y) \oplus z$	*Associativity of $\oplus$*
c.	$x \wedge (y \oplus z) = (x \wedge y) \oplus (x \wedge z)$	*Distributivity of $\oplus$*
d.	$x \oplus x' = 1$	*Complementation for $\oplus$*
e.	$x \oplus x = 0$	*Idempotency for $\oplus$*
f.	$x \oplus 0 = x$	*Identity element of $\oplus$*
g.	$x \oplus 1 = x'$	*Inverse element of $\oplus$*
h.	$(x \oplus y)' = x \oplus y' = x' \oplus y$	*De Morgan's law for $\oplus$*

Figure 5.3.11.: Properties of XOR.

▶ *Do Exercises 5.17-20.*

Exercises

Exercise 5.1. Research into the reasons why the following connectives, which can all be added to O_{L1}, are said to be non-classical:

1. $\Box^1$ (read: *necessarily*) in modal logic.

2. $\Diamond^1$ (read: *possibly*) in modal logic.

3. $\sim^1$ in intuitionistic logic.

---

[13] Additionally, from the viewpoint of efficiency these gates are generally faster.

5. The language of classical logic

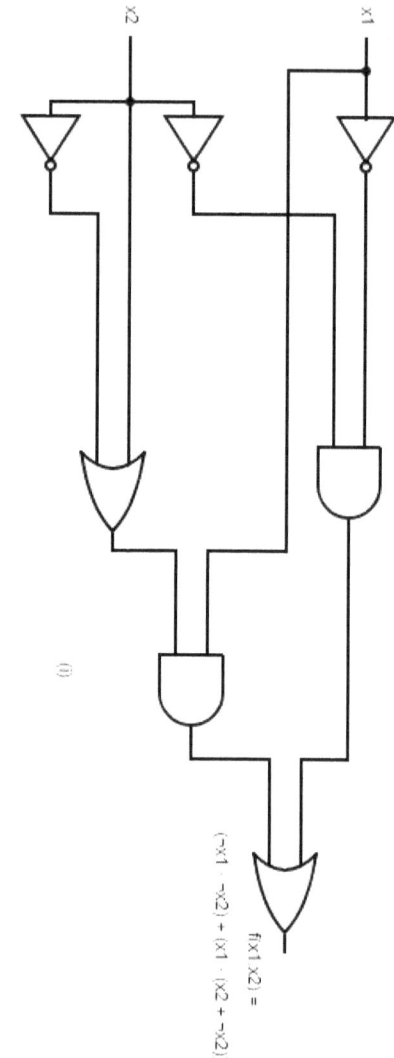

Figure 5.3.9.: Two functionally equivalent logic circuits.

5.3. Applications of L1

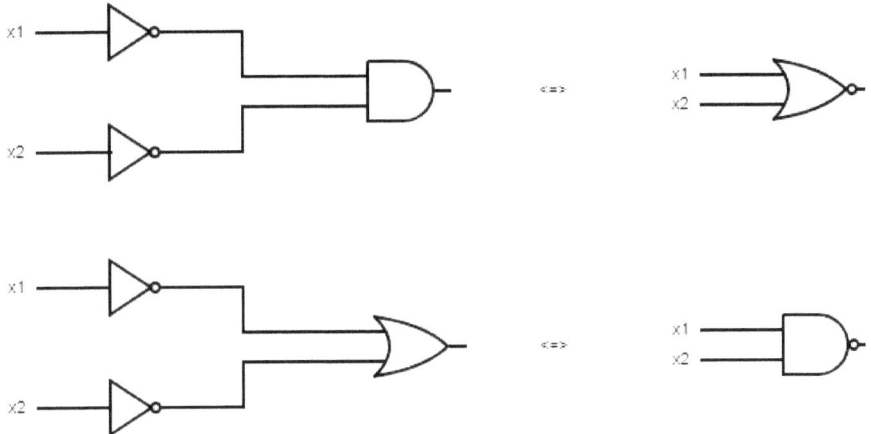

Figure 5.3.10.: The De Morgan's laws and the NOR and NAND gates.

4. O^1 (read: *it is obligatory*) in deontic logic.

5. $\&^2$, known as *strong conjunction* in fuzzy logic.

6. ∇^2, known as *strong disjunction* in fuzzy logic.

Exercise 5.2. Simplify the following formulas of L0, if possible, by removing superfluous parentheses:

1. $(\neg((\neg(\neg(Q \wedge R))) \to (\neg(P \leftrightarrow Q))) \wedge (\neg(\neg P)))$.
2. $(\neg((\neg(A) \to (B \vee C)) \leftrightarrow (\neg(A \wedge (\neg B)))))$.
3. $((A \to B) \to (\neg(\neg(A \leftrightarrow B))))$.

Exercise 5.3. Find a shorter formula than that of the proof of Proposition 5.9 with the Sheffer stroke and equivalent to $A \leftrightarrow B$.

Exercise 5.4. Rewrite the formulas of Exercise 2.9 by using solely the Sheffer stroke.

Exercise 5.5. Show that the subset $O'_L = \{\bot^0, \to^2\}$ is functionally complete.

Exercise 5.6. Do the following with respect to the indicated Propositions:

1. Prove Propositions 5.7 and 5.10.
2. Complete the proof of Proposition 5.9.

5. The language of classical logic

Exercise 5.7. Prove the following equivalences with respect to the set $Q_{L1} = \{\forall, \exists\}$:

1. $\neg \forall x \, (\phi) \equiv \exists x \, (\neg \phi)$
2. $\neg \exists x \, (\phi) \equiv \forall x \, (\neg \phi)$
3. $\neg \forall x \, (\neg \phi) \equiv \exists x \, (\phi)$
4. $\neg \exists x \, (\neg \phi) \equiv \forall x \, (\phi)$

Exercise 5.8. Draw two intersecting circles (a Venn diagram) and represent the following propositions by means of shading minterms with no members and marking areas with at least one member with a cross:

1. It is false that all X are Y.
2. It is false that all X are not Y.
3. It is false that some X are Y.
4. It is false that no X is Y.
5. No X are non-Y.
6. All X are non-Y.
7. Some X are not non-Y.
8. Some X are non-Y.
9. Some non-Y are non-X.
10. All non-Y are non-X.
11. No non-Y are non-X.
12. Some non-Y are not non-X.

Exercise 5.9. Check the validity of the following syllogisms by means of Venn and Euler diagrams:

1. All caterpillars are worms. All larvae are caterpillars. Hence, all larvae are worms.

2. No snakes have spectacles. All boa constrictors are snakes. Therefore, no boa constrictors have spectacles.

3. No logic exercise is fun. Some homework is logic exercises. As a conclusion, some homework is not fun.

4. All dinosaurs are extinct. All velociraptors are dinosaurs. Thus, some velociraptors are extinct.

5. All crickets are musical. No praying mantas are musical. Ergo, some praying mantas are not crickets.

6. Some cats have no fur. All cats are mammals. Therefore, some mammals have no fur.

7. All doodles are deedle. All goglers are doodles. Hence, all goglers are deedle.

8. No butterfly is repellent. Some insects are butterflies. Thus, some insects are not repellent.

Exercise 5.10. Research into the topic of *existential fallacy*.

1. Explain in what this consists.

2. Give the eight inference forms (i.e. one premise followed by the conclusion) of the existential fallacy.

Exercise 5.11. With respect to the program *Fatherhood* (Example 5.24), ask the following queries in the correct format:

1. Who is Sara's father?

2. Is Rick Carl's father?

3. Who are John's children?

4. Who are John's sons?

5. Who are John's daughters?

6. Who is Mary's father?

7. Has Rick got any children?

8. Are Louis, Mary, and Jane all Harry's children?

Exercise 5.12. With respect to the program *Fatherhood*:

5. The language of classical logic

1. Create a rule that determines that being a father is a predicate of a male.

2. Create a rule that allows us to find if one of the males is a grandfather. By means of this rule, find whether Harry, John, or Rick are grandfathers.

3. Create a rule that allows us to determine who a grandfather's grandchildren are.

Exercise 5.13. Add rules to the program *Fatherhood* so that this becomes the program *Parenthood*.

Exercise 5.14. Consider the following two rules:
nearby (X, Y) ← connected (X, Y, W).
nearby (X, Y) ← connected (X, Z, W), connected (Z, Y, W).

1. Write a program *European_borders* for the countries Spain, Portugal, France, Italy, Germany, Switzerland.

2. Instantiate the variable W to other values to produce different programs.

3. Add one or more rules to allow the program *European_borders* to give "true" replies to queries such as nearby (France, Italy)? and nearby (Italy, France)?.

Exercise 5.15. Consider the following Prolog program *Addition*.
plus (0, X, X) ← natural_number(X).
plus (s (X), Y, s (Z)) ← plus (X, Y, Z).
natural_number (0).
natural_number (s (X)) ← natural_number (X).

1. What kind of goals can be obtained from it?

Exercise 5.16. With respect to the argument in Exercise 2.13:

1. Write the premises of the argument as a Prolog program.

2. Formulate the following as a query:

   a) What country is Daffy an enemy of?

5.3. Applications of L1

b) Who is a native of country C?

c) What does the West sell to Daffy?

d) Who is hostile?

e) Is M1 a weapon?

f) What does Daffy own?

g) Does the West sell a missile to Daffy?

h) Does Daffy own a M1 missile?

Exercise 5.17. Show by perfect induction (i.e. by substituting truth values for variables) that the following properties hold for $x, y, z \in A$, $\mathfrak{B} = (A, \neg, \wedge, \vee)$ is a Boolean algebra:

1. $(x \wedge y) \vee (x \wedge \neg y) = x$

2. $(x \vee y) \wedge (x \vee \neg y) = x$

3. $x \vee (\neg x \wedge y) = x \vee y$

4. $x \wedge (\neg x \vee y) = x \wedge y$

5. $(x \wedge y) \vee (y \wedge z) \vee (\neg x \wedge z) = (x \wedge y) \vee (\neg x \wedge z)$

6. $(x \vee y) \wedge (y \vee z) \wedge (\neg x \vee z) = (x \vee y) \wedge (\neg x \vee z)$

Exercise 5.18. Prove the properties of $\oplus$ in Figure 5.3.11.

Exercise 5.19. Show that the following functions are equivalent:

1. $f(x, y) = (x \wedge y) \vee (\neg x \wedge \neg y) \vee (\neg x \wedge y)$; $g(x, y) = \neg x \vee y$

2. $f(x_1, x_2, x_3) = \overline{x}_1 \overline{x}_2 x_3 + x_1 \overline{x}_2 \overline{x}_3 + x_1 \overline{x}_2 x_3 + x_1 x_2 \overline{x}_3$;
   $g(x_1, x_2, x_3) = x_1 \overline{x}_3 + \overline{x}_2 x_3$

3. $f(x_1, x_2, x_3) = (x_1 + x_2 + x_3)(x_1 + x_2 + \overline{x}_3)(\overline{x}_1 + x_2 + \overline{x}_3)$;
   $g(x_1, x_2, x_3) = x_2 + x_1 \overline{x}_3$

4. $f(x_1, x_2) = (\neg x_1 \wedge x_2) \vee (x_1 \wedge \neg x_2)$; $g(x_1, x_2) = x_1 \nleftrightarrow x_2$

5. $f(x, y) = (xy) + (\overline{xy})$; $g(x, y) = \overline{(x \leftrightarrow y)}$

Exercise 5.20. Determine the functions implemented in the circuit diagrams of Figures 5.3.12.(i)-(iii) and minimize the circuits whenever possible.

173

5. The language of classical logic

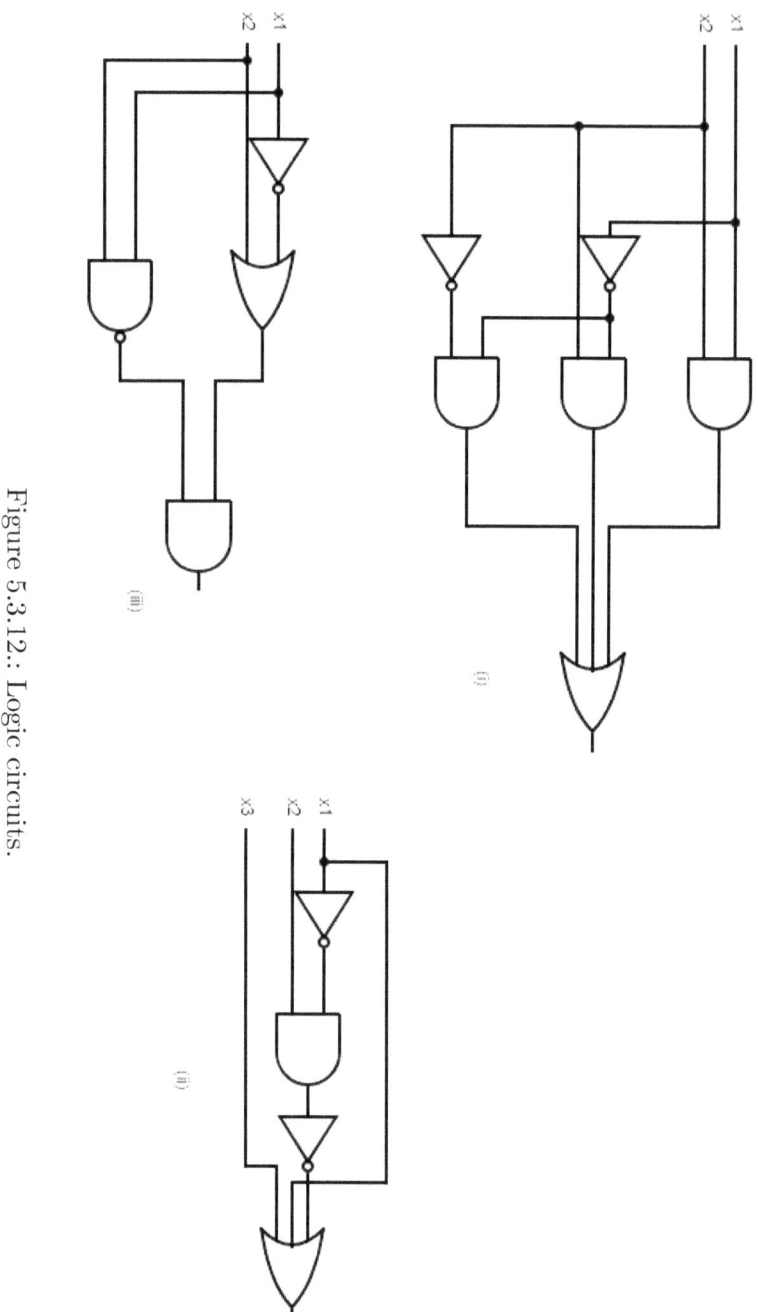

Figure 5.3.12.: Logic circuits.

6. Classical logical consequence

As stated above, logical consequence is the central notion for a logical system L or a logic **L**. In Chapter 4, we elaborated on this topic from a general viewpoint; in this Chapter, we approach it from the perspective of classical logic. Here, we discuss solely the essential aspects; a comprehensive discussion of classical logical consequence can be found in Augusto (2017).

6.1. Classical $\heartsuit$-consequences

To define the k_i-ary operations $\heartsuit_i^k$ on a set of formulas F_L, $0 \leq k \leq 2$, equates with imposing specific conditions on the consequence relation/operation in presence of the $\heartsuit_i^k$. We thus speak of $\heartsuit$-*consequences*. By this means, we define not only negation, conjunction, disjunction, etc., but also their *behavior* (e.g., classical, modal, paraconsistent, etc.). We first focus on the classical $\heartsuit$-consequences with generality in mind in this Chapter; below, in Chapter 12, we specify this behavior for the calculi $\mathcal{LK}$ (cf. the logical rules) and $\mathcal{NK}$ with inference rules in mind.

We consider the language L with $O_\mathsf{L} = \{\top, \bot, \neg, \wedge, \vee, \rightarrow\}$ the set of connectives; $\top$ and $\bot$ are taken as 0-ary connectives and the connective $\leftrightarrow$ can be defined in terms of the other members of O_L in the usual ways (cf. Example 5.6.3).

With generality in mind, we first introduce the $\heartsuit$-consequence operations, and then separately approach the syntactical and semantical $\heartsuit$-relations. With respect to the $\heartsuit$-consequence operations $\heartsuit_{Cn}$, they are spoken of as *Tarski-style conditions for Cn* if they involve exactly one connective of the language at hand.

Definition 6.1. The following are the Tarski-style conditions for the *classical consequence operation* Cn_L for a (possibly empty) consistent set $X \subseteq F_\mathsf{L}$ and any formulas $\phi, \psi \in F_\mathsf{L}$:[1]

1. ($\top_{Cn}$) $\top \in Cn(X)$ and $\phi \in Cn(X, \top)$

---

[1] Not all of them were considered by Tarski, despite this label; in fact, Tarski (1930) considered solely $\neg_{Cn}$ and $\rightarrow_{Cn}$. We obtain $\wedge_{Cn}$ and $\vee_{Cn}$ by defining the connectives $\wedge$ and $\vee$ in terms of $O'_\mathsf{L} = \{\neg, \rightarrow\}$ (cf. Example 5.6.1-2).

6. Classical logical consequence

2. ($\perp_{Cn}$) $\perp \notin Cn(X)$ but $\phi \in Cn(X, \perp)$

3. ($\neg_{Cn}$) $\phi \in Cn(X)$ iff $Cn(X, \neg\phi) = F_L$

4. ($\wedge_{Cn}$) $Cn(X, \phi \wedge \psi) = Cn(X, \phi, \psi)$

5. ($\vee_{Cn}$) $Cn(X, \phi \vee \psi) = Cn(X, \phi) \cap Cn(X, \psi)$

6. ($\rightarrow_{Cn}$) $\phi \rightarrow \psi \in Cn(X)$ iff $\psi \in Cn(X, \phi)$

Informally, Definition 6.1.1 tells us that ⊤ is a classical consequence of any empty or consistent set X and any formula ϕ is a consequence of $X \cup \{\top\}$.[2] Definition 6.1.2 formulates the fact that, if X is a consistent set, then $\perp$ is not one of its consequences, but any formula whatsoever is a consequence of $X \cup \{\perp\}$, a principle known as *ex falso* (or *contradictione*) *quodlibet*, abbreviated as EFQ (ECQ, respectively). In 6.1.3, Cn is defined with relation to $\neg$ in terms of consistency: an arbitrary formula ϕ is a consequence of X iff $X \cup \{\neg\phi\}$ is inconsistent. Definition 6.1.4 tells us that the consequences of a set X and of a formula $\phi \wedge \psi$ are exactly the consequences of $X \cup \{\phi\} \cup \{\psi\}$, whereas Definition 6.1.5 tells us that the consequences of a set X and of a formula $\phi \vee \psi$ are the intersection of the consequences of $X \cup \{\phi\}$ and of $X \cup \{\psi\}$. Finally, Definition 6.1.6 defines the $\rightarrow$-consequence in terms of DDT (cf. Exercise 4.16), which is made clearer in Definition 6.5.6 below.

Because the Tarski-style conditions for the connectives are in fact rules (of inference), we can also determine Cn with respect to the rules MP and DT:[3]

Remark 6.2. MP_{Cn} holds iff MP is a rule of Cn, and DT_{Cn} holds if DT is a rule of Cn.

6.1.1. Classical syntactical ♡-consequences

Given the inter-definition with respect to consequence relations and operations (cf. Prop. 4.6), we can now obtain the classical syntactical ♡-consequence relations from the Tarski-style definitions above of the classical ♡-consequence operations. We use the symbols $\Rightarrow, \Leftrightarrow \in O^{\mu^L}$ for conditional and iff-sentences, respectively, at the metalanguage level.

Definition 6.3. A consequence relation $\vdash$ over the language L, for $X \subseteq F_L$ and all $\phi, \psi, \theta \in F_L$, is

---

[2] We henceforth omit in this passage the adjective "classical," in order to avoid repetition.

[3] Clearly, DT can be seen as a rule of inference.

6.1. Classical ♡-consequences

1. ⊤-*classical* iff
    a) $\vdash \top$
    b) $X, \top \vdash \phi \Rightarrow X \vdash \phi$

2. ⊥-*classical* iff $\bot \vdash \phi$

3. ¬-*classical* iff
    a) $(X, \phi \vdash \psi$ and $X, \neg\phi \vdash \psi) \Rightarrow X \vdash \psi$ and
    b) $\phi, \neg\phi \vdash \psi$

4. ∧-*classical* iff $X, \phi \wedge \psi \vdash \theta \Leftrightarrow X, \phi, \psi \vdash \theta$

5. ∨-*classical* iff $X, \phi \vee \psi \vdash \theta \Leftrightarrow X, \phi \vdash \theta$ and $X, \psi \vdash \theta$

6. →-*classical* iff $X, \phi \vdash \psi \Leftrightarrow X \vdash \phi \to \psi$

Although essentially the same as in Definition 6.1, the above Definition 6.3 allows us to specify a few more properties of the classical logic system. Definition 6.3.1.b) expresses *monotonicity* (M; a weak version thereof, actually). With respect to Definition 6.3.3.a), this is known as *redundancy* (abbr.: RD) and 6.3.3.b) can also be called the *principle of explosion* (PE). Further characterizations of classicality for the connectives, some involving more than one connective, are as follows.[4]

Definition 6.4. For a set of formulas $X \subseteq F_L$ and for all $\phi, \psi, \theta \in F_L$, if $\vdash$ is

1. ¬-classical, then $X, \neg\phi \vdash \phi \Rightarrow X \vdash \phi$ (*reductio ad absurdum*, strong version, RAstr) and $\phi \dashv\vdash \neg\neg\phi$ (*involution*–or *double negation–law*, DN).

2. ∧-classical, then
    a) $\phi, \psi \vdash \phi \wedge \psi$ (*adjunction*, AD) and
    b) $\begin{cases} \phi \wedge \psi \vdash \phi \\ \phi \wedge \psi \vdash \psi \end{cases}$ (*simplification*, SP).

3. ∨-classical, then
    a) $\begin{cases} \phi \vdash \phi \vee \psi \\ \phi \vdash \psi \vee \phi \end{cases}$ (*addition*, AT) and

---

[4]Whenever it is adequate to do so, we omit X in the premises and abbreviate (e.g., written in full, AD (Def. 6.4.2.a) is: if $X \vdash \phi$ and $X \vdash \psi$, then $X \vdash \phi \wedge \psi$).

b) $(X, \phi \vdash \theta$ and $X, \psi \vdash \theta \Rightarrow X, \phi \vee \psi \vdash \theta)$ (*summation*, SM).

4. $\wedge$- *and* $\neg$-*classical, then* $X, \phi \vdash \psi \wedge \neg \psi$ *implies* $X \vdash \neg \phi$ (*principle of non-contradiction*, PNC).[5]

5. $\vee$- *and* $\neg$-*classical, then* $\vdash \phi \vee \neg \phi$ (*tertium non datur* or *principle of excluded middle*, PEM).

The above Definitions can be added to the argument forms of Figure 2.3.1 whenever there is more than one premise. These, as well as some of the Definitions above, can be formulated as classical tautologies (see below Prop. 7.1).

We can now characterize the classical syntactical $\heartsuit$-consequence operations in terms of their associated relations as follows:

Proposition 6.5. Classical Cn satisfies

1. $\top_{Cn}$ *iff it satisfies* M.

2. $\bot_{Cn}$ *iff it satisfies* EFQ.

3. $\neg_{Cn}$ *iff it satisfies* RD, PE, RAstr, PNC, PEM, *and* DN.

4. $\wedge_{Cn}$ *iff it satisfies* AD, SP, *and* PNC.

5. $\vee_{Cn}$ *iff it satisfies* AT, SM, *and* PEM.

6. $\rightarrow_{Cn}$ *iff it satisfies both* MP_{Cn} *and* DT_{Cn}.

6.1.2. Classical semantical $\heartsuit$-consequences

The logical system $\mathrm{CL} = (\mathsf{L}, \Vdash_{\mathrm{CL}})$, known as classical logic, is a deductive system (cf. Chapter 4). We can define deduction from the semantical viewpoint as *preservation of truth* (or *truth-preservation*), so that we say that an argument is deductively valid iff, if the premises are true, then the conclusion must be true, too. For CPL, it suffices to define the classical semantical $\heartsuit$-consequence relations in Boolean terms, i.e. in terms of their truth tables (see Example 3.19; cf. Prop. 3.23.1-6).

Proposition 6.6. For all $\phi, \psi \in F_{\mathsf{L}}$, a valuation val is

1. $\top$-Boolean iff $val(\top) = \mathtt{t}$

2. $\bot$-Boolean iff $val(\bot) = \mathtt{f}$

---

[5]Or, more correctly perhaps, *proof by contradiction* or *reductio ad impossibilem* (see Section 1.3.2).

6.1. Classical ♡-consequences

3. ¬-Boolean iff $val(\neg\phi) = \mathtt{t} \Leftrightarrow val(\phi) = \mathtt{f}$
4. ∧- Boolean iff $val(\phi \wedge \psi) = \mathtt{t} \Leftrightarrow val(\phi) = \mathtt{t}$ and $val(\psi) = \mathtt{t}$
5. ∨- Boolean iff $val(\phi \vee \psi) = \mathtt{t} \Leftrightarrow val(\phi) = \mathtt{t}$ or $val(\psi) = \mathtt{t}$
6. →-Boolean iff $val(\phi \to \psi) = \mathtt{t} \Leftrightarrow val(\phi) = \mathtt{f}$ or $val(\psi) = \mathtt{t}$

Remark 6.7. In order to extend the above results to CFOL an interpretation $\mathcal{I} = (\mathscr{D}, \Theta, \varpi)$ (cf. Definition 3.25) is required. In particular, the Boolean valuations are now constrained by val_ϖ, i.e. a valuation under the variable assignment ϖ, which assigns to a variable x a constant a in the domain of discourse $\mathscr{D}$ (denoted by $\varpi(x) = a$).

Definition 6.8. For all $\phi, \psi \in F_{\mathsf{L1}}$, given an interpretation $\mathcal{I} = (\mathscr{D}, \Theta, \varpi)$ a valuation $val_\mathcal{I}$ is

1. ⊤-Boolean iff $val_\mathcal{I} = \mathtt{t}$
2. ⊥-Boolean iff $val_\mathcal{I} = \mathtt{f}$
3. ¬-Boolean iff $val_{\mathcal{I},\varpi}(\neg\phi) = \mathtt{t} \Leftrightarrow val_{\mathcal{I},\varpi}(\phi) = \mathtt{f}$
4. ∧- Boolean iff

$$val_{\mathcal{I},\varpi}(\phi \wedge \psi) = \mathtt{t} \Leftrightarrow val_{\mathcal{I},\varpi}(\phi) = \mathtt{t} \text{ and } val_{\mathcal{I},\varpi}(\psi) = \mathtt{t}$$

5. ∨- Boolean iff

$$val_{\mathcal{I},\varpi}(\phi \vee \psi) = \mathtt{t} \Leftrightarrow val_{\mathcal{I},\varpi}(\phi) = \mathtt{t} \text{ or } val_{\mathcal{I},\varpi}(\psi) = \mathtt{t}$$

6. →-Boolean iff

$$val_{\mathcal{I},\varpi}(\phi \to \psi) = \mathtt{t} \Leftrightarrow val_{\mathcal{I},\varpi}(\phi) = \mathtt{f} \text{ or } val_{\mathcal{I},\varpi}(\psi) = \mathtt{t}$$

We now define the classical conditions for the semantical ♡-consequence relations by means of the connectives of L1 (cf. Prop. 3.23).

Definition 6.9. Let $\mathcal{I} = (\mathscr{D}, \Theta, \varpi)$ be an interpretation for L1. We have

1. $\models_{\mathcal{I},\varpi} \top$ is classical
2. $\not\models_{\mathcal{I},\varpi} \bot$ is classical
3. $\models_{\mathcal{I},\varpi} \neg\phi$ is classical iff $\not\models_{\mathcal{I},\varpi} \phi$

6. Classical logical consequence

4. $\models_{I,\varpi} \phi \wedge \psi$ is classical iff $\models_{I,\varpi} \phi$ and $\models_{I,\varpi} \psi$

5. $\models_{I,\varpi} \phi \vee \psi$ is classical iff $\models_{I,\varpi} \phi$ or $\models_{I,\varpi} \psi$

6. $\models_{I,\varpi} \phi \to \psi$ is classical iff $\not\models_{I,\varpi} \phi$ or $\models_{I,\varpi} \psi$

Given the adequateness of classical logic, the classical semantical ♡-consequence relations can be *mutatis mutandis* further specified as in Definitions 6.3-4 for $\models$, as in fact the conditions for the classical consequence operation in Proposition 6.5 hold also in a semantical perspective.

▶ *Do Exercises 6.1-3.*

6.2. Classical ♦-consequences

The extension from L0 to L1 can be quite problematic in philosophical terms insofar as the introduction of the quantifiers is concerned. In particular, Boolean truth valuations are missing with respect to the quantifiers, and this motivates the suspicion of arbitrariness with respect to the choice of–and restriction to–the quantifiers ∀ and ∃. Nevertheless, completeness for classical FOL has been proven (see below), and from a mathematical perspective things are (more) clearly defined, namely in regard to the definitions of the consequence relations with respect to the quantifiers.

The syntactical classicality conditions for the ∀- and ∃-consequence relations can be formulated in terms of rules of inference. More specifically, these classicality conditions are formulated in terms of the rules for the introduction and elimination of the quantifiers (see Chapter 12 below).

Definition 6.10. For the language L1, the consequence relation $\vdash$ is *classical* iff

1. for ∀ it is the case that

    a) $\phi(a) \vdash \forall x \phi(x)$, provided a is arbitrary;
    b) $\forall x \phi(x) \vdash \phi(a)$.

2. for ∃ it is the case that

a) $\phi(a) \vdash \exists x \phi(x)$;

b) $\exists x \phi(x) \vdash \phi(a)$, provided a has a merely auxiliary role.

The semantical counterpart of the classical ∀- and ∃-consequence relations originated in the work of Tarski (1935). The next definition stipulates the *Tarskian truth conditions* for the quantifiers:

Definition 6.11. (Tarski, 1935) Given an assignment $\varpi : Vi \to \mathscr{D}$,

1. $\forall x \phi(x)$ is true under ϖ iff every a in $\mathscr{D}$ satisfies $\phi(x)$ under ϖ.

2. $\exists x \phi(x)$ is true under ϖ iff at least one a in $\mathscr{D}$ satisfies $\phi(x)$ under ϖ.

It is easy to see that this corresponds to the following definitions in terms of the semantical ♦-consequence relations:

Definition 6.12. For the language L1, given an interpretation $\mathcal{I} = (\mathscr{D}, \Theta, \varpi)$ the consequence relation $\models$ is classical iff

1. $\models_{\mathcal{I},\varpi} \forall x \phi(x)$ iff for every $a \in \mathscr{D}$ it is the case that $\models_{\mathcal{I}_a^x} \phi(a)$.

2. $\models_{\mathcal{I},\varpi} \exists x \phi(x)$ iff for some $a \in \mathscr{D}$ it is the case that $\models_{\mathcal{I}_a^x} \phi(a)$.

And the corresponding semantical ♦-consequence operation is defined as follows:

Definition 6.13. For the language L1, given an interpretation $\mathcal{I} = (\mathscr{D}, \Theta, \varpi)$ the consequence operation Cn is classical iff

1. $\phi[x/a] \in Cn_{\mathcal{I},\varpi}(\forall x \phi(x))$ if every $a \in \mathscr{D}$ is substitutable for x in $\phi(x)$ under ϖ.

2. $\phi[x/a] \in Cn_{\mathcal{I},\varpi}(\exists x \phi(x))$ if some $a \in \mathscr{D}$ is substitutable for x in $\phi(x)$ under ϖ.

▶ *Do Exercise 6.4.*

6. Classical logical consequence

Exercises

Exercise 6.1. Define $\leftrightarrow_{Cn}$ and formalize the syntactical and semantical classicality conditions for the $\leftrightarrow$-consequence.

Exercise 6.2. Could we have

$$(\neg^*_{Cn}) \qquad \neg\phi \in Cn(X) \text{ iff } Cn(X,\phi) = F_L$$

as a Tarski-style condition for classical $\neg_{Cn}$ instead of Definition 6.1.3? Why (not)?

Exercise 6.3. Let $\phi, \psi, \chi \in F_{L0}$ be arbitrary formulas. Indicate whether the following statements are true or false. Account for your answer, with counter-examples if necessary.

1. $\phi \Vdash \psi$ and $\phi \Vdash \chi \;\;\Rightarrow\;\; \psi \Vdash \chi$.
2. $\phi \Vdash \psi$ and $\psi \Vdash \chi \;\;\Rightarrow\;\; \phi \Vdash \chi$.
3. $\phi \Vdash \psi \vee \chi \;\;\Rightarrow\;\; \phi \Vdash \psi$.
4. $\phi \equiv \psi \vee \chi \;\;\Rightarrow\;\; \psi \Vdash \phi$.

Exercise 6.4. Comment on the following condition for the classical $\forall$-consequence operation first stated by Tarski (1935):

$$(C\forall) \qquad \forall x \phi(x) \in Cn(\phi(x)).$$

7. CL and extensions

7.1. The logic CL

Given the Tarski-style conditions above, the deductive system CL generates tautologies that characterize it uniquely as a logic.

Proposition 7.1. *The following are some of the valid sentences of the logic* **CL**:

⊩1	$A \vee \neg A$	PEM
⊩2	$\neg(A \wedge \neg A)$	PNC
⊩3	$A \rightarrow A$	Law of identity
⊩4	$\neg\neg A \leftrightarrow A$	DN
⊩5	$((A \rightarrow B) \wedge A) \rightarrow B$	MP
⊩6	$((A \rightarrow B) \wedge \neg B) \rightarrow \neg A$	MT
⊩7	$(A \rightarrow B) \leftrightarrow (\neg B \rightarrow \neg A)$	Law of contraposition
⊩8	$((A \rightarrow B) \wedge (B \rightarrow C)) \rightarrow (A \rightarrow C)$	HS
⊩9	$(\neg A \rightarrow (B \wedge \neg B)) \rightarrow A$	*Reductio ad absurdum*
⊩10	$\neg(A \wedge B) \leftrightarrow (\neg A \vee \neg B)$	De Morgan's law (DeM$_\wedge$)
⊩11	$\neg(A \vee B) \leftrightarrow (\neg A \wedge \neg B)$	De Morgan's law (DeM$_\vee$)

Proof: The proof is by examination of the truth tables or by producing the proofs in a calculus. **QED**

Proposition 7.2. *The negation of any tautology of CL is a contradiction, and the negation of a contradiction of CL is a tautology, i.e.* $\neg \top \equiv \bot$ *and* $\neg \bot \equiv \top$.

Proof: Obvious, given the Boolean nature of CL. **QED**

Definition 7.3. In particular, and denoting a classical contradiction or falsity by the symbol $\bot$, we have in CL

$$(\bot_{df}) \qquad A \wedge \neg A.$$

183

7. CL and extensions

Note that we can equivalently define a classical contradiction as $\Box := A \wedge \neg A$ (see Chapter 14).

Proposition 7.4. *Given any formula A, in CL we have*

$$A \wedge \bot = \bot.$$

Proof: Obvious, given $\bot_{df}$ and the classical truth table for $\wedge$. **QED**

We began this Section by speaking of unique characterizations with respect to a deductive system. By *characterizing uniquely*, we mean that given different conditions for the connectives of a given logical language, this typically gives rise to different logics. This unique characterization is particularly important in the case of **CL**, as it stands as the reference logic that other logics either restrict or extend. If a deductive system S does not generate at least one of the tautologies of the logic **CL**, or does not derive at least one of the theorems of **CL**, then we say that S is *non*-classical and we speak of the logic **S** as being a *non*-classical logic. For instance, intuitionistic logic does not generate or derive ⊩ 1 above. If a deductive system S has a many-valued semantics, then $A \wedge \neg A$ may not be a contradiction, and thus ⊨2 does not belong to **S**. These non-classical logics are thus subsets of **CL**, which makes them–arguably–less expressive logics. But the opposite is true of logics whose sets of tautologies/theorems are actually bigger than **CL**; examples of these are the modal logics.[1]

In some cases, there is only partial acceptance or rejection of some of the formulas in Proposition 7.1. For instance, PEM (⊩1) is not eliminated *tout court* in intuitionistic logic, but rather in (the FO formalization of) mathematical theories involving infinity. For instance, let $A = \forall x \exists y\, (P(y, x))$; if $x, y \in \mathbb{Z}^+$ and $P(y, x)$ stands for the ordering relation $y > x$, then we should have A or $\neg A$, but this is rejected in intuitionistic logic on the grounds that it does not appear possible to provide a proof of either.[2] Still with respect to intuitionistic logic, only the ←-direction of ⊩ 4 is a valid sentence of this logic.

---

[1] For this, usually extensions of **L1** are required. For instance, in the case of the modal logics, we have the extended set of connectives $O_M = \{\neg^1, \Box^1, \Diamond^1, \wedge^2, \vee^2, \rightarrow^2\}$ ($\leftrightarrow$ is defined as usually by means of the other connectives).

[2] This requirement of intuitionistic logic is known as the *disjunction property*, according to which, given some (possibly empty) set of formulas X and some logical system J, we have

$$X \vdash_J (\phi \vee \psi) \quad \Rightarrow \quad X \vdash_J \phi \text{ or } X \vdash_J \psi$$

for two formulas $\phi, \psi \in F_{L1}$, where $\vdash_J (\phi \vee \psi)$ actually denotes that *there is a proof* (from X) in J of either ϕ or ψ from which *construction*, in turn, $X \vdash_J \phi$

Take now ⊩ 2; clearly, this is a tautology iff $A \land \neg A$ is a contradiction. The reason to reject contradictions in CL is that, if accepted, then CL would be a trivial deductive system, as anything whatsoever follows classically from a contradiction (the principle of explosion; cf. Def. 6.3.3.b). However, some logics–the dialetheic logics–are founded on the stance that there are indeed true contradictions, and some other–the paraconsistent logics–"tolerate" them in view of the fact that they do not necessarily lead to trivial theories.[3]

We concentrated the discussion in the last paragraphs on ⊩1-3.[4] In effect, a logic in which any or all of these do not hold is indeed a non-classical logic. In fact, ⊩1-3 are often spoken of as the pillars of CL. Historically, they can all be traced back to Aristotle, not without reason called the father of CL. We give here the Aristotelian passages where these classical logical *laws* were first conceived, not as historical curiosities, but as food for thought on the very core of CL.[5]

> [⊩ 1]   But on the other hand there cannot be an intermediate between contradictories, but of one subject we must either affirm or deny any one predicate. This is clear, in the first place, if we define what the true and the false are. To say of what is that it is not, or of what is not that it is, is false, while to say of what is that it is, and of what is not that it is not, is true; so that he who says of anything that it is, or that it is not, will say either what is true or what is false; but neither what is nor what is not is said to be or not to be. (Aristotle, *Metaphysics*, Book IV, Part 7)
>
> [⊩ 2]   [I]t is impossible, then, that "being a man" should mean precisely "not being a man", if "man" not only signifies something about one subject but also has one significance

---

or $X \vdash_J \psi$ denote that there is a proof, in this constructive sense, of either ϕ or ψ (from X). Contrast this *constructivist* notion of proof with the classical one in Section 4.1.2 above.

[3] See Chapter 4 of Augusto (2017) for a discussion on non-classical deductive consequences.

[4] Note that ⊩ 1 ≡ ⊩ 3, as we have

$$A \to A \equiv \neg A \lor A.$$

This formal identity, however, does not necessarily entail equivalent linguistic accounts of the two laws. In fact, ⊩ 3 is often formalized by employing the symbol for equality, i.e. $A = A$. But the employment of the symbol for equality in CL is not without issues (cf. Section 7.2 below), and so we prefer the formalization above.

[5] Traditionally, the first law to be presented is ⊩ 3. However, the order is irrelevant given ⊩ 1 ≡ ⊩ 3.

7. CL and extensions

> ... And it will not be possible to be and not to be the same thing, except in virtue of an ambiguity, just as if one whom we call "man", others were to call "not-man"; but the point in question is not this, whether the same thing can at the same time be and not be a man in name, but whether it can be in fact. (Aristotle, *Metaphysics*, Book IV, Part 4)

> [⊩ 3]   First then this at least is obviously true, that the word "be" or "not be" has a definite meaning, so that not everything will be "so and not so". Again, if "man" has one meaning, let this be "two-footed animal"; by having one meaning I understand this:—if "man" means "X", then if A is a man "X" will be what "being a man" means for him. (It makes no difference even if one were to say a word has several meanings, if only they are limited in number; for to each definition there might be assigned a different word. For instance, we might say that "man" has not one meaning but several, one of which would have one definition, viz. "two-footed animal", while there might be also several other definitions if only they were limited in number; for a peculiar name might be assigned to each of the definitions. If, however, they were not limited but one were to say that the word has an infinite number of meanings, obviously reasoning would be impossible; for not to have one meaning is to have no meaning, and if words have no meaning our reasoning with one another, and indeed with ourselves, has been annihilated; for it is impossible to think of anything if we do not think of one thing; but if this is possible, one name might be assigned to this thing.) (Aristotle, *Metaphysics*, Book IV, Part 4)

▶ *Do Exercises 7.1-2.*

7.2. The extension CL$^=$: CL with equality

So far, other than some brief discussions in Chapter 2, we have been mostly silent on the topic of *equality* in CL. The reason for this silence is that this relation happens to have more to it than first meets the eye. Put simply, it "complicates" CFOL if we introduce it in the alphabet of L1 by means of a new symbol, typically "=". However, equality is a property that is quite important, especially for the formalization of mathematical theories, and many authors see it as actually indispensable. As it is often

7.2. The extension CL$^=$: CL with equality

put intuitively, equality is a predicate that is meaningful regardless of the domain of discourse.

In Example 2.25, we avoided the introduction of the symbol for equality by means of a predicate E of arity n, but equality in a logical language is a *logical* symbol rather than a parameter. In particular, it allows the expression of two fundamental laws:

(SubP) $\quad \forall x \forall y\, [(x = y) \to \forall P\, (P(x) \leftrightarrow P(y))]$

and

(IdI) $\quad \forall x \forall y\, [\forall P\, (P(x) \leftrightarrow P(y)) \to (x = y)]$

known as *substitution principle* and (Leibniz's principle of) *identity of indiscernibles*, respectively. These two laws, in turn, allow us to express Leibniz's law:

(LL) $\quad \forall x \forall y\, [\forall P\, (P(x) \leftrightarrow P(y)) \leftrightarrow (x = y)].$

It is obvious, however, that the three laws above are *not* expressible in FOL, as universal quantification over predicates is required. In the above paragraph we wrote *complicate* between inverted commas to express the fact that we use this term somehow reluctantly. In effect, to be rigorous, from the formal point of view, a FO language cannot fully express equality. This can easily be verified in the following axiomatization in the language L1$^=$, i.e. L1 augmented with the equality symbol.

Proposition 7.5. *The following is an axiomatization of equality as a logical symbol:*

($\mathcal{E}$1) $\quad \forall x\, (x = x)$

($\mathcal{E}$2) $\quad \forall x \forall y\, ((x = y) \to (y = x))$

($\mathcal{E}$3) $\quad \forall x \forall y \forall z\, [((x = y) \land (y = z)) \to (x = z)]$

($\mathcal{E}$4) $\quad \forall f \forall x \forall y\, [(x = y) \to (f(x) = f(y))]$

($\mathcal{E}$5) $\quad \forall P \forall x \forall y\, [(x = y) \to (P(x) = P(y))]$

Proof: Left as an exercise (Hint: Use SubP, IdI, and $\mathcal{E}$1 to prove $\mathcal{E}$2-5).

7. CL and extensions

To be more precise, in axioms $\mathcal{E}$4-5 we must introduce sequences of variables $\vec{u} = u_1, ..., u_n$ and $\vec{z} = z_1, ..., z_m$, for $n, m \geq 0$, such that we have

$$(\mathcal{E}4) \quad \forall f \forall x \forall y \left[(x = y) \to (f(\vec{u}, x, \vec{z}) = f(\vec{u}, y, \vec{z})) \right]$$

and

$$(\mathcal{E}5) \quad \forall P \forall x \forall y \left[(x = y) \to (P(\vec{u}, x, \vec{z}) = P(\vec{u}, y, \vec{z})) \right].$$

Again, it will be easily verified that this axiomatization of equality is actually a *second-order* axiomatization (cf. axioms $\mathcal{E}$4-5). This means that the quantifiers are allowed to quantify over functions and predicates, and this entails that we are no longer using **L1**, but rather **L2**$^=$, i.e. a second-order extension of **L1** with the equality symbol. It is thus obvious that the above theory

$$\Theta_{Id} = \{\text{SubP}\} \cup \{\text{IdI}\} \cup \{\text{LL}\} \cup \{AX_{Eq}\}$$

where AX_{Eq} is the set of the five axioms of Proposition 7.5 and which is sometimes called *theory of logical identity* (e.g., Tarski, 1994), is not a FO theory.

Is there some way to introduce *de facto* the symbol "=" in CFOL? Just as in the case of all other logical symbols in a truth-preserving logical system, we require what can be called *a semantics of the symbol "="*. In effect, the fact that CFOL is not capable of expressing the theory of equality Θ_{Id} does not entail that we cannot extend it by including "=" in its set of logical symbols.

Definition 7.6. Given an interpretation $\mathcal{I} = (\mathcal{D}, \Theta, \varpi)$, we say that a model $\mathcal{M}$ is *normal* if = denotes the equality relation on $\mathcal{D}$.

By this simple means, we have in fact obtained an extension of **CL** that we shall denote by **CL**$^=$ and speak of as *classical FO logic with equality*. Similarly, the language **L1** is now **L1**$^=$.

Definition 7.7. Let us denote the semantical consequence relation in **CL**$^=$ by $\models_=$. Then, given a set of formulas $X \subseteq$ **L1**$^=$ and a formula $\phi \in$ **L1**$^=$, we write

$$X \models_= \phi$$

provided that $val_\mathcal{I}(\phi) = \mathbf{t}$ in every normal model in which $val_\mathcal{I}(\chi_i) = \mathbf{t}$ for all the $\chi_i \in X$.

We then have the following trivial proposition, where we denote the classical semantical consequence relation simply by $\models$:

7.2. The extension CL=: CL with equality

Proposition 7.8. *If $X \models \phi$, then $X \models_= \phi$.*

Proof: Left as an exercise.

Example 7.9. We formalize the argument of Exercise 2.10.4 in L1=. Axioms 1-3 are the proper axioms of the group theory $\Theta_{\mathcal{G}}$ for a group a pair $\mathcal{G} = (G, \star)$. Note that we did not add the logical axioms of equality, nor did we include any of SubP, IdI, or LL. Note also that we chose to write the predicate $\star$ in infix notation, as is more common in mathematics.

$$
\begin{array}{ll}
(1) & \forall x \forall y \forall z \, ((x \star y) \star z = x \star (y \star z)) \\
(2) & \forall x \exists w \, (x \star w = x \land w \star x = x) \\
(3) & \forall x \exists y \exists w \, (x \star y = w \land y \star x = w) \\
\hline
(4) & \forall x \forall y \forall z \, ((x \star z = y \star z) \to x = y)
\end{array}
$$

We can prove the correctness of the argument in Example 7.9 by applying directly some proof calculus that includes equality. Does this mean that we can do without Θ_{Id} when working in L1=? Not really, as a rigorous formal FO theory of equality cannot dispense with Θ_{Id}, totally or in part. However, we might not need it if we use a proof calculus that has its own way to tackle equality. This is the case of the resolution calculus, which has a proof technique called paramodulation (see Section 14.3). The Gentzen calculus $\mathcal{NK}$ has an extension $\mathcal{NK}^=$ for equality with two rules for the introduction and elimination of the symbol $=$ (cf. Section 12.1.3). In other cases, there might be a specific equality predicate; for instance, Prolog has a built-in predicate for this relation.

Of course, one may wonder whether these correspond to the equality defined in Θ_{Id}. Fitting (1996) gives a formal means to integrate the axioms of AX_{Eq} in CFOL via function and predicate *replacement axioms*.

Definition 7.10. Let $f \in Fun$ and $P \in Pred$. Then,

1. a replacement axiom $\mathcal{F}_{repl} \in Fun_{repl}$ for f is

$$\forall v_1 ... \forall v_n \forall w_1 ... \forall w_n [((v_1 = w_1) \land ... \land (v_n = w_n)) \to$$
$$(f(v_1, ..., v_n) = f(w_1, ..., w_n))].$$

7. CL and extensions

2. a replacement axiom $\mathcal{P}_{repl} \in Pred_{repl}$ for P is

$$\forall v_1...\forall v_n \forall w_1...\forall w_n[((v_1 = w_1) \wedge ... \wedge (v_n = w_n)) \rightarrow$$
$$(P(v_1, ..., v_n) = P(w_1, ..., w_n))].$$

Example 7.11. Let us consider $\star$ as a binary function symbol. Let there be given the replacement axiom

$$(\mathcal{F}_{repl}1) \quad \forall x \forall y \forall z \forall w \left[((x = z) \wedge (y = w)) \rightarrow (x \star y = z \star w)\right].$$

Let now $c \in Cons$. Then we have

$$\{\mathcal{F}_{repl}1\} \models \forall x \forall z \left[((x = z) \wedge (c = c)) \rightarrow (x \star c = z \star c)\right]$$

and

$$\{\mathcal{F}_{repl}1\} \cup \{\mathcal{E}1\} \models \forall x \forall z \left((x = z) \rightarrow (x \star c = z \star c)\right).$$

Let $AX_{repl} = Fun_{repl} \cup Pred_{repl}$ be the set of the replacement axioms. Then we have the following result:

Theorem 7.12. *For a set of FOL formulas X and a FO formula ϕ,*

$$X \models_= \phi \quad iff \quad X \cup \{\mathcal{E}1\} \cup \{AX_{repl}\} \models \phi.$$

Proof: Left as an exercise.

Rigorously, we now have

$$\mathbf{CL}^= = \mathbf{CL} \cup \{\mathcal{E}1\} \cup \{AX_{repl}\}.$$

However, just as we may dispense with Θ_{Id} in some cases, we may also relax this and simply consider $\mathbf{CL}^=$ as classical FO logic with equality without going all the formal way above. Were our main concern mathematical proofs, we would not so easily get away with this relaxation; but we are concerned here with a general notion of proof, and thus feel this relaxation is permissible. The reader not willing to condescend with this relaxation can benefit from Fitting (1996), which integrates equality in a rigorous way in both resolution and analytic tableaux.

In the Chapters that follow, we consider mostly CL *without* equality. Whenever this is considered, namely by means of the extension $\mathbf{CL}^=$, we shall make it clear; otherwise, in presence of the symbol "=" the reader may safely assume that we are at the metalanguage level.[6]

---

[6] Given our restricted and localized use of the symbol "=" at the object-language

7.2. The extension CL=: CL with equality

▶ **Do Exercises 7.3-12.**

Exercises

Exercise 7.1. Comment on Aristotle's passages above on the classical sentences ⊩-1-3.

Exercise 7.2. Research into further rationales to expand or restrict the set of sentences of **CL**.

Exercise 7.3. Let the binary relation $R \subseteq A \times A$ over A be an *equivalence relation* if R is reflexive, symmetric, and transitive, i.e. if for every $x, y, z \in A$ we have respectively

- xRx
- $xRy \Rightarrow yRx$
- $(xRy \wedge yRz) \Rightarrow xRz$

Show that $\mathcal{E}$1-3 define an equivalence relation.

Exercise 7.4. Formalize the argument in Example 2.25 over L1=.

Exercise 7.5. The substitution principle (SubP) is often also called *principle of extensionality*. Account for this synonymy.

Exercise 7.6. Sometimes the following axioms are given for **CL**= instead of $\mathcal{E}$4 and $\mathcal{E}$5, respectively:

$$\forall x \forall y \forall z \left[((f(x) = z) \wedge (x = y)) \to (f(y) = z) \right]$$

$$\forall x \forall y \left[(P(x) \wedge (x = y)) \to P(y) \right].$$

Are these equivalent to $\mathcal{E}$4 and $\mathcal{E}$5, respectively?

Exercise 7.7. Produce an example showing that the converse of Proposition 7.8 is not true.

Exercise 7.8. Consider $<$ as a binary predicate symbol and give an example of a replacement axiom for it.

Exercise 7.9. Explain formally why **CL**= is an extension of **CL**.

Exercise 7.10. Show that the compactness theorem carries over to **CL**=.

---

level, we see no need to distinguish this from equality at the metalanguage level by means of a different symbol. In particular, we see no need to distinguish *here* mathematical *equality* from logical *identity*, as, for instance, Tarski (1994) advocates.

7. CL and extensions

Exercise 7.11. Prove Propositions 7.5 and 7.8.

Exercise 7.12. Prove Theorem 7.12.

8. Classical FO theories and the adequateness of CFOL

The reader might have noticed that in Section 7.1 above we approached the logic **CL** almost exclusively from the propositional perspective. In effect, in a strict sense we may say that classical FO logic (CFOL) has no tautologies proper, as any open or closed formula $\phi(x_1,...,x_n) = \phi(\vec{x})$ requires an interpretation of the form $\mathcal{I}_{\vec{a}}^{\vec{x}}$ for $\vec{x} \in Vi(\phi)$ and $\vec{a} \in \mathscr{D}$, which can entail arbitrarily finitely or infinitely many models. Thus, rather than tautologies, in CFOL we must speak of *instances* of FO tautologies. On the syntactical side, we may consider $\neg \forall x (\phi) \leftrightarrow \exists x (\neg \phi)$ and $\neg \exists x (\phi) \leftrightarrow \forall x (\neg \phi)$ (cf. Def. 2.32.4), as well as $\forall x (\phi \wedge \psi) \leftrightarrow (\forall x (\phi) \wedge \forall x (\psi))$ and $\exists x (\phi \vee \psi) \leftrightarrow (\exists x (\phi) \vee \exists x (\psi))$, as theorems, but this is quite unnecessary, as these–and other examples–can be taken as equivalence-based rules of inference proper. Nevertheless, in a broader sense, we may say that there are FO tautologies and theorems, which allows us to say, for instance, that the set of (instances of) classical FO tautologies/theorems is recursively enumerable. Hence, semi-decidability is assured for CFOL theories.

Besides (semi-)decidability, we also often require adequateness of the deductive systems in which we treat our theories, i.e. we want to be assured that we only prove tautologies or valid formulas, and that all the valid formulas in the theories are provable (see Section 4.1.4). While adequateness of the deductive systems has to be evaluated with specific theories in mind,[1] we can give some general results. In this perspective, CFOL is an adequate deductive system for many theories, in particular so for many mathematical theories. The proof of the soundness of CFOL is essentially the same as that given for Theorem 4.48, but the reader is referred to, for instance, Enderton (2001) for a complete proof.

The proof of CFOL completeness, firstly given by Gödel (1930), is not such a simple matter, being, in mathematical lingo, a *deeper result*. In Section 4.4.2, we discussed the topic of FO completeness with two main aspects in mind: generality and decidability. That is, we were concerned with FO languages in general and how to tame their expressiveness so

---

[1] For instance, any FO theory containing elementary arithmetic is incomplete. This is the famous incompleteness result first published in Gödel (1931).

8. Classical FO theories and the adequateness of CFOL

as to allow for increased decidability in logical systems based on them. Given these two aspects, all the results (including the restrictions) in the mentioned Section hold for CFOL. However, it is mandatory in a book on the classical formalization of logic to give the proof of completeness for CFOL. This we do below, based on the proof given by L. Henkin (1949), in the present Section. However, the proof of completeness of CFOL–Gödel's or anyone else's[2]–is convoluted, as befits a deep result, and we provide only some of the main aspects thereof. This we do also with the objective of leaving the details as exercises for the reader, who is referred to Henkin's original proof, as well as to Enderton (2001) and Mendelson (2015) for different presentations of this celebrated result.

As said above, although we can prove the adequateness of a deductive system, this is often done with specific theories in mind. We shall now write $\vdash_\Theta \phi$ instead of $\Theta \vdash \phi$ to emphasize that Θ is not just a set of formulas, actually including axioms and/or rules of inference; in other words, Θ may be simply a calculus.[3] The whole point of the completeness theorem for CFOL is that, given some theory Θ, we have $\vdash_\Theta \phi$ if we have $\models_\Theta \phi$, so a theory Θ can also be defined semantically as a closed set of formulas for which, given an interpretation $\mathcal{I}$, there is some model $\mathcal{M}$, i.e. there is an interpretation in which all the axioms of Θ are valuated to **true**. In other words, the validity of the formulas of Θ is preserved by the rules of inference of Θ, and Θ is said to be *complete* in the sense that no more rules of inference are necessary to prove all the tautologies of Θ.

Although somewhat unorthodoxly coined, the following theories will play an important role in the proof of completeness of the classical FO predicate calculus.

Definition 8.1. Let Θ be a theory in a FO language. Then, Θ is a *scapegoat theory* if for any formula $\phi(x)$ that has x as its only free variable there is a ground term t such that

$$(\Upsilon) \qquad \vdash_\Theta \exists x \, (\neg \phi(x)) \to \neg \phi(t).$$

Note that Υ is actually an application of DT to $\exists x \, (\neg \phi(x)) \vdash_\Theta \neg \phi(t)$, and this is equivalent to $\neg \forall x \phi(x) \vdash_\Theta \neg \phi(t)$. Clearly, only a language with the required resources–such as L1–can produce a scapegoat theory.

In the following fundamental theorem, the symbols $\vdash, \models$ without any

---

[2] E.g., Beth (1960). As it is easy to guess, completeness of CPL is a far less complex affair to prove; see, e.g., Quine (1938).

[3] See Chapters 11 and 12 below for an elaboration on axiomatic and rule-based direct proof systems. However, we may wish to distinguish the proper axioms from some calculus proper $\mathcal{P}$, in which case we write $\Theta \vdash_\mathcal{P} \phi$.

subscripts denote the classical FO consequence relations for a FO language with countably many symbols and formulas–i.e. **L1**.

Theorem 8.2. (*The completeness theorem; Gödel, 1930*) *The following statements are equivalent:*

1. *In any FO theory Θ, the theorems of Θ are precisely the valid formulas of Θ, i.e.*

$$\text{if } \models_\Theta \phi, \text{ then } \vdash_\Theta \phi.$$

2. *Any consistent FO theory has a model.*

Proof: First, we have to prove that statements 1 and 2 are equivalent. This done, it suffices to prove either of the statements. We leave the first part as an exercise, and choose to prove statement 2. We begin by preparing the "setting" (step I) and then proceed to (the sketch of) the proof proper (step II).

I. Let Θ be a consistent set of FO formulas of the countable language **L1**. Henkin (1949) considers a theory Θ,[4] with MP and GEN[5] as inference rules, comprising the following axioms 1-5 (with restrictions indicated by *) and the classical FO theorems 6-10:

1. $C \to (B \to C)$
2. $(A \to B) \to ((A \to (B \to C)) \to (A \to C))$
3. $\neg\neg A \to A$
4. $\forall x (A \to B) \to (A \to \forall x (B))$ (*)
5. $\forall x (A) \to B[x/y]$ (*)
6. $DT_\vdash$ (Theorem 4.17; version for $\vdash$)
7. $\vdash \neg B \to (B \to C)$
8. $\vdash B \to (\neg C \to \neg (B \to C))$
9. $\vdash \forall x (\neg A) \to \neg \exists x (A)$
10. $\vdash \neg \forall x (B) \to \exists x (\neg B)$

The restrictions are as follows: In 4, x does not occur freely in A; in 5, y is any symbol replacing each free occurrence of x in A and no free occurrence of x in A occurs in a sub-formula of A of the form $\forall y (C)$.

---
[4] Actually, Henkin considers a *class* of deductive systems. This is obviously the class of classical FO logics, or CFOL for short. We work with a theory for reasons that should become clear in the proof below.

[5] This rule, called *generalization rule*, states that one can infer $\forall x (A)$ from A. See below Proposition 11.11.

8. Classical FO theories and the adequateness of CFOL

Axiom 1 is so also in the axiom system $\mathcal{L}$ of Chapter 11. Axiom 3 is our DN, i.e. involution or double negation law (cf. Prop. 2.32.3). The theorems have already been given in some form or can easily be verified to be theorems of CFOL. We give here Henkin's formal system, even though we do not strictly follow his original proof, and the reader is free to add axioms and/or theorems at will as long as they are well established as classical and can actually be invoked in this proof of FOL completeness. In particular, we shall invoke the following axiom in our theory Θ:

$$(\dagger) \qquad \forall x_i \phi(x_i) \to \phi(t)$$

which holds if $\phi(x_i) \in F_{L1}$ and t is a term free for x_i in $\phi(x_i)$. In fact, it may be the case that $x_i = t$, so that we have the axiom

$$(\dagger') \qquad \forall x_i \phi(x_i) \to \phi(x_i).$$

We shall also invoke a particularization of †, denoted by ‡, which states that if t is free for x in $\phi(x)$, then

$$(\ddagger) \qquad \forall x \phi(x) \vdash \phi(t).$$

As x is free for x in $\phi(x)$, a special form of ‡ is

$$(\ddagger') \qquad \forall x \phi(x) \vdash \phi(x).$$

Besides the above rules of inference, Henkin considers an additional rule:

11. Let X be a set of formulas no one of which contains a free occurrence of the individual symbol c. Let further a formula B be obtained from a formula A by replacing each free occurrence of c by the individual symbol x, it being the case that none of these occurrences of x is bound in B. Then, if $X \vdash A$, also $X \vdash B$.

With respect to rule GEN, we have the following specification: Given $X \Vdash \phi$, then we can obtain $X \Vdash \forall x \phi [c/x]$ if $c \notin X$ is a constant symbol and x does not occur in ϕ; furthermore, there is a deduction $X \Vdash \forall x \phi [c/x]$ in which c does not occur. We call this specification *constant generalization* (abbr.: CGEN).

II. Expand L1 with a countably infinite set of new constants $Cons^*$, so that the extension Θ^* remains consistent in the new language $L1^*$ (1. How?). For every $\phi \in L1^*$, for any variable x, and any constant $c \in Cons^*$, we let $A \in \Theta^*$,

$$A = \neg \forall x \phi(x) \to \neg \phi[x/c]$$

(2. Why?). Let now

$$\theta_1 = \neg \forall x_1 \phi(x_1) \to \neg \phi_1 [x_1/c_1]$$

$$\vdots$$

$$\theta_n = \neg \forall x_n \phi(x_n) \to \neg \phi_n [x_n/c_n]$$

where c_i is the i-th new constant not occurring in ϕ_i or in θ_j for $j < i$. Let

$$\Theta^* - \Theta = \bigcup_{i=1}^{n} \theta_i.$$

Then, $\Theta \cup \Theta^*$ is consistent. (3. Proof?) We extend (4. How?) $\Theta \cup \Theta^*$ to a consistent set Θ^∞; Θ^∞ is maximal, i.e. for any closed formula $\psi \in \mathsf{L1}^*$, either $\psi \in \Theta^\infty$ or $\neg \psi \in \Theta^\infty$, and Θ^∞ is deductively closed. In effect, we have

$$\vdash_{\Theta^\infty} \psi \;\Rightarrow\; \nvdash_{\Theta^\infty} \neg \psi$$

$$\Rightarrow \neg \psi \notin \Theta^\infty$$

$$\Rightarrow \psi \in \Theta^\infty$$

From Θ^∞, we now construct (5. How?) an interpretation $\mathcal{I} = (\mathcal{D}, \Theta, \varpi)$ for $\mathsf{L1}^*$ where $\mathcal{D}$ is the set of ground terms of Θ^∞ and for every predicate symbol P of $\mathsf{L1}^*$ we have

$$val_\mathcal{I}(P(t_1, ..., t_n)) = \{(t_1, ..., t_n) \mid \vdash_{\Theta^\infty} P(t_1, ..., t_n)\}.$$

In other words, the closed formula $P(t_1, ..., t_n)$ is *true* in $\mathcal{I}$ iff it is provable in Θ^∞, so that, given the consistency of Θ^∞, we must have

(♣) $\quad \vdash_{\Theta^\infty} P(t_1, ..., t_n) \;\Rightarrow\; (t_1, ..., t_n) \in \mathcal{I}(P)$

(♠) $\quad \vdash_{\Theta^\infty} \neg P(t_1, ..., t_n) \;\Rightarrow\; (t_1, ..., t_n) \notin \mathcal{I}(P)$

So, we need to prove that, given $\mathcal{I}$, there is a model $\mathcal{M}$ such that if $\vdash_{\Theta^\infty} \psi$, then $\models_\mathcal{M} \psi$ for a formula ψ. This we do by induction on atomic ψ.

(i) Let $\psi = P(t_1, ..., t_n)$. By ♣, we have $\models_\mathcal{M} P(t_1, ..., t_n)$, and so $\models_\mathcal{M} \psi$.

(ii) Let now $\psi = \neg \chi$, so that we have $\models_\mathcal{M} \neg \chi$.

(iii) Let now $\psi = \chi \to \omega$ and let $val_\mathcal{I}(\chi \to \omega) = \mathtt{f}$ to apply *reductio ad absurdum*, so that we have $\vdash_{\Theta^\infty} \neg \psi$ (6. How?) and Θ^∞ is inconsistent. But this is a contradiction.

8. Classical FO theories and the adequateness of CFOL

(iv) Make now $\psi = \forall x_i \phi(x_i)$. If x_i does not occur free in ϕ, ϕ is closed, and by the induction hypothesis, if $\vdash_{\Theta^\infty} \phi$, then $\models_\mathcal{M} \phi$. If $\vdash_{\Theta^\infty} \phi$, then $\vdash_{\Theta^\infty} \forall x_i \phi$, and $\models_\mathcal{M} \forall x_i \phi$ iff $\models_\mathcal{M} \phi$ (7. How?). Hence, we have $\models_\mathcal{M} \psi$ iff $\vdash_{\Theta^\infty} \psi$. On the other hand, if x_i occurs free in $\phi(x_i)$, then it must be the only free variable, given that $\psi = \forall x_i \phi(x_i)$ is closed. Thus, it must be the case that $\phi(x_i) = \theta_i \in \Theta^*$. Let it be θ_k. Then, $\psi = \forall x_k \phi(x_k)$. Suppose now $\vdash_{\Theta^\infty} \psi$ but $\not\models_\mathcal{M} \psi$. Then, there is a sequence that does not satisfy $\forall x_k \phi(x_k)$ and so neither does it satisfy $\phi(x_k)$. But then neither does it satisfy $\phi[x_k/c_k]$ and we have $\models_\mathcal{M} \neg\phi(c_k)$ (8. How?). But $\vdash_{\Theta^\infty} \forall x_k \phi(x_k) \to \phi(c_k)$, from which by MP we have $\vdash_{\Theta^\infty} \phi(c_k)$; by the induction hypothesis, we have $\models_\mathcal{M} \phi(c_k)$, and we reach a contradiction.

Thus, all the formulas in Θ^* have a common model $\mathcal{M}$ in $\mathcal{I}$ iff they are derivable in Θ^∞, and we proved that any consistent FOL theory has a model. **QED**

▶ *Do Exercises 8.1-10.*

Exercises

Exercise 8.1. Give an intuitive account of a scapegoat theory. (Hint: Find the "scapegoat" in Υ.)

Exercise 8.2. For Theorem 8.2, prove the equivalence of statements 1 and 2.

Exercise 8.3. In the proof of Theorem 8.2, give answers to the questions posed in it. The following are hints to some of these questions:

1. Invoke constant generalization.

2. Focus on $\neg$.

3. The proof is by RA. Assume that $\Theta \cup \left(\Theta^* = \{\theta_1, ..., \theta_k, \theta_{k+1}\}\right)$, $k \geq 0$, is inconsistent. Let $\theta_{k+1} = A$ and take the least k to obtain

$$\vdash_{\Theta \cup \{\theta_1,...,\theta_k\}} \begin{cases} \neg \forall x \phi(x) \\ \phi[x/c] \end{cases}$$

so that we have both $\neg \forall x \phi$ and $\forall x \phi$, the latter by $\forall y\, (\phi[y/x])$.

4. Let $\Theta^\infty = \Theta \cup \Theta^* \cup \Xi$, Ξ is the set of axioms for L1*, and let $\Theta^\infty = \{\psi | val(\psi) = t\}$. What are the *additional* axioms of Θ^∞?

5. Define $\mathcal{I}$ for every constant symbol c, every function symbol f, and every predicate symbol P of L1*.

Exercise 8.4. The proof of Theorem 8.2 is given for countable languages. Indicate, in general terms, what changes might be required to this proof to accommodate languages with a higher cardinality.

Exercise 8.5. Let the following lemma be given:

Lemma 8.3. *Every consistent theory Θ has a consistent extension Θ^* that contains denumerably many ground terms and is a scapegoat theory.*

1. Where was this lemma employed in the proof of Theorem 8.2?
2. Prove it in detail.

Exercise 8.6. Let the following lemma be given:

Lemma 8.4. *Let Θ be a complete scapegoat theory. Then Θ has a model $\mathcal{M}$ whose domain is the set of ground terms of Θ.*

1. Where was this lemma employed in the proof of Theorem 8.2?
2. Prove it in detail.

Exercise 8.7. Where in the proof of Theorem 8.2 is Lindenbaum's Theorem (Theorem 4.62) implicitly invoked?

Exercise 8.8. Which of the following statements, given any FOL theory Θ, can be a (multiple-statement) corollary to the completeness theorem?

1. For $\phi \in \Theta$ and $X \subseteq \Theta$, if $\phi \in Cn(X)$, then we have $X \vdash_\Theta \phi$.
2. ϕ is true in every denumerable model of Θ iff $\vdash_\Theta \phi$.
3. The set of terms of a language L1 is denumerable.
4. If, in every model of Θ, every sequence that satisfies all $\chi_i \in X$ also satisfies ϕ, then $X \vdash_\Theta \phi$.
5. For $\phi, \chi \in \Theta$, if $\phi \in Cn(\chi)$, then $\chi \vdash_\Theta \phi$.
6. If Θ is consistent, then it has a denumerable model.

Exercise 8.9. In Section 4.2.2.1, we made the completeness theorem for FOL follow from the FO-compactness theorem. Given the completeness proof in this Section, how is the reverse obtained, i.e. how do we obtain the FO-compactness theorem from the completeness proof?

Exercise 8.10. Prove Theorem 8.2 for **CL**$^=$.

Part III.

Classical Models

9. Three formal semantics for classical logic

In Chapter 2 above, we concentrated on form, or structure, alone with respect to a logical language. That is to say, we learned how to arrange and combine symbols of a logical alphabet into well-formed formulas and into formalized arguments. We were not concerned with *meaning* at all, but it is obvious that structure alone does not suffice if we want to move from *well-formed* strings to *valid* formulas and up to *valid* arguments. This little step that is actually a giant leap is made possible, as seen in Chapter 3, by the introduction of a *semantics*. This is said to be *classical* if it is based on bivalent valuations, i.e. if the semantical correlates of the formulas are restricted to the set $G = \{0, 1\}$ or their denotations are limited to *truth* and *falsity*. Furthermore, a classical semantics has to satisfy all the conditions for classicality defined in Chapter 6, namely with respect to the ♡- and ♦-consequence relations/operations.

We now elaborate on the formal semantics that are commonly invoked for classical logic. These are Tarski and Herbrand semantics, as well as the algebraic semantics provided by Boolean algebras. We call these semantics *formal* in the strictest sense: by the expression "formal semantics" we mean to capture the import of form, or structure, to meaning. In effect, in common these three semantics have the fact that the meaning of a proposition is (i) its truth value and (ii) a function of the meaning of its constituents. In spite of these commonalities, these semantics are quite different from each other, and this difference is mirrored in their distinct applications: Tarski semantics is typically employed in adequate proof systems such as Hilbert and Gentzen systems (Chapters 11 and 12), Herbrand semantics is vastly required in the field of automated theorem proving (see Part V of this book), and Boolean algebras are prominently so in mathematical and computational treatments of CFOL (e.g., Section 5.3.3). In the following Sections, we give the main points of these formal semantics.[1]

---

[1] With but only a few notable exceptions (e.g., van Fraassen, 1971), the expression "formal semantics" is rarely used in logic, being, on the contrary, frequently to be found in linguistics and programming. By employing it here, we mean to emphasize that our approach to logic is formal through and through, as expounded

203

9. Three formal semantics for classical logic

9.1. Tarskian semantics

Most of the semantical notions discussed above were so from the viewpoint of what can be called *Tarskian semantics*. This despite the fact that this semantics can be captured by the term *denotational semantics*, which, in turn, is actually a Fregean achievement. Recall the principle of extensionality or of the compositionality of meaning, formalized in Proposition 3.12, in particular in item 2 of this proposition; this principle states that the meaning of a proposition (its semantical correlate or, more generally, its truth value) is a function of the meaning of its components. We already know from Section 3.3 that this lies at the root of the concept of truth-functionality, but it can be further specified in terms of *denotation* as follows: as seen in the aforementioned Section, when we formalize natural language assertions by means of the alphabet of a logical language, we do so by assigning denotations to its symbols; these denotations are in fact the predicates, functions, and constants (or objects) of the logical language at hand.[2] Then, the meaning of an atomic formula is considered in terms of what the expressions thereof denote, and the meaning of compound formulas is the rule-based composition of the meaning of their atomic formulas.

But by *denotation* of an expression at a metalogical level we mean actually the truth value assigned to it by means of an interpretation, more precisely so a valuation (cf. Def. 3.25), so that in fact meaning and denotation coincide with the semantical correlates, which can be *logical truth values*, such as `true` and `false`, but also *algebraic truth values*, as seen in Definition 3.14. In either case, we are not speaking here of the natural-language adjectives "true" and "false," and it is this specification of "truth" in a formal(ized) language that earned A. Tarski the honor of having this denotational semantics named after him (cf. Tarski, 1935).

---

on in the Preface and in Part I. The following account largely characterizes our own perspective:

> In formal semantics, we deal with a class of structures called *(formal) languages*; they are called languages because they are believed to provide rational reconstructions of (parts of) natural languages and, indeed, adequate reconstructions relative to certain purposes. A logical system is considered correct for a language if it provides a catalogue of the valid inferences in that language. (van Fraassen, 1971)

[2]More properly, the denotations are so of predicates and functions when their terms are replaced or specified by constants, i.e. objects of the domain of discourse, so that Tarskian semantics can be seen as an interpretation of constants.

9.1. Tarskian semantics

Above, we spoke of rule-based composition of meaning. This immediately suggests adequateness of a logical system, in particular in the sense that one can speak of the *truth* or *falsity* of a rule of inference. For instance, an intuitive explanation for a logic-programming rule of the characteristic form

$$(\mathcal{C}_P) \qquad A \leftarrow B_1, ..., B_n$$

(cf. Def. 5.15) is as follows: if every B_i is true, then A is true, or A can be proved by proving all the B_i. This truth- or provability-preservation actually implicitly entails that rule $\mathcal{C}_P$, concretized in the connective $\leftarrow$ (essentially a reversed $\rightarrow$), is itself *true*. Without this *truth* of the inference rule at hand there simply is no way to assure us that the rule "holds," but this *truth* is a function of $\leftarrow$ with respect to the meanings of A and the B_i. As it is, $\mathcal{C}_P$ coincides with a Horn clause, a formula of the form

$$A \vee \neg B_1 \vee ... \vee \neg B_n$$

which is valuated as `true` if at least one of its sub-formulas is so valuated. But this valuation depends on the denotation of the expressions constituting each of the sub-formulas. For instance, the atomic Prolog formula `father(john, sara)` is valuated to `true` iff there is an interpretation $\mathcal{I} = (\mathscr{D}, \Theta, \varpi)$ such that, given the predicate `father(X, Y)` and the variable assignments $\varpi(\texttt{X}) = \texttt{john}$ and $\varpi(\texttt{Y}) = \texttt{sara}$, we have

$$val_\mathcal{I}(\texttt{father(john, sara)}) = \Theta(\texttt{father})(val_\mathcal{I}(\texttt{john}), val_\mathcal{I}(\texttt{sara})) = \texttt{t}.$$

Although this might be terrain for philosophical debates, from a formal perspective we talk here of the soundness and completeness of a rule of inference, it being the case that this adequateness holds in virtue of the denotational semantics that determines that the meaning of a classical FO formula is the meaning, or the denotation, of its constituting parts, which meaning in turn is dependent on a rule of inference concretized by some logical operation. Even though a deductive system might be given such a Tarskian semantics only implicitly–as is the case in logic programming–, this provides it with a warranty that whatever the logic generated by it is, it is a *rule-based truth-preserving logic*, and this is what classical logic is, after all, (mostly, if not entirely) about.

In effect, although resolution and analytic tableaux are essentially proof calculi, their ability to prove a set of formulas inconsistent is determined against a Tarskian semantics, making unsatisfiability out of this inconsistency. To be sure, especially in the case of the resolution calculus–but also not negligibly so in the analytic tableaux calculus–, it

205

9. Three formal semantics for classical logic

is of Herbrand (un)satisfiability that we speak, mainly due to Herbrand's Theorem (cf. Section 13.3), but Herbrand semantics is employed at a for some very high cost: the *propositionalization* of CFOL.[3] And this is precisely so because, as elaborated on in the next Section, Herbrand semantics is not denotational, treating predicates, functions, and constants at a fixed-interpretation level.

▶ *Do Exercise 9.1.*

9.2. Herbrand semantics

The formal semantics for CFOL that we approach now was conceived by the French mathematician J. Herbrand in his doctoral dissertation (Herbrand, 1930). As a matter of fact, Herbrand's dissertation was first and foremost in the proof theory of CFOL, but it so happens that from this work a semantics for CFOL can be extracted. We call it *Herbrand semantics*.

From a computational viewpoint, Herbrand semantics exhibits many advantages over the Tarskian semantics elaborated on in (especially) the Section immediately above, and in particular so with respect to automated deduction. This is so mostly because Herbrand semantics provides a fixed interpretation for CFOL, thus allowing for a purely syntactical manipulation of symbols in FO formulas while at the same time providing these symbols with a bivalent semantics. We thank this feature to Herbrand's concern with finitistic approaches to CFOL, in line with Hilbert's formalist-finitist program, and in particular to his dislike–or outright rejection–of infinite models.

The use of Herbrand semantics in automated deduction is, however, sanctioned mostly by the fact that in this we are mainly concerned with satisfiability. Indeed, Herbrand semantics is solely satisfiability-preserving, being incomplete and not validity-preserving.

Contrary to Tarskian semantics, which is diffused in different parts of this book, we concentrate our discussion of Herbrand semantics in this and in Section 13.3 below. In the present Section, we introduce the basic elements of this semantics, as well as the main results concerning it, and below we elaborate on it from the viewpoint of satisfiability testing.

Although our discussion of Herbrand semantics serves mainly our treatment of resolution, which is clause-based, this semantics is equally relevant to the analytic tableaux calculus; the reason for our concentration in resolution lies in the fact that discussing this semantics in a

---
[3] Cf. Section 4.2.2.1.

9.2. Herbrand semantics

clause-based formalism is particularly adequate. In effect, although we restrict our discussion here to sets of clauses, i.e. finite disjunctions of literals, Herbrand's results hold for any set X of FOL formulas.

Let C be a set of clauses. Then the following definitions hold.

Definition 9.1. The *Herbrand universe* of C, denoted by H_C, is the set of all ground terms built up from the constants and functions of C in the following way:[4]

1. if C contains no function, then $H_C = H_0$ is the set of constants occurring in C; if no constant occurs in C, then $H_C = H_0$ consists of a single arbitrary constant, say, $H_C = \{a\}$, i.e.

$$H_0 = \begin{cases} Cons\,(C) & \text{if } Cons\,(C) \neq \emptyset \\ \{a\} & \text{if } Cons\,(C) = \emptyset \end{cases} ;$$

2. if C contains a function, then $H_C = \bigcup_{i=0}^{\infty} H_i$, $H_i = H_{i-1} \cup Fun$,

$$Fun = \{f\,(t_1, ..., t_n)\,|\,f \in Fun\,(C)\,, (t_1, ..., t_n) \in H_{i-1}, n \in \mathbb{N}\}.$$

H_i, for $1 \leq i \leq \infty$, is called the *i-level constant set of C*. Clearly, H_C is infinite iff $Fun\,(C) \neq \emptyset$.

Example 9.2. We exemplify the above:

- Let $C_1 = \{P(b), \neg P(x) \vee Q(y)\}$. Then $H_{C_1} = H_0 = \{b\}$.
- Let $C_2 = \{P(x) \vee Q(x), R(z), T(y) \vee \neg S(y)\}$. Then we let $H_{C_2} = H_0 = \{a\}$.
- Let $C_3 = \{P(f(x)), Q\,(a)\,, R\,(g(y), b)\}$. Then

$H_0 = \{a, b\}$
$H_1 = \{a, b, f(a), f(b), g(a), g(b)\}$
$H_2 = \{a, b, f(a), f(b), g(a), g(b), f\,(f\,(a))\,, f\,(f\,(b))\,, f\,(g\,(a))\,, f\,(g\,(b))\,, g\,(f\,(a))\,, g\,(f\,(b))\,, g\,(g\,(a))\,, g\,(g\,(b))\}$
$\vdots$
$H_{C_3} = \{a, b, f(a), f(b), g(a), g(b), f\,(f\,(a))\,, f\,(f\,(b))\,, f\,(g\,(a))\,, f\,(g\,(b))\,, g\,(f\,(a))\,, g\,(f\,(b))\,, g\,(g\,(a))\,, g\,(g\,(b))\,, ...\}$

---
[4] In other words, the Herbrand universe of C is the set of all ground terms definable over $\Upsilon_C - Pred\,(C)$. See Definition 2.13.

9. Three formal semantics for classical logic

Definition 9.3. A *ground instance* of a clause $\mathcal{C}$ of C is a clause obtained by replacing variables in $\mathcal{C}$ by members of H_C. A *Herbrand instance (H-instance)* of $\mathcal{C}$ is a ground instance $\mathcal{C}\theta$ of $\mathcal{C}$ such that θ is based on C. The *Herbrand base* of C, denoted by $H(C)$, is the set of all Herbrand instances of atoms occurring in clauses of C.

Example 9.4. The Herbrand base of the clauses in Example 9.2 is as follows:

- $H(C_1) = \{P(b), Q(b)\}$
- $H(C_2) = \{P(a), Q(a), R(a), T(a), S(a)\}$
- $H(C_3) = \{P(a), Q(a), R(a,a), R(a,b), P(b), Q(b), R(b,a), ...\}$

Obviously, $H(C)$ is finite iff H_C is finite.

Definition 9.5. A *Herbrand interpretation (H-interpretation)* for C, denoted by $H\mathcal{I}_C$, is a triple (H_C, Θ, ϖ) (cf. Def.s 3.24-5) such that

1. $\Theta(c) = c$ for every $c \in Cons(C)$;
2. $\Theta(f)(t_1, ..., t_n) = f(t_1, ..., t_n)$ for all $t_1, ..., t_n \in H_C$, if $f \in Fun(C)$.

$H\mathcal{I}_C$ provides a *fixed* interpretation, as every constant symbol is interpreted as itself (i.e., $H\mathcal{I}_C$ maps every constant to itself), and every function symbol is interpreted as a term builder over H_C, or, in other words, as the function that applies it (i.e. $H\mathcal{I}_C$ maps every function symbol $f \in Fun(C)$ with arity > 0 to the n-ary function that maps every n-tuple $(t_1, ..., t_n)$ of terms $t_1, ..., t_n \in H_C$ to the term $f(t_1, ..., t_n)$). Moreover, because clauses are interpreted as closed formulas, ϖ is irrelevant in $H\mathcal{I}_C$.

All this entails that we end up with a purely syntactical interpretation, being meant by this that the symbols in a set of clauses are interpreted independently of any domain.

Definition 9.6. A H-interpretation $H\mathcal{I}_C$ is a subset $H'(C)$ of $H(C)$ such that the truth value **t** is assigned to all elements of $H\mathcal{I}_C$ and the truth value **f** is assigned to all atoms in $H(C) - H\mathcal{I}_C$. The subset $H'(C)$ is in fact a *Herbrand model (H-model)* $H\mathcal{M}_C$ of C, because for an interpretation $H\mathcal{I}_C$ and some $P \in Pred(C)$ we have

$$\Theta(P)(t_1, ..., t_n) = val_{H\mathcal{I}}(P(t_1, ..., t_n))$$

9.2. Herbrand semantics

so that for $t_i \in H(C)$ we have

$$HM_C = \{P(t_1, ..., t_n) \,|\, \Theta(P)(t_1, ..., t_n) = \mathtt{t}\}.$$

Example 9.7. Let $C = \{P(x) \vee Q(x), R(f(y))\}$. Then,

$$H_C = \{a, f(a), f(f(a)), ...\}$$

and

$$H(C) = \{P(a), Q(a), R(a), P(f(a)), Q(f(a)), R(f(a)), ...\}$$

The following are H-interpretations for C:

$HI_{C_1} = \{P(a), Q(a), R(a), P(f(a)), Q(f(a)), R(f(a)), ...\}$
$HI_{C_2} = \{\neg P(a), \neg Q(a), \neg R(a), \neg P(f(a)), \neg Q(f(a)), \neg R(f(a)), ...\}$
$HI_{C_3} = \{P(a), Q(a), \neg R(a), P(f(a)), Q(f(a)), \neg R(f(a)), ...\}$

It is easy to see that C is satisfied by HI_{C_1}, but falsified by HI_{C_2} and HI_{C_3}. Thus, only HI_{C_1} is a H-model HM_C.

As said above, Herbrand semantics applies to any set of FOL formulas, and not only to sets of clauses. The following shows this by means of the inductive definition of the satisfiability relation in Herbrand semantics.

Definition 9.8. Let $P(t_1, ..., t_n)$, $n \geq 1$, be an atomic ground formula and let ϕ, ψ be (atomic) ground formulas. Then, given a Herbrand interpretation HI, the *Herbrand satisfiability (H-satisfiability) relation* is defined inductively as follows:

1. $\models_{HI} \top$
2. $\not\models_{HI} \bot$
3. $\models_{HI} P(t_1, ..., t_n)$ iff $P(t_1, ..., t_n) \in HM$
4. $\models_{HI} \neg \phi$ iff $\not\models_{HI} \phi$
5. $\models_{HI} \phi \wedge \psi$ iff $\models_{HI} \phi$ and $\models_{HI} \psi$
6. $\models_{HI} \phi \vee \psi$ iff $\models_{HI} \phi$ or $\models_{HI} \psi$
7. $\models_{HI} \phi \rightarrow \psi$ iff $\not\models_{HI} \phi$ or $\models_{HI} \psi$
8. $\models_{HI} \forall x \phi(x)$ iff $\models_{HI} \phi(t)$ for every ground term $t \in H_\phi$

9. Three formal semantics for classical logic

9. $\models_{HI} \exists x \phi(x)$ iff $\models_{HI} \phi(t)$ for some ground term $t \in H_\phi$

A comparison of Definition 9.8 with the Tarski-style semantical classicality conditions in Chapter 6 reveals the essentially classical character of Herbrand semantics. However, the condition above that an atomic formula be of the form $P(x_1, ..., x_n)$ for $n \geq 1$ makes this semantics uninteresting to CPL itself. In effect, the main pay-off of employing the less-studied Herbrand semantics is that of "simulating" a classical propositional calculus for a given FO language, which has attached advantages such as finiteness of models and thus the existence of algorithmic decision procedures. In fact, the "simplification" that Herbrand semantics entails with respect to Tarskian semantics is to be found in Definition 9.8.3; in the latter semantics, and according to Definition 3.25.3, this condition would be formulated as

$$\models_\mathcal{I} P(t_1, ..., t_n) \text{ iff } \Theta(P)(val_\mathcal{I}(t_1), ..., val_\mathcal{I}(t_n)) = \mathbf{t}$$

for every $t_1, ..., t_n$ in the alphabet Σ of some FO language. This means that in Tarskian semantics every term t_i in P is valuated in an interpretation $\mathcal{I} = (\mathcal{D}, \Theta, \varpi)$ and we have $val_\mathcal{I}(P(t_1, ..., t_n)) = \mathbf{t}$ iff every t_i is valuated to $\mathbf{t}$. Given an infinite domain, we may have infinite models, i.e. interpretations in which $val_\mathcal{I}(P(t_1, ..., t_n)) = \mathbf{t}$. Compare with Definition 9.6: a H-model of $P(t_1, ..., t_n)$ just is a subset of the Herbrand base in which for every t_i in this base it is the case that $val_{HI}(P(t_1, ..., t_n)) = \mathbf{t}$, it being the case that Definition 9.5 eliminates the domain of discourse, replacing it by the Herbrand universe, and thus renders the variable assignment ϖ wholly superfluous. If we reduce the alphabet of a FO language to $\Sigma = \{P, a\}$ then there is only one model for the atomic formula $P(a)$, and that is $\{P(a)\}$. Increase finitely the alphabet with the constants $b, c, d, ...$ and one still has a finite number of models for $P(a)$: $\{P(a), P(b)\}$, $\{P(a), P(b), P(c)\}$, etc.

In terms of the relation between satisfiability and validity, the results obtained in Tarskian semantics (cf. Prop. 4.37) hold in Herbrand semantics. But this entails an advantage of the latter over the former in CFOL: under this, CFOL is generally undecidable (at best semi-decidable), but it becomes decidable under Herbrand semantics, even if only given some restrictions to a classical signature.

Theorem 9.9. *Given the language* L1, *let* $\Upsilon_{L1} = (Pred, Fun, Cons)$ *where* $Fun = \emptyset$ *and* $|Cons|$ *is finite. Then, the problem of validity in* L1 *is decidable under Herbrand semantics.*

Proof: For the given Υ_{L1}, the Herbrand universe H_F for a finite set of formulas $F \subseteq$ L1 is finite by Definition 9.1.1. Hence, $H(F)$, the set of ground instances of atoms of F, is also finite, and the models $H\mathcal{M}_F$ are necessarily in finite number, too. Assign to each ground atom in $H(F)$ a propositional symbol: you have a propositional rewriting or transformation of the FO set of formulas F. (Cf. Section 4.2.2.1 for the details of this transformation.) Thus, the validity problem in L1 can be reduced by means of grounding to the validity problem in L0, which is known to be decidable. **QED**

Let us denote the above *function-free* fragment of L1 by L1$_{ff}$ and let us denote the Herbrand semantics above by $H\mathfrak{S}$. Then we have the following obvious result for some set $X \subseteq \mathsf{L1}_{ff}$ and some formula $\phi \in \mathsf{L1}_{ff}$.

Corollary 9.10. $X \models_{H\mathfrak{S}} \phi$ *is decidable.*

Proof: L1$_{ff}$ generates solely (sets of) ground formulas that are substitutable by propositional formulas of L0, for which there is a semantical decision procedure, namely the truth-table construction. **QED**

In the following two theorems, we consider some *Herbrand axiomatization*, and we denote it by $H\mathcal{P}$.[5]

Theorem 9.11. *Given a set of formulas $X \subseteq F_{\mathsf{L1}}$ and a formula $\phi \in F_{\mathsf{L1}}$, if $X \vdash_{H\mathcal{P}} \phi$, then $X \models_{H\mathfrak{S}} \phi$.*

Proof: (Idea) A Herbrand axiomatization of L1 only proves (sets of) formulas of L1 that have a model in Herbrand semantics, both proofs and models being finite. (Of course, you have to show that finite models are finitely axiomatizable.) **QED**

Theorem 9.12. *There exist a set of formulas $X \subseteq F_{\mathsf{L1}}$ and a formula $\phi \in F_{\mathsf{L1}}$ such that $X \models_{H\mathfrak{S}} \phi$ but $X \nvdash_{H\mathcal{P}} \phi$.*

Proof: (Idea) Some formulas derivable in $H\mathfrak{S}$ require infinite proofs, but by definition proofs are finite objects. (Hint: Compactness is missing in $H\mathfrak{S}$. See Exercise 9.4) **QED**

▶ *Do Exercises 9.2-9.*

---

[5]Such an axiomatization may well be the (extended) $\mathcal{NK}$ calculus (see Section 12.1 below). Herbrand (1930) does indeed provide his own proof calculus for CFOL, but it does not differ significantly from this calculus, although it considers the elimination of the rule of inference MP.

9. Three formal semantics for classical logic

9.3. Algebraic semantics: Boolean algebras

By *algebraic semantics*, we mean here the employment of algebraic structures to give meaning to logical formulas. As a matter of fact, models are algebraic structures, and we have been employing them so far in association to interpretations and logical consequence, in both Tarskian and Herbrand semantics.[6] In both these semantics, a FO formula ϕ is said to be satisfiable iff, given an interpretation $\mathcal{I} = (\mathcal{D}, \Theta, \varpi)$ (a H-interpretation $H\mathcal{I} = (H_\phi, \Theta, \varpi)$), there is a model $\mathcal{M}$ (a H-model $H\mathcal{M}$) such that we have $\models_\mathcal{M} \phi$ ($\models_{H\mathcal{M}} \phi$, respectively). But this is not quite what we mean here by *algebraic semantics*; by this label, we intend to capture the fact that logical theories and calculi obey the laws of some (classes of) algebras. In particular, the theories or calculi obey the *logical laws* that are in fact exactly the *algebraic laws* of some given (class of) algebras. In this case, we say that the given (classes of) algebras *characterize* a certain logic. For instance, the Boolean algebras are known to characterize CPL, and Heyting algebras do so for propositional intuitionistic logic.

In an algebraic semantics, a logic or a logical system is provided with models that are in fact order-theoretical structures, such as lattices: the meaning of some proposition $p \in F_L$, where L is a logical language, is identified with an element of a specific lattice, and the connectives $\heartsuit_i \in O_L$ are interpreted as lattice-operations thereof.[7] We then speak not only of a model $\mathcal{M}$, but of an *algebraic model* $\mathcal{M} = \mathcal{U}_\mathcal{L}$, for a given lattice $\mathcal{L}$. If a logic L is characterizable by some algebra $\mathfrak{A}$, we then write

$$\models_\mathcal{M} p \quad \Leftrightarrow \quad \leq_{\mathcal{U}_\mathcal{L}} [p]$$

where $\mathcal{U}_\mathcal{L}$ denotes a filter of $\mathcal{L}$ and $[p]$ denotes the equivalence class of p. More precisely, we have $p \in hull(h)$ for h a Boolean homomorphism $h : \mathfrak{L} \longrightarrow 2$ where $\mathfrak{L} \supseteq \{p\}$ is an algebra of (propositional) formulas

---

[6]Strictly defined, a model $\mathcal{M}$ is an algebraic structure

$$\mathcal{M} = (U, R)$$

where U is a non-empty set and R is a relation over U. In order to contrast this algebraic structure with an algebra (cf. Section 3.2), note that an algebraic structure is a triple $\mathfrak{Z} = (U, O, R)$; thus, an algebra is the special case of an algebraic structure $\mathfrak{A} = (U, O, \emptyset)$, and a model is the special case of $\mathcal{M} = (U, \emptyset, R)$. For example, $(\mathbb{Z}, +)$ is an algebra, and $(\mathbb{Z}, \leq)$ is a model.

[7]Thus, the syntax of some logical language can be defined as an algebra, too. In particular, a propositional language L0 can be seen as an algebra of formulas $\mathfrak{L}$ freely generated by the set $V \subseteq F_{L0}$ (cf. Def. 3.16 and, below, Def. 9.14).

and 2 is the two-element Boolean algebra.[8] In other words, $\mathcal{U}_\mathcal{L}$ is an ultrafilter.[9]

The symbol $\Leftrightarrow$ denotes here a "bridge" between the logic **L** and an algebra $\mathfrak{A}$ established by a Lindenbaum-Tarski algebra. We give below the basic details of how this "bridge" is obtained between CPL and a Boolean algebra, but this construction carries over to many classes of algebras and many logics.

Definition 9.13. Given the language LO and the set of all formulas F_{L0}, we define the equivalence relation $\sim$ over F_{L0} as[10]

$$\phi \sim \psi \quad \text{iff} \quad \Theta \Vdash \phi \leftrightarrow \psi$$

for a consistent theory $\Theta \subseteq F_{\text{L0}}$. The set B of all equivalence classes defined as $[\phi] = \{\psi | \Vdash \phi \leftrightarrow \psi\}$ is a Boolean algebra under the operations $\vee, \wedge, ', 1$, and 0, in the following ways:

$$[\phi] \vee [\psi] = [\phi \vee \psi]$$
$$[\phi] \wedge [\psi] = [\phi \wedge \psi]$$
$$[\phi]' = [\neg \phi]$$
$$1 = [\phi \vee \neg \phi]$$
$$0 = [\phi \wedge \neg \phi]$$

This algebra is called the *Lindenbaum-Tarski algebra of the language* LO. This algebra can be extended to L1 by considering its set T of all terms. Then, it can be shown that

$$\bigvee_{t \in T} [\phi(t)] := [\exists x \phi(x)]$$

---

[8] Let h be a Boolean homomorphism. Then, the *hull* of h is the set

$$\text{hull}(h) = h^{-1}[\{1\}]$$

where h^{-1} denotes an *inverse homomorphism*. The dual of a hull is a *kernel*.

[9] Let $\mathcal{L}$ be a lattice and A be a set of elements of $\mathcal{L}$. Then, A is said to be a *filter* of $\mathcal{L}$ if (i) $x, y \in A$ implies that $(x \cap y) \in A$ and (ii) $x \in \mathcal{L}, y \in A$ and $x \geq y$ imply that $x \in A$. A is said to be a *proper filter* of $\mathcal{L}$ if $A \neq \mathcal{L}$, and a proper filter A is called an *ultrafilter* if for any filter A it is the case that

$$A \subseteq X \subseteq \mathcal{L} \quad \Rightarrow \quad A = X \text{ or } X = \mathcal{L}.$$

The dual of a filter is an *ideal*.

[10] Cf. Exercise 7.3 for an equivalence relation.

9. Three formal semantics for classical logic

and
$$\bigwedge_{t \in T} [\phi(t)] := [\forall x \phi(x)].$$

Definition 9.14. Let the set $B = \mathsf{L0}/\sim$ of propositional formulas of $\mathsf{L0}$ *modulo equivalence* be the set of equivalence classes. Then, the algebra $\mathfrak{B} = (B, ', \vee, \wedge, 0, 1)$ is a Boolean algebra. Moreover, this is a free Boolean algebra freely generated by the set of propositional variables V_{L0}.

Note that what we have here is a means of investigating sets of CFOL formulas by investigating their associated Lindenbaum-Tarski algebras instead, and as these can be shown to be Boolean algebras,[11] we may investigate these sets of formulas by investigating Boolean algebras instead.

With the two above-discussed semantics, Herbrand and Tarskian, we were able to prove the completeness of CFOL (see Section 4.1.4 and Chapter 8); in other words, we related the logical language L1 to both model-theoretical and proof-theoretical structures, to wit, models and proofs. Because logical languages, models, and proofs are all objects of logic, we did not leave the terrain of logic proper. But now, having shown that classical logical theories coincide with Boolean algebras, we have a much easier way to prove the completeness of CFOL via a *representation theorem*, i.e. a theorem that puts families or classes of mathematical structures in relation with one of its proper sub-families or sub-classes. In fact, the completeness theorem can be considered a corollary of the representation theorem for Boolean algebras.

Theorem 9.15. *(Representation theorem; Stone, 1936) Let A be the set of two-valued homomorphisms on a Boolean algebra $\mathfrak{B}$. Then, $\mathfrak{B}$ is embeddable into 2^A via the mapping defined by*

$$f(p) = \{h \in A | h(p) = 1\}$$

for every $p \in \mathfrak{B}$.

Proof: A Boolean algebra $\mathfrak{B} = (A, ', \wedge, \vee, 0, 1)$ has the operations of complementation, meet, and join, denoted by $', \wedge, \vee$, respectively. Recall

---

[11] More specifically, Lindenbaum-Tarski algebras of propositional languages coincide with free Boolean algebras, and Lindenbaum-Tarski algebras of FO languages coincide with polyadic Boolean algebras. We eschew any discussion of the latter algebras for several reasons, not the least of which is that our interest here in algebraic semantics is mostly connected to the SAT, which can be naturally generalized to CFOL and adequately tackled then with Herbrand semantics (see Part V).

9.3. Algebraic semantics: Boolean algebras

that an embedding, or monomorphism, is a one-to-one homomorphism (cf. footnotes 6 and 9 in Sections 3.2 and 3.3, respectively). Let $p, q \in \mathfrak{B}$. We show that h is a homomorphism:

$$\begin{aligned} h(p \vee q) &= \{h \in A | h(p \vee q) = 1\} \\ &= \{h \in A | h(p) \vee h(q) = 1\} \\ &= \{h \in A | h(p) = 1 \text{ or } h(q) = 1\} \\ &= \{h \in A | h(p) = 1\} \cup \{h \in A | h(q) = 1\} \\ &= h(p) \cup h(q) \end{aligned}$$

Similarly, we have

$$\begin{aligned} h(p') &= \{h \in A | h(p') = 1\} \\ &= \{h \in A | h(p)' = 1\} \\ &= \{h \in A | h(p) = 0\} \\ &= \{h \in A | h(p) = 1\}' \\ &= h(p)' \end{aligned}$$

The meet-operation can be defined by means of the join and complementation operations $(p \wedge q = (p' \vee q')')$, and we have shown that h is a homomorphism. Now, we have to show that h is one-to-one, an easy thing to do in algebra (hint: kernel!). **QED**

It should be easy to see that in the above homomorphism $h : \mathfrak{B} \longrightarrow 2^A$, A can be seen as the set of ultrafilters of $\mathfrak{B}$.

Corollary 9.16. *Every Boolean algebra is isomorphic to a field of sets.*

Proof: (Idea) A *field* (or *algebra*) *of sets with unit* U is a structure $\mathfrak{F} = (\Psi, -, \cap, \cup, \emptyset, U)$ where $\Psi \supseteq 2^U$ and Ψ is closed under the operations $-, \cap, \cup$. Moreover, Let $\mathfrak{A} = (U, O), \mathfrak{B} = (A, O)$ be similar algebras; if $h : \mathfrak{A} \longrightarrow \mathfrak{B}$ is a homomorphism such that $h(U) = A$, then h is an *epimorphism*. An epimorphism that is also a monomorphism is an *isomorphism* of $\mathfrak{A}$ and $\mathfrak{B}$. **QED**

▶ **Do Exercises 9.10-21.**[12]

---
[12] The exercises in this Section are all geared to a fuller grasp of the characterization of CPL by means of Boolean algebras. Although some of these are likely the most difficult exercises in this book, solving them will provide a deeper understanding of the importance of Boolean algebra for many applications of CPL (and, by extension, of CFOL), such as in logic circuits and logic programming.

9. Three formal semantics for classical logic

Exercises

Exercise 9.1. Reflect on the following passages by A. Tarski on the concept of truth.

1. We regard the truth of a sentence as its "correspondence with reality". This rather vague phrase, which can certainly lead to various misunderstandings and has often done so in the past, is interpreted as follows. We shall regard as valid all such statements as:

the sentence "it is snowing" is true if and only if it is snowing; the sentence "the world war will begin in the year 1963" is true if and only if the world war will begin in 1963.

Quite generally we shall accept as valid any sentence of the form

the sentence x is true if and only if p

where "p" is to be replaced by any sentence of the language under investigation and "x" by any individual name of that sentence provided this name occurs in the metalanguage. (In colloquial language such names are usually formed by means of quotation marks.) Statements of this form can be regarded as partial definitions of the concept of truth. They explain in a precise way, and in conformity with common usage, the sense of all special expressions of the type: *the sentence x is true.* Now, if we succeed in introducing the term "true" into the metalanguage in such a way that every statement of the form discussed can be proved on the basis of the axioms and rules of inference of the metalanguage, then we shall say that the way of using the concept of truth which has thus been established is *materially adequate.* In particular, if we succeed in introducing such a concept of truth by means of a definition, then we shall also say that the corresponding definition is materially adequate. We can apply an analogous method to any other semantical concepts as well. (Tarski, 1935/1956)

2. Let us consider the following sentence:

if 1 is a positive number and $1 < 2$, then 1 is a positive number.

This sentence is obviously true, it contains exclusively constants belonging to the domain of logic and arithmetic, and yet the idea of listing this sentence as a special theorem in a textbook of mathematics would not occur to anybody. If one reflects why this is so, one comes to the conclusion that this sentence is completely uninteresting from the standpoint of arithmetic; it fails to enrich our knowledge about numbers in any way, since its truth does not depend at all upon the content of the arithmetical terms occurring within it, but only upon the meaning of the words "*and*", "*if*", "*then*". (Tarski, 1994)[13]

3. [W]hen dealing with the problem of truth, we are concerned with relating expressions (in this case, sentences) and the objects to which the expressions refer, or, which they "talk about". (Tarski, 1994)

Exercise 9.2. For each of the following sets of clauses C, find the Herbrand universe H_C and the Herbrand base $H(C)$:

1. $C = \{P(x, y) \vee \neg Q(b), \neg P(a, x) \vee Q(b)\}$
2. $C = \{P(x, f(y)), P(z, g(z))\}$
3. $C = \{P(a, f(x, y)), P(b, f(x, y))\}$
4. $C = \{(R(g(a, f(b))))\}$

Exercise 9.3. Interpret formula χ of Example 3.28 under Herbrand semantics.

Exercise 9.4. Herbrand semantics is not FO-compact.

1. Show that there are infinite sets of FO formulas that are unsatisfiable while every finite subset thereof is satisfiable.

2. Say how this impacts on satisfiability testing.

Exercise 9.5. Show by means of an example that Skolemization does not preserve satisfiability in Herbrand semantics.

Exercise 9.6. Herbrand semantics allows for more expressive power of a logic in comparison to Tarskian semantics. Comment on this.

---

[13]This is an English translation of a text originally published in Polish.

9. Three formal semantics for classical logic

Exercise 9.7. CFOL with Herbrand semantics loses semi-decidability. Why?

Exercise 9.8. Complete the proof of Lemma 4.72. (Hint: Restate the lemma in terms of a model $H\mathcal{M}$.)

Exercise 9.9. Prove Theorems 9.11-2 by following the given ideas and hint.

Exercise 9.10. Show that all the properties of a Boolean algebra (Fig. 3.5.1) hold in CPL.

Exercise 9.11. Let Θ be a propositional theory and $\mathcal{M}_\Theta$ be a set of Θ-models such that for every model $\mathcal{M}$ satisfying all the formulas in Θ there is a model $\mathcal{M}' \in \mathcal{M}_\Theta$ that is elementarily equivalent to $\mathcal{M}$. Let $\mathcal{M}_\phi$ denote the set of models in $\mathcal{M}_\Theta$ that satisfy a given formula $\phi \in \Theta$. Recall now the definition of a field of sets (cf. proof of Corolary 9.16). Show that the $\mathcal{M}'_\phi$ form a field of sets with unit universe $\mathcal{M}_\Theta$.

Exercise 9.12. Denote the above field of sets by $\mathfrak{F}_\Theta$. Then, we say that the algebraization of CPL is the class $Iso(\mathfrak{F}_\Theta)$ of algebras, where Iso denotes isomorphism (cf. proof of Corolary 9.16) and Θ varies over the possible theories on the possible propositional languages. Show that $Iso(\mathfrak{F}_\Theta)$ of CPL coincides with the isomorphic closure of $\mathfrak{F}_\Theta$.

Exercise 9.13. A class of algebras is said to be a *variety* if it can be axiomatized by a set of equations.

1. Show that $Iso(\mathfrak{F}_\Theta)$ is a variety and that it is the class of Boolean algebras.

2. Say informally how this result can be seen as a completeness theorem for CPL.

Exercise 9.14. Let us denote the Lindenbaum-Tarski algebra for a propositional theory Θ by $\mathfrak{B}_\Theta$. Let $Iso(\mathfrak{B}_\Theta)$ denote the algebraization of CPL based on a given proof system.

1. Show that $\mathfrak{B}_\Theta$ and $\mathfrak{F}_\Theta$ are isomorphic.

2. Explain how this result can be considered a completeness theorem for CPL.

Exercise 9.15. Match the expressions/symbols in column A (Boolean algebras) with the expressions in column B (propositional logics):

9.3. Algebraic semantics: Boolean algebras

	A	B	
1.	Boolean filter	Taut	a.
2.	$\mathfrak{B}_\emptyset \longrightarrow \mathscr{D}$	complete logical theory	b.
3.	Boolean ultrafilter	CPL is complete	c.
4.	$\leq$	propositional model	d.
5.	Boolean algebras are a variety	logical theory	e.
6.	1	$\models$	f.

Exercise 9.16. The *Ultrafilter theorem* states that every filter is included in an ultrafilter. Establish the relation for propositional logic between this and Lindenbaum's Theorem (Theorem 4.62).

Exercise 9.17. Let $\mathfrak{B}$ be a Lindenbaum-Tarski algebra. Show that the following statements are equivalent for a set of formulas $X \in F_\mathsf{L}$:

1. X is consistent.

2. X is satisfiable.

3. The set $\{[x] \mid x \in X\}$ has the finite-meet property in $\mathfrak{B}$.[14]

Exercise 9.18. Show that every Boolean function can be expressed as a Boolean formula.

Exercise 9.19. Show that for all Boolean functions f and g the following statements are equivalent:

1. $f \leq g$

2. $f \vee g = g$

3. $\bar{f} \vee g = 1$

4. $f \wedge g = f$

5. $f \wedge \bar{g} = 0$

Exercise 9.20. Complete the proof of Theorem 9.15.

Exercise 9.21. Prove Corollary 9.16.

---

[14] A subset X of a lattice $\mathcal{L}$ is said to have the *finite-meet property* if whenever $x_1, ..., x_n \in X$, then $\bigwedge_{i=1}^{n} x_i \neq 0$.

Part IV.
Classical Proofs I: Direct Proofs

10. The validity problem, or *VAL*

The decidability of theories, and particularly so of theories over L1, is one of the major concerns of formal logic. As seen in Section 4.2.2, given a theory Θ and a formula ϕ over a logical language L, the decision problem $DP(\Theta, \phi)$ consists in finding out whether we have $\Theta \models \phi$ in all interpretations, or in at least one interpretation, over some semantics $\mathfrak{S}$. We focus here on the former, known as the *validity problem* (abbr.: *VAL*), and discuss the latter (the satisfiability problem) in Chapter 13.

Although current computational applications of CL favor methods for unsatisfiability testing, or indirect proofs, validity-proving, or direct proof, methods are still crucial for a satisfactory understanding of logical proof. Moreover, the VAL constitutes a historical scenario that, going from D. Hilbert to A. Turing, provided to modern logic some of its most impacting results. For these pedagogical and historical reasons, we offer here a short elaboration on this problem and below, in Chapters 11 and 12, elaborate at length on the most significant validity-proving systems. Our discussion of the VAL falls on three essential components of this problem, to wit, its original formulation by D. Hilbert and A. Turing's celebrated negative answer (cf. Section 10.1), direct proofs (Section 10.2), and the complexity of *VAL* (Section 10.3).[1]

10.1. The *Entscheidungsproblem* and Turing's negative answer

As is well known, the decision problem for CPL is solvable: A truth table is in general a decision procedure for some propositional formula or theory. Whether this is an efficient procedure, that is altogether another question, as the complexity of a propositional formula grows

---

[1] Recall that a decision/computational problem can be formulated as a language. Strictly conceived, the validity problem is a language, namely the language *VAL* (see below); broadly conceived, it is the framework of the (sub-)problem(s) and proposed solutions. Accordingly, we write "*VAL*" whenever we mean the language, and "the VAL" as an abbreviation for the validity problem taken broadly. We proceed in the same way for the satisfiability problem (see Chapter 13).

10. The validity problem, or VAL

exponentially with the number n of atoms, so that a logical problem that is decidable may turn out to be in fact non-computable. Be it as it may, we consider the decision problem for CPL "uninteresting," and by referring to this problem we more often than not mean *the decision problem for CFOL*. This focus is made quite clear in D. Hilbert's original formulation of the *Entscheidungsproblem* as arising in the context of (the completeness) of CFOL:

> Whether the axiom system is complete, at least in the sense that all logical formulas that are correct for every domain of individuals can be derived from it, is still an unresolved question. (Hilbert & Ackermann, 1928)

He actually considered this to be the main problem of mathematical logic, and briefly stated the need for–and thereby formulated the notion of–an algorithmic solution in the following terms:[2]

> The decision problem (*Entscheidungsproblem*) is solved if we know a procedure with a finite number of operations that determines the validity or satisfiability of any given logical expression. (Hilbert & Ackermann, 1928)

To be sure, D. Hilbert speaks here of of validity or satisfiability, but the *Entscheidungsproblem* came to be associated to the validity problem, an association clearly established by A. Church:

> By the *Entscheidungsproblem* of a system of symbolic logic is here understood the problem to find an effective method by which, given any expression Q in the notation of the system, it can be determined whether or not Q is provable in the system. Hilbert and Ackerman (1928) understand the *Entscheidungsproblem* of the *engere Funktionenkalkül* in a slightly different sense. But the two senses are equivalent in view of the proof by Kurt Gödel of the completeness of the *engere Funktionenkalkül* (Gödel, 1930). (Church, 1936b; slightly changed)

Although there are some fragments of CFOL in which the decision problem is decidable (cf. Section 4.2.2), the general negative answer to the *Entscheidungsproblem* remains as one of the great feats of modern logic. This answer was given independently by A. Church (Church, 1936a, b)

---

[2] By "any given logical expression," Hilbert means a formula of what he called the "restricted function calculus" (*engere Funktionenkalkül* in German), i.e. FO predicate logic. See Section 1.2.2 for a brief account of the expression "function calculus."

10.1. The Entscheidungsproblem and Turing's negative answer

and A. Turing (Turing, 1936-7), being thus known as the *Church-Turing Theorem*. Essentially, this theorem states that there is no general algorithm to decide whether a FOL formula is provable in a FOL system L from the axioms by using the rules of L. Although Church's λ-calculus and the Turing machine are equivalent algorithmic constructs, the latter remains to this day the best known of both, not only because of its more intuitive character, but also because it was an inspiration for the conception of the digital computer.[3]

We present this negative result as Turing's Theorem and then give a sketch of Turing's proof, leaving many of its details as (research) exercises.

Theorem 10.1. *(Turing's Theorem; Turing, 1936-7)* The Entscheidungsproblem *has no solution.*

Proof: (Sketch) The core of the proof is as follows: Turing shows that given an arbitrary FO formula ϕ there is no general process to determine whether ϕ is provable by showing that there can be *no* automatic computing machine that, given ϕ as input, will eventually decide whether ϕ is provable. Let us imagine that there is an automatic machine whose tape contains the axioms of the Hilbert FO calculus (cf. Chapter 11). Let us call this machine K, abbreviating thus the German word *Kalkül*, the English equivalent of *calculation* or *calculus*. Given the additional axioms for supplementary predicates, i.e. the proper axioms of some theory, the machine automatically implements the rules of inference. Recall that DT allows us to express any reasoning instance $\Theta \vdash \phi$, where Θ is as in Definition 4.56, as the single theorem $\vdash \Theta \rightarrow \phi$, so that this is actually what might be the initial input given to K.[4]

Let us consider the following proper axioms:

P1. $\exists u \, (N(u))$
P2. $\forall x \, (N(x) \rightarrow \exists y \, (F(x, y)))$
P3. $F(x, y) \rightarrow N(y)$

We consider now some valuation val such that $val(N(x)) = 1$ if x is a non-negative integer and $val(F(x, y)) = 1$ if $y = x + 1$, i.e. y is the successor if x. These are actually axioms from *Peano arithmetic*, more

---

[3] See Augusto (2020b, Introduction).
[4] However, the rules of inference need not be part of Θ, as they can be implemented by means of the transition functions of K.

10. The validity problem, or VAL

specifically axioms for the successor function over N.[5] We can join the three axioms above in the single formula

$$\phi = (\exists u\, (N(u))) \wedge (\forall x\, (N(x) \to \exists y\, (F(x,y)))) \wedge (F(x,y) \to N(y))$$

Formula ϕ defines the sequence α in the sense that $\neg \phi$ is not provable and, for each n, one of the following A_n or B_n is provable:

$$(A_n) \qquad \left(\phi \wedge F^{(n)}\right) \to G_\alpha\left(u^{(n)}\right)$$

$$(B_n) \qquad \left(\phi \wedge F^{(n)}\right) \to \neg G_\alpha\left(u^{(n)}\right)$$

where $F^{(n)}$ stands for

$$F(u, u') \wedge F(u', u'') \wedge \ldots \wedge F\left(u^{(n-1)}, u^{(n)}\right)$$

and $G_\alpha(x)$ and $\neg G_\alpha(x)$ denote the propositions "The x-th figure of α is 1" and "The x-th figure of α is 0," respectively.

Example 10.2. Let $\alpha = \{0, 1, 2, 3, \ldots\} = \mathbb{N}$ and $G(x)$ be the unary predicate "x is even." Then, where u denotes the element "0" in the sequence α, we have the following provable formulas for $n = 5$:[6]

$$A_0 \qquad \left(\phi \wedge F^{(0)}\right) \to G_\alpha(u)$$

$$B_1 \qquad \left(\phi \wedge F^{(1)}\right) \to \neg G_\alpha(u')$$

$$A_2 \qquad \left(\phi \wedge F^{(2)}\right) \to G_\alpha(u'')$$

$$B_3 \qquad \left(\phi \wedge F^{(3)}\right) \to \neg G_\alpha(u''')$$

---

[5]This set of axioms does not, in fact, establish two important facts of Peano arithmetic, to wit, the existence of 0 and the uniqueness of a successor. This is accomplished in the following axioms for the successor function s (corresponding predicate symbol: S) in Hilbert & Bernays (1934):

1. $\forall x \exists y\, (S(x,y))$
2. $\exists x \forall y \neg (S(y,x))$
3. $\forall x \forall y \forall r \forall s\, ((S(x,r) \wedge S(y,r) \wedge S(s,x)) \to S(s,y))$

[6]Because Turing did not establish the fact that 0 is not the successor of any other natural number we need to define $F^{(0)}$ as the special case $F(u,u)$, and then count $n+1$ where n stands for prime marks (e.g. u''' has three primes or three prime marks). In any case, we may, as Turing did, start with $F^{(1)}$ for $F(u,u')$.

$$A_4 \quad \left(\phi \wedge F^{(4)}\right) \to G_\alpha\left(u^{(4)}\right)$$

$$B_5 \quad \left(\phi \wedge F^{(5)}\right) \to \neg G_\alpha\left(u^{(5)}\right)$$

This gives us the sequence $\alpha_G = 1, 0, 1, 0, 1, 0$.

Proof (cont.): Recall the notion of decidability or computability (cf. Section 1.4). Clearly, the sequence α is decidable, or computable. In effect, Let K_α be the automatic machine K that computes α; then, K_α computes the characteristic function for any $x \in \alpha$:

$$\chi_{G_\alpha}(x) = \begin{cases} 1 & \text{if we have formula } A_x \\ 0 & \text{if we have formula } B_x \end{cases}.$$

Furthermore, defined in this way via ϕ (literally: "ϕ defines α"), α includes all the computable numbers.[7] But these do not include all the *definable numbers*. In effect, the reals can be clearly defined, but they are not computable.[8]

The "morale" is: One may have a perfectly unequivocal formal definition of a problem, namely in FO predicate logic, and yet be confronted with the fact that there is no solution thereto, or, better put, there is no automatic machine that can solve it.

▶ *Do Exercises 10.1-5.*

10.2. *VAL* and direct proofs

The general validity problem can be defined as a formal language in the following way:

Definition 10.3. The language

$$VAL = \{\langle \alpha \rangle \,|\, \alpha \text{ is valid}\}$$

---

[7] It is in this seminal paper, to wit, Turing (1936-7), that Turing first introduces the Turing machine (see Section 1.4). K_α is the Turing machine computing sequence α. We leave it as an exercise to describe the functioning of this Turing machine.

[8] As Turing puts it:

> It is (so far as we know at present) possible that any assigned number of figures of [the sequence of reals] δ can be calculated, but not by a uniform process. When sufficiently many figures of δ have been calculated, an essentially new method is necessary in order to obtain more figures. (Turing, 1936-7)

The proof is by application of the *diagonal process*. We leave this topic as an exercise.

10. The validity problem, or VAL

for a given argument α and $\langle\alpha\rangle$ an adequate encoding thereof, is called the *validity problem*, or, abbreviated, *VAL*.

We shall consider that an adequate encoding $\langle\alpha\rangle$ of $\alpha = A_1, ..., A_n/B$ just is a pair (X, ϕ) where X is a set of FO formalizations of $A_1, ..., A_n$ and ϕ is the FO formalization of B.

As seen in the Section above, the *Entscheidungsproblem* asks whether, given a complete logical system, there is an algorithm to decide on the validity of an arbitrary FO *formula*. Given our definition of validity via the notion of semantical logical consequence, we can reformulate Definition 10.3 as follows:

Definition 10.4. The language

$$VAL = \{\phi \mid \models \phi\}$$

for a given FO formula ϕ is called the validity problem, or, abbreviated, *VAL*.

As seen in Section 4.2.2, we can generalize this problem to logical theories as

$$VAL = \{(\Theta, \phi) \mid \Theta \models \phi\}$$

for Θ a FO theory and ϕ a FO formula, given that by DT we have

$$VAL = \{(\Theta, \phi) \mid \models \Theta \to \phi\}.$$

Recall that, given an adequate logical system, we can use some proof system $\mathcal{P} = (\mathsf{L}, AX, RI)$ to solve the validity problem, as given an adequate logical system L we have (Prop. 4.52)

$$\Theta \vdash_{\mathsf{L},\mathcal{P}} \phi \quad \text{iff} \quad \Theta \models_{\mathsf{L},\mathfrak{S}} \phi$$

for a theory Θ and a proposition ϕ (or, more generally, for a–possibly empty–set of formulas X and some formula ϕ). As seen in Chapter 4, CL is an adequate system. Hence, if, given some theory $\Theta = \{\theta_1, ..., \theta_k\}$ and some formula ϕ over L1, we wish to know whether we have $\Theta \models \phi$ in all interpretations, we may do so by producing a direct proof in some calculus $\mathcal{P}$ such that $\Theta \vdash_{\mathcal{P}} \phi$. Recall the general definition of a logical proof in Definition 4.16; the following definitions specify the case of a direct proof, so coined because the conclusion is derived *directly* from the premises and/or from constituents of the calculus at hand such as axioms and rules.

Definition 10.5. A *direct proof* ■ $\in \mathcal{P}$ for some proof system $\mathcal{P}$ and

10.2. VAL and direct proofs

some decision problem $DP(\Theta, \phi)$ where $\Theta = \{\theta_1, ..., \theta_k\}$ is a set of FO formulas and ϕ is a FO formula consists in a sequence

$$
\begin{array}{ll}
1. & \theta_1 \\
2. & \theta_2 \\
\vdots & \vdots \\
k. & \theta_k \\
(k+1). & \theta_{k+1} \\
\vdots & \vdots \\
k_m. = (n-1). & \theta_m \\
n. & \vdash \theta_n = \phi
\end{array}
$$

where $\theta_i \in (F_L \cup AX \cup RI)$ for $1 \leq i \leq m$, or $\theta_i \in Cn(\{\theta_{j-l}, ..., \theta_{j-1}\})$ for some $1 < i \leq m$ such that $j - l \leq j - 1 < i$, and $\phi = \theta_n$.

1. In case we have, for $1 \leq i \leq k$, $[(F_L - (AX \cup RI)) \neq \emptyset] \supseteq \Theta$, then we have a theory proper whose (all or some) θ_i are premises in the proof of ϕ.

2. If $\{\theta_i\}_{i=1}^{k} \subseteq (AX \cup RI)$, AX is a set of logical axioms, then we have a calculus and ■ is the proof of a theorem.

With respect to Definition 10.5, note the following specification:

Definition 10.6. In a direct proof $\Theta \vdash \phi$ where $\Theta = \{\theta_1, ..., \theta_k\}$, the given $\theta_i \in [\Theta \subseteq (F_L \cup AX \cup RI)]$, $1 \leq i \leq k$, are the *premises*. A formula $\theta_l \in F_L$ for $k < l \leq m$ (where possibly $k = 0$) assumed (to be *true*) in an ad-hoc manner in the proof at hand or in a sub-proof thereof (cf. Def. 12.4 below) is an *assumption* (or a *hypothesis*).[9]

Example 10.7. When proving a theorem, one may start from one or more axioms in a Hilbert system (cf. Def. 10.5.2) or from arbitrary formulas in a Gentzen system (cf. Def. 10.6). In Example 11.2, the first two steps of the proof of the theorem $\vdash_\mathcal{L} \phi \to \phi$ are constituted by axioms of the axiom system $\mathcal{L}p$. In the proof of Lemma 12.16.1 in the Gentzen system $\mathcal{NK}$, i.e. of the theorem $\vdash_{\mathcal{NK}} \phi \leftrightarrow \neg\neg\phi$, the assumptions in the first steps are arbitrary formulas. Additionally, in a Hilbert-style system, one may also start a proof from assumptions.

Remark 10.8. By DT, (cf. Theorem 4.25), given the decision problem $DP(\Theta, \phi)$ in which Θ may be empty, one may opt to prove $\vdash \Theta \to \phi$, in

---

[9] We remark that this terminological convention is not fixed in the literature, with *premise*, *assumption*, and *hypothesis* being often considered synonyms.

10. The validity problem, or VAL

which case Step 1 in the sequence in Definition 10.5 must be an axiom (possibly followed by other axioms) or an assumption (possibly followed by other assumptions). Still by DT, for non-empty $\Theta = \{\theta_1, ..., \theta_k\}$, one may prove $\Theta' \vdash \varphi$ for $\Theta' = \{\theta_1, ..., \theta_r\}$ given some $r < k$ such that $\Theta - \Theta' = \{\theta_{k-1}, ..., \theta_{k-r}\}$, and $\varphi = \left(\bigwedge_{j=k-1}^{k-r} \theta_j\right) \to \phi$, or, equivalently, $\varphi = \theta_{k-1} \to (... \to (\theta_{k-r} \to \phi))$ (cf. Theorem 4.25 and Prop. 4.27).

Example 10.9. Care should be taken when applying Remark 10.8, as one may complicate a proof rather than simplify it. A particularly good example is provided by an argument of the form $\{\phi \lor \psi, \neg\phi\} \vdash \psi$, known as *modus tollendo ponens* (TP; cf. Fig. 2.3.1): the proof $\{\phi \lor \psi\} \vdash \neg\phi \to \psi$ is complete, as it corresponds to the classical definition of the connective $\lor$ by means of the functionally complete set $O'_L = \{\neg, \to\}$ (cf. Example 5.6.1), whereas the proof $\{\neg\phi\} \vdash (\phi \lor \psi) \to \psi$ is more complex.

Remark 10.10. A direct proof may in fact be a hybrid with an indirect proof if it has the form of a proof by contradiction (cf. Def. 1.6). In such a proof, for some $l \geq k+1$ where possibly $k = 0$ one assumes $\theta_l = \neg\phi$, from which assumption it follows that some two formulas θ_i, θ_j, $i, j \neq l, 1 \leq i \leq m-1$ and $1 \leq j \leq m-1$, such that $\theta_i \land \theta_j \equiv \bot$ can be derived, from which, by PNC, $\neg\neg\phi = \theta_m$ is derived in Step $n-1$ of the proof. By DN, we obtain $\phi = \theta_n$, having thus the proof $\Theta \vdash \phi$. Although providing actually a proof by contradiction, if the proof has the form in Definition 10.5, then it is still a direct proof, namely because there must be a (derived) rule in the calculus at hand that, from $\phi \vdash \bot$, stipulates the derivation of $\neg\phi$.

Example 10.11. See Example 12.13 below for a direct proof in the form of a proof by contradiction. The rule invoked for the production of the proof by contradiction is inference rule 7 ($\neg I$) of the $\mathcal{NK}$ calculus (cf. Prop. 12.1)

Remark 10.12. There are variations for the form of a direct proof as in Definition 10.5: The order of the steps may be inverted (i.e. one starts with Step n and ends the proof in Step 1) and/or Steps 1 to m may be constituted by formulas θ_{jl} denoting the j-th formula in the l-th branch. When the latter is the case, we speak of a proof as a *(labeled) tree*.

Example 10.13. In the sequent calculus $\mathcal{LK}$, the proofs have both properties detailed in Remark 10.12: A direct proof in this calculus is an upwards-growing labeled tree in which the root is labeled with the formula ϕ to be proven or even with $\Theta \vdash \phi$ (see, for example, Figure 12.2.1).

In this Section, we elaborated on the VAL from the viewpoints of its definition as a formal language and the notion of a direct proof. Strictly conceived, we speak of *(classical) validity* when, given some argument, it is the case that it is not possible to derive a false conclusion if the premises are all valuated to truth. In spite of this condition and the formal definitions above, however, validity remains a problematic notion in CL. We leave some of the main problems–not the least of which is the relation between validity and logical form–as (research) exercises.

▶ *Do Exercises 10.6-8.*

10.3. The complexity of *VAL*

Recall now the complexity and tractability aspects briefly discussed in Section 1.4. Proving *VAL* might be no tougher than proving *SAT* (cf. Chapter 13). Indeed, *VAL* is considered a co-**NP**-complete problem, it being the case that co-**NP** is the complexity class of the problems whose complements are in the **NP** class, or, formally put for decision problems defined as languages,[10]

$$\text{co-}\mathbf{NP} = \left\{ L | \overline{L} \in \mathbf{NP} \right\}$$

and, in turn, a co-**NP**-complete problem is one whose complement is **NP**-complete. In effect, the complement of *VAL*, also denoted $\overline{VAL}$, is *SAT*, which was the first problem to be classified as **NP**-complete. Just as in the case of **NP**-completeness, a computational problem A is said to be co-**NP**-complete iff (i) $A \in$ co-**NP** and (ii) for every other computational problem $A_i \in$ co-**NP** it is the case that $A_i \preceq_P A$. (Cf. Def. 1.18.)

Theorem 10.14. *VAL is co-**NP**-complete.*

Proof: (Idea) Let $L \in$ co-**NP**, where L is a decision or computational problem defined as a language. Then, we aim to show that $L \preceq_P VAL$. **QED**

Given that co-**NP**-complete problems are not thought to be harder than **NP**-complete ones, the question arises why *SAT* has all but obliterated *VAL* in many fields in which CFOL is applied. In other words, whenever there is sought a decision concerning whether we have $\Theta \models \phi$

---
[10] Another, equivalent way, to put this is to say that the co-**NP** class contains the decision problems for which the "No" instances can be accepted in polynomial time by a non-deterministic Turing machine.

10. The validity problem, or VAL

for a given FO theory Θ and a FO formula ϕ, the problem is more often than not defined in terms of SAT. As seen, the *Entscheidungsproblem* asks more specifically whether there is some algorithm Ψ that constitutes a decision procedure for the question $\phi \in ?VAL$ for a FO formula ϕ. For CPL, a truth table is such an algorithm Ψ; however, this is only the case for a relatively small number of atoms. More importantly, truth tables are generally not appropriate for CFOL. In other words, and as proven above (cf. Turing's Theorem), there is *no* general decision procedure for VAL for CFOL.

One of the reasons that explains the overwhelming preference for SAT is the amenability of satisfiability-testing methods to full automation, with calculi such as resolution and analytic tableaux providing refutation-based decision procedures, or indirect proof procedures, for which there are several provers.[11] These are *de-facto* decision procedures, i.e. algorithmic proof methods that indeed give us a "Yes/No" answer after a finite number of steps, to a great extent because Herbrand semantics applied to a decision problem formulated in terms of SAT over L1 gives to it the highly desired property of being decidable (see Sections 9.2 and 13.3).

Differently from this, the direct proof $\Theta \vdash_\blacksquare \phi$ (cf. Def. 10.5) is in general *not* an algorithmic procedure, mostly because of the fact that often we have to make assumptions, i.e. extra, *ad-hoc*, formulas (assumed to be *true*). This gives to proofs a character of arbitrariness in the choice of the assumptions that casts doubt on the notion of logical proof; as a matter of fact, we may assume whatever we think will help us to obtain ϕ as a logical consequence of some (possibly empty) set X. In the case of elements from the set $(AX \cup RI)$, the choice of these may not be trivial. These aspects entail that automation is in principle not feasible–or, more simply, not efficient–for direct proof methods.

▶ *Do Exercises 10.9-10.*

Exercises

Exercise 10.1. Consider the Turing machine K_α that computes sequence α given above.

1. Give its formal description.

2. Describe its (possible) behavior.

---

[11] A *prover* is a fully automated proving software. A partially automated proving software is called an *assistant*.

10.3. The complexity of VAL

3. Compare your answers with the contents in Turing (1936-7).

Exercise 10.2. A *circle-free* Turing machine is one that indefinitely goes on printing 0s and 1s; its opposite is a *circular* Turing machine. (This terminology is from Turing (1936-7).) What might be the interpretation of the following short citation from this paper?

> K_α is a circle-free machine; α is a computable sequence. (Turing, 1936-7)

Exercise 10.3. The proof of the *non-computability* of the set of the real numbers equates with showing that they are *uncountable*, i.e. $\mathbb{R}$ is neither finite nor denumerable (or countably infinite). This is said to be a proof by the *diagonal process*.

1. Research into this proof.

2. We can actually speak of *Cantor diagonalization argument*. Find out why.

3. Relate the citation above in Exercise 10.2 with this particular proof.

4. Find other proofs of mathematical theorems by the diagonal process.

Exercise 10.4. Comment on the following statement from Turing (1936-7) by focusing on the terms in italics:

> I shall show that there is no general method which tells whether a given formula ϕ is *provable* in **K** [i.e. CFOL], or, what comes to the same, whether the system consisting of **K** with $\neg\phi$ adjoined as an extra axiom is *consistent*. (Turing, 1936-7; our italics; slightly changed)

Exercise 10.5. Research into A. Church's answer to the *Entscheidungsproblem* and give its main aspects.

Exercise 10.6. With respect to validity, comment, by providing examples, on the following facts:

1. An argument with false premises and a false conclusion can be valid.

2. True premises and a true conclusion do not establish validity of an argument.

10. The validity problem, or VAL

Exercise 10.7. Give some thought to the following statements with respect to *(logical) validity*:

1. We should like to comment on the two words, *"valid"* and *"true"*. The term *"true sentence"* ... and its opposite, *"false sentence"*, have been the subject of many philosophical discussions, but they are used in this book in a straightforward manner: with reference to sentences of elementary mathematical theories or to simple examples from everyday language. The term *"valid"*, on the other hand, is broader in scope (and it has not given rise to such controversies). It is often used in mathematics in a similar way as *"true"*. Thus, *"valid"* may mean *"true"* when referring to a special context (in particular, when referring to chosen models and interpretations ...). In fact, situations sometimes arise when the two words are used interchangeably. There is, moreover, a certain preference for using *"valid"* when one deals with sentential functions ... Then *"valid"* means about the same as *"satisfied under every substitution of variables"*. Furthermore, sometimes *"valid"* means the same as *"proved"* or *"provable"*. (Tarski, 1994)

2. $\exists x \, (Q(x) \to \forall x \, (Q(x)))$. This is a strange–but valid–sentence. (Enderton, 2001; adapted notation)

3. In contrast to the finitary procedure for [propositional] tautologies, suppose that you want to know whether or not a well-formed formula ϕ (of our first-order language) is valid. The definition requires that you consider every structure $(\mathscr{D}, \Theta)$. (In particular this requires using every nonempty set, of which there are a great many.) For each of these structures, you then must consider each function ϖ from the set Vi of variables into $\mathscr{D}$. And for each given $(\mathscr{D}, \Theta)$ and ϖ, you must determine whether or not $(\mathscr{D}, \Theta)$ satisfies ϕ with ϖ. When $\mathscr{D}$ is infinite, this is a complicated notion in itself. In view of these complications, it is not surprising that the set of valid formulas fails to be decidable. What is surprising is that the concept of validity turns out to be equivalent to another concept (deducibility) whose definition is much closer to being finitary. (Enderton, 2001; slightly abridged)

4. When it comes to validity ... we now have two goals on the table. One is to find a precise analysis of validity. ... The other is to find a method of assessing arguments for validity that is both (1.) foolproof: it can be followed in a straightforward, routine way, without recourse to intuition or imagination–and it always gives

10.3. The complexity of VAL

the right answer; and (2.) general: it can be applied to any argument. ... It is the fact that validity can be assessed on the basis of form, in abstraction from the specific content of the propositions involved in an argument (i.e., the specific claims about the world–what ways, exactly, the propositions that make up the argument are representing the world to be), that will bring this goal [2.] within reach. (Smith, 2012)

Exercise 10.8. Consider the following passage on validity:

To say that an inference is valid is to say that its conclusion is a logical consequence of the initial premise(s) of the inference. There are inferences whose validity arises solely from the meanings of certain expressions occurring in them. ... [W]e say such inferences are analytically valid. To illustrate, consider the following valid inference.

$$\frac{(Sister(paige, kelly) \land Sister(paige, shannon))}{\therefore Sister(paige, kelly)}$$

[T]he validity of the inference arises solely from the meaning of "$\land$" ... as specified by, say, a truth table. (McKeon, 2010; adapted notation)

1. Comment on the view exposed above, known as *analytical validity*, that an inference that is permissible by a meaning-determined rule is sufficient for inferential validity.

2. Research into A. Prior's "tonk" argument against analytical validity (see Prior, 1960).

Exercise 10.9. Give examples of problems known to be co-**NP**-complete.

Exercise 10.10. Research into the complete proof of Theorem 10.14.

11. Hilbert-style systems

In this Chapter, we elaborate on *axiom systems proper*.[1] Proofs in these systems are carried out by applying axiom schemata to the formulas (of arguments) at hand, with a few inference rules providing a mostly auxiliary role; in other words, the set AX plays in axiom systems proper a more important role than the set RI. Accordingly, an axiom system proper is a triple $\mathcal{A} = (F_L, AX, RI)$ where $|AX| \geq |RI|$. The cardinality of the set AX can vary greatly from system to system, with some systems having as few as a single axiom and others more than ten axioms, but it is important to check the *independence* of the axioms from each other; if one can derive an axiom from other axioms of the same system, then one has a superfluous axiom. Other than independence, *consistency* and *sufficiency* are also important properties of an axiom system proper: no inconsistent formula should be derivable from the set of axioms, and these should suffice to prove every theorem of the logic in consideration.

Proving theoremhood in an axiom system proper $\mathcal{A} = (F_L, AX, RI)$ consists in making sure that every line of the proof is a theorem. These proofs, especially in the case of the classical tautologies, typically have no assumptions or hypotheses. Given a set of premises (or assumptions) $X = \{\chi_1, ..., \chi_n\}$, we say that a conclusion ϕ *is derivable from X in* $\mathcal{A}$, and write $X \vdash_{\mathcal{A}} \phi$, if there is a chain or sequence of formulas that are instances of the axiom schemata in AX or that are obtained from previous instances by means of a few rules of inference, it being the case that typically there is one single such rule in an axiom system proper, to wit, MP.[2] Importantly, rule SUB (cf. Def. 4.18) is often tacitly considered as an inference rule; indeed, the very definition of axiom schema (cf. Def. 2.26) implicitly takes SUB to be an essential rule of axiom systems proper.

We call axiom systems proper *Hilbert systems*, or *Hilbert-style systems*, though D. Hilbert was not the first to conceive one such system.[3] There

---

[1] This specification is necessary, because it is often the case that proof systems are called axiom systems regardless of whether they are axiom- or rule-base proof systems.
[2] If X comprises only assumptions, then it is important to state that ϕ *is derivable from the assumptions* $\chi_1, ..., \chi_n$.
[3] G. Frege was (cf. Frege, 1879). By doing so, we simply follow a tradition respected

237

11. Hilbert-style systems

is actually an abundance of Hilbert-style systems, but we focus on a particularly well-studied axiom system known as Frege-Łukasiewicz's. This we do in Section 11.1. The proof of the completeness of this system actually holds for the class of the Hilbert systems based on the subset $O'_L = \{\neg, \rightarrow\}$, of which we give some main examples in Section 11.2.1.

11.1. The axiom system $\mathcal{L}$

We elaborate here on the Frege-Łukasiewicz axiom system, thus coined on account of its being greatly a simplification carried out by the Polish logician J. Łukasiewicz of the original six-axiom system conceived by G. Frege (see Section 11.2.1).[4] We denote it generally by $\mathcal{L}$, but specify its propositional restriction as $\mathcal{L}p$ (where p stands for "propositional") and its FO extension as $\mathcal{L}q$ (where q stands for "quantification" or "quantified"). These latter specifications are, however, used solely for disambiguation.

11.1.1. The propositional system $\mathcal{L}$

Proposition 11.1. *The following axiom schemata $\mathcal{L}$1-3, together with rule MP, constitute a proof system for classical propositional logic known as the* Frege-Łukasiewicz axiom system $\mathcal{L}p = (F_L, \{\mathcal{L}1, \mathcal{L}2, \mathcal{L}3\}, \{MP\})$:

$(\mathcal{L}1) \quad A \rightarrow (B \rightarrow A)$
$(\mathcal{L}2) \quad (A \rightarrow (B \rightarrow C)) \rightarrow ((A \rightarrow B) \rightarrow (A \rightarrow C))$
$(\mathcal{L}3) \quad (\neg A \rightarrow \neg B) \rightarrow (B \rightarrow A)$

Example 11.2. Figure 11.1.1 shows the proof $\vdash_\mathcal{L} P \rightarrow P$. On the right, the axiom schemata and/or rule used in each step of the derivation are indicated. For instance, "MP (1, 2)" denotes the application of rule MP on the previous steps 1 and 2. In the application of the schemata, we indicate the substitutions carried out.

---

by some reference texts in this topic (e.g., Troelstra & Schwichtenberg, 2000), and thus contribute to keeping terminological consistency. But because G. Frege was actually the conceiver of the first axiom system proper the label "Frege system" is also to be often found in the literature.

[4]This was not a one-shot conception, there being actually at least four simplifications. See, for instance, Łukasiewicz (1934).

11.1. The axiom system $\mathcal{L}$

1. $\vdash_\mathcal{L} (P \to ((P \to P) \to P)) \to ((P \to (P \to P)) \to (P \to P))$ $\mathcal{L}2$ with $A/P; B/P \to P; C/P$
2. $\vdash_\mathcal{L} P \to ((P \to P) \to P)$ $\mathcal{L}1$ with $A/P; B/P \to P$
3. $\vdash_\mathcal{L} (P \to (P \to P)) \to (P \to P)$ MP (1, 2)
4. $\vdash_\mathcal{L} P \to (P \to P)$ $\mathcal{L}1$ with $A/P; B/P$
5. $\vdash_\mathcal{L} P \to P$ MP (3, 4)

Figure 11.1.1.: Proof of $\vdash_\mathcal{L} P \to P$.

11. Hilbert-style systems

The proof in Example 11.2 is in a linear format, but this is not a requirement; actually, it is often the case that proofs in Hilbert systems are implemented by trees.

As can be easily seen, this axiom system is based on the functionally complete subset $O'_L = \{\neg, \rightarrow\}$. Hence, formulas over the complete set of connectives O_L must first be rewritten using only $O'_L = \{\neg, \rightarrow\}$. The following proposition, whose (easy) proof we leave as an exercise, can assist us in this task.

Proposition 11.3. *The following are classical equivalences over* L:

1.
$$A \wedge B \equiv \neg(A \rightarrow \neg B)$$

2.
$$A \wedge \neg B \equiv \neg(A \rightarrow B)$$

3.
$$\neg A \wedge B \equiv \neg(B \rightarrow A)$$

4.
$$\neg A \wedge \neg B \equiv \neg(\neg A \rightarrow B)$$

5.
$$A \vee B \equiv \neg A \rightarrow B$$

6.
$$\neg A \vee B \equiv A \rightarrow B$$

7.
$$A \vee \neg B \equiv \neg A \rightarrow \neg B$$

8.
$$\neg A \vee \neg B \equiv A \rightarrow \neg B$$

In Example 11.2, we remain at the metalanguage level, and thus employ the symbol $\vdash_{\mathcal{L}}$ on every line of the proof, namely to make it clear that every line is a theorem. However, this is not the case in a regular proof, as the next example shows. Moreover, in Example 11.2 we prove a theorem, reason why there are no premises or assumptions; this is obviously not the case for proofs of arguments.

The frugality of the sets $AX_{\mathcal{L}}$ and $RI_{\mathcal{L}}$ should make it obvious that proofs in $\mathcal{L}$ can be extremely cumbersome if we restrict ourselves to

11.1. The axiom system $\mathcal{L}$

these sets. More often than not, we resort to the application of any of the classical tautologies. Assumptions, or hypotheses, in a proof in an axiom system proper can take many forms, from *ad-hoc* formulas, as in the proof of the following example, to lemmas and theorems, or tautologies in general. In a certain sense, an assumption is an additional premise that is not a sentence in the theory, and thus can be spoken of as a hypothesis.

Example 11.4. Figure 11.1.2 shows a proof in $\mathcal{L}$ of the argument

$$\{(P \vee R) \to \neg(Q \wedge T), \neg S \to Q, P \wedge T\} \vdash S.$$

We opted for the employment of an equivalence, to wit, $\neg Q \to S \equiv \neg S \to Q$, without which the proof would be far more convoluted. In this example, it is noteworthy that none of the axioms of $\mathcal{L}$ was actually employed, with MP the single element of $\mathcal{L}$. Nonetheless, this is considered a proof in $\mathcal{L}$. (We leave it as an exercise for the reader to prove this argument in $\mathcal{L}$ by actually employing the axioms of $\mathcal{L}$.) It is also noteworthy in this proof that a single premise of the set of three premises was actually employed in it, but this is merely the property of classical logic known as monotonicity at play (cf. Section 4.1.1). Indeed, let $X = \{(P \vee R) \to \neg(Q \wedge T), \neg S \to Q, P \wedge T\}$ and $X' = \{\neg S \to Q\}$. Then, we also have the property of compactness at play (cf. Def. 4.11), a property that contributes in a fundamental way to the characterization of a logical system as a deductive system.

1.	$(P \vee R) \to \neg(Q \wedge T)$	Premise
2.	$\neg S \to Q$	Premise
3.	$P \wedge T$	Premise
4.	$\neg Q \to S$	$\equiv (2)$
5.	$\neg Q$	Assumption
6.	S	MP (4, 5)

Figure 11.1.2.: Proof of an argument in $\mathcal{L}p$.

Because $\mathcal{L}$ is an adequate system, theoremhood in the proofs above, signaled by the syntactical consequence relation symbol $\vdash$, can be substituted by validity, signaled by the semantical consequence relation symbol $\models$. Thus, though $\mathcal{L}$ is a proof system, and thus a system for proving theoremhood, a feature denoted by $\vdash_\mathcal{L} \phi$, it is also a system for proving validity.

11. Hilbert-style systems

The proof of the soundness of $\mathcal{L}p$ is an easy matter and we leave it largely as an exercise. We give the idea for the proof for weak soundness and leave the proof for strong soundness as an exercise.

Theorem 11.5. *(Soundness of $\mathcal{L}p$) For a propositional formula ϕ,*

$$\text{if } \vdash_{\mathcal{L}p} \phi, \text{ then } \models \phi.$$

Proof: (Idea) We first prove that all the axioms of $\mathcal{L}p$ are tautologies. The only inference rule in $\mathcal{L}p$ is MP; we then show that this rule preserves tautologousness. **QED**

As usually, however, the proof of completeness of $\mathcal{L}p$ is not such an easy matter. When proving the completeness of $\mathcal{L}p$, it is useful to prove simultaneously the completeness of the particular class of propositional Hilbert-style systems, denoted simply by $\mathcal{H}$, to which $\mathcal{L}p$ belongs. To prove the completeness of this class $\mathcal{H}$, we isolate a subset of classical propositional axioms–which is actually $AX_{\mathcal{L}p} = \{\mathcal{L}1, \mathcal{L}2, \mathcal{L}3\}$–and call this set of axiom schemata and derived theorems together with rule MP the H propositional axiom system. We consider $AX_{\mathcal{H}p} = \{\mathcal{H}1, \mathcal{H}2, \mathcal{H}3\}$ to be the axioms and theorems $\mathcal{H}4\text{-}9$ can all be proved in $\mathcal{H}p/\mathcal{L}p$ (see Example 11.2 for the proof of $\mathcal{H}4$ as $\vdash_{\mathcal{L}p} \phi \to \phi$; the remaining proofs are left as an exercise), so that they can be applied in proofs as *derived axiom schemata*.

Proposition 11.6. *The following axiom schemata $\mathcal{H}1\text{-}3$ and theorems $\mathcal{H}4\text{-}9$, together with rule MP, constitute a proof system for classical propositional logic known as the H propositional axiom system $\mathcal{H}p = (F_L, \{\mathcal{H}1, \mathcal{H}2, \mathcal{H}3\} \cup \{\mathcal{H}4, ..., \mathcal{H}9\}, \{MP\})$:*

$$
\begin{array}{ll}
(\mathcal{H}1 = \mathcal{L}1) & A \to (B \to A) \\
(\mathcal{H}2 = \mathcal{L}2) & (A \to (B \to C)) \to ((A \to B) \to (A \to \chi)) \\
(\mathcal{H}3 = \mathcal{L}3) & (\neg A \to \neg B) \to (B \to A) \\
(\mathcal{H}4) & A \to A \\
(\mathcal{H}5) & A \to \neg\neg A \\
(\mathcal{H}6) & \neg A \to (A \to B) \\
(\mathcal{H}7) & A \to (\neg B \to \neg(A \to B)) \\
(\mathcal{H}8) & (A \to B) \to ((\neg A \to B) \to B) \\
(\mathcal{H}9) & (\neg A \to A) \to A
\end{array}
$$

For the proof of completeness of $\mathcal{H}p$, we draw on Kalmár (1935), a proof that became so to say the standard proof for Hilbert-style systems.[5] We assume that $\mathcal{H}p$ is sound (as is easily proven). As in the case of the proof of soundness, we give the proof of weak completeness, and leave the proof of strong completeness as an exercise. The following contents will be required in the sketch we give here.

Definition 11.7. Let $\phi(p_1, ..., p_n)$ be a propositional formula over L. We define, for a valuation function $val : (F_L \supseteq V_L) \longrightarrow \{0,1\}$,

$$\phi' = \begin{cases} \phi & \text{if } val(\phi) = 1 \\ \neg \phi & \text{if } val(\phi) = 0 \end{cases}$$

and

$$p_i' = \begin{cases} p_i & \text{if } val(p_i) = 1 \\ \neg p_i & \text{if } val(p_i) = 0 \end{cases}.$$

Example 11.8. Let $\phi = \phi(p_1, p_2) = \neg P \to Q$. Applying Definition 11.7 to ϕ, we have $p_1 = P$ and $p_2 = Q$. Then we have the following truth table for ϕ and $\neg \phi$:

P	Q	$\neg P \to Q$	$\neg(\neg P \to Q)$
1	1	1	0
1	0	1	0
0	1	1	0
0	0	0	1

Then, we have, say,

$$\neg P, Q \vdash (\neg P \to Q),$$

$$P, \neg Q \vdash \neg\neg(\neg P \to Q),$$

and

$$\neg P, \neg Q \vdash \neg(\neg P \to Q).$$

Lemma 11.9. *For a propositional formula $\phi = \phi(p_1, ..., p_n)$, if $\phi', p_1', ..., p_n'$ are defined as in Definition 11.7, then*

$$(\odot) \qquad p_1', ..., p_n' \vdash \phi'.$$

Proof: (Sketch) Let n denote the number of occurrences in ϕ of the connectives of the set $O_L' = \{\neg, \to\}$. The proof is by induction on n. If

---
[5] See below, Section 11.2, for Kalmár's own axiom system.

11. Hilbert-style systems

$n = 0$, then ϕ just is a propositional atom; hence, ϕ consists of a single propositional variable p, and we have $P \vdash P$ and $\neg P \vdash \neg P$, and $\odot$ holds in both cases. Let us now assume that the lemma holds for all $j < n$. Then, there are two cases to consider. Let $\phi = \neg \psi$; we call this Case 1. Then, ψ has less than n occurrences of elements in $O'_L = \{\neg, \rightarrow\}$. We now have two subcases:

- (Subcase 1a) Let $val(\psi) = 1$, so that $val(\phi) = 0$. We thus have $\psi = \psi'$ and $\phi' = \neg\phi = \neg\neg\psi$. Applying the inductive hypothesis to ψ, we have

$$(\odot) \quad p'_1, ..., p'_n \vdash \psi$$

By $\mathcal{H}_5$, we have $\vdash \psi \rightarrow \neg\neg\psi$, and by monotonicity of $\vdash$, we have $p'_1, ..., p'_n \vdash \psi \rightarrow \neg\neg\psi$; by $\odot$ and MP we get $p'_1, ..., p'_n \vdash \neg\neg\psi$, and hence $\odot$, and the lemma is proven for this subcase.

- (Subcase 1b) Let $val(\psi) = 0$, so that $val(\phi) = 1$. Then, $\psi = \neg\phi$ and $\phi' = \phi$. By the inductive hypothesis,

$$p'_1, ..., p'_n \vdash \neg\psi$$

but $\neg\psi = \phi'$ and so $\odot$ holds, and the lemma is proven for this subcase.

In the remaining Case 2, we let ϕ be $\phi_1 \rightarrow \phi_2$. We leave the proof as an exercise. (Hint: Consider 3 subcases.) **QED**

Given Definition 11.7 and the complete proof of Lemma 11.9, it should be obvious that this lemma describes a way to establish a correspondence between the classical semantics and deducibility via the axiom system $\mathcal{H}p$, so that if we have $X \models \psi$, then we have $X \vdash_{\mathcal{H}p} \psi$. We now have to turn this description into a theorem for the completeness of $\mathcal{H}p$. As said above, we shall concern ourselves with the weak version of this completeness result.

Theorem 11.10. *(Completeness of $\mathcal{H}p$) For a propositional formula ϕ,*

$$\text{if } \models \phi, \text{ then } \vdash_{\mathcal{H}p} \phi.$$

Proof: Assume $\models \phi$, where $\phi = \phi(p_1, ..., p_n)$ and the only connectives in ϕ are from the set $O'_L = \{\neg, \rightarrow\}$. For any truth assignment, by Lemma 11.9 we have $p'_1, ..., p'_n \vdash \phi$, namely because ϕ is always valuated to 1 and hence ϕ' is ϕ. Consider the propositional variable p_n; if $val(p_n) = 1$ we have $p'_1, ..., p'_{n-1}, p_n \vdash \phi$, and if $val(p_n) = 0$ we have $p'_1, ..., p'_{n-1}, \neg p_n \vdash \phi$. By DT, we get $p'_1, ..., p'_{n-1} \vdash p_n \rightarrow \phi$ and $p'_1, ..., p'_{n-1} \vdash \neg p_n \rightarrow \phi$. By

244

11.1. The axiom system $\mathcal{L}$

$\mathcal{H}8$ and MP, we get $p'_1, ..., p'_{n-1} \vdash \phi$. Repeat for $p'_1, ..., p'_{n-1}$; after n such steps, we have $\vdash \phi$, and because we applied $\mathcal{H}p$ (both in this particular proof and in the proofs above),[6] we actually have $\vdash_{\mathcal{H}p} \phi$. **QED**

Theorem 11.10 can be proven for any propositional axiom system $\mathcal{A} = (F_L, AX, RI)$ by applying the corresponding axioms, theorems/derived axiom schemata, and/or rules of inference. In particular, and given Proposition 11.6, we proved the completeness of the class $\mathcal{H}$ of propositional axiom systems.

▶ *Do Exercises 11.1-8.*

11.1.2. The FO system $\mathcal{L}$

Proposition 11.11. *In order to obtain the axiom system $\mathcal{L}$ for L1, denoted by $\mathcal{L}q$, we add to $\mathcal{L}$1-3 further axiom schemata–the* quantifier axioms $Q1$-2, *which assume* A^x_t–*and an additional inference rule, known as* generalization rule *(GEN):*[7]

$$(Q1) \quad \forall x A(x) \to A(t)$$
$$(Q2) \quad A(t) \to \exists x A(x)$$

$$(\text{GEN}) \quad \frac{A(x)}{\forall y A(y)}; \quad \frac{A}{\forall x A}$$

Example 11.12. We prove in $\mathcal{L}q$ the following valid syllogism:

1.	All big cats are finicky.	$\forall x (B(x) \to F(x))$
2.	All black panthers are big cats.	$\forall x (P(x) \to B(x))$
3.	All black panthers are finicky.	$\forall x (P(x) \to F(x))$

---
[6] This includes the proof of $DT_\vdash$, which is left as an exercise.
[7] We give two versions of GEN.

11. Hilbert-style systems

1.	$\forall x\,(B(x) \to F(x))$	Premise
2.	$\forall x\,(P(x) \to B(x))$	Premise
3.	$B(a) \to F(a)$	$\mathcal{Q}1$ (1)
4.	$P(a) \to B(a)$	$\mathcal{Q}1$ (2)
5.	$P(a) \to (B(a) \to F(a))$	Assumption
6.	$(P(a) \to)(B(a) \to F(a)) \to$ $((P(a) \to B(a)) \to (P(a) \to F(a)))$	$\mathcal{L}2$
7.	$(P(a) \to B(a)) \to (P(a) \to F(a))$	MP (5, 6)
8.	$P(a) \to F(a)$	MP (4, 7)
9.	$\forall x\,(P(x) \to F(x))$	GEN (8)

Figure 11.1.3.: Proof in $\mathcal{L}q$ of a valid syllogism.

Note in the proof above (cf. Fig. 11.1.3) that we had to make an assumption (Step 5), without which we would not have been able to complete the proof.

Just as in the case of $\mathcal{L}p$, formulas over L1 containing the connectives $O'_\mathsf{L} = \{\land, \lor, \leftrightarrow\}$ must be rewritten using solely the connectives $\neg$ and $\to$. Further equivalences and inter-definitions involving the quantifiers might also be relevant for proofs in $\mathcal{L}q$.

The soundness of $\mathcal{L}q$ is proven in a similar way as for $\mathcal{L}p$. With respect to completeness, the proof given for Theorem 8.2 can easily be adapted to this particular FO axiom system. Both proofs are left as exercises.

▶ *Do Exercises 11.9-12.*

11.2. Further Hilbert-style systems

In this Section, we give the essentials of some relevant propositional axiom systems for classical logic. As above, we write the axiom schemata with the uppercase Roman letters A, B, C, ... to stand for formulas where they might have been originally written with Roman or Greek lowercase letters. Moreover, for the sake of intratextual consistency we keep to our own notation and fonts.

11.2.1. The class $\mathscr{H}$

As seen above, $\mathscr{H}$ is the class of Hilbert systems whose axiom schemata range over the functionally complete set $O'_\mathsf{L} = \{\neg, \to\}$. We give here the

11.2. Further Hilbert-style systems

main axiom systems of this class. All these systems have MP as single inference rule.

Proposition 11.13. *The original Frege axiom system (cf. Frege, 1879), denoted by $\mathcal{F}$, had the following six axiom schemata:*

$(\mathcal{F}1)$ $\quad A \to (B \to A)$
$(\mathcal{F}2)$ $\quad (A \to (B \to C)) \to ((A \to B) \to (A \to C))$
$(\mathcal{F}3)$ $\quad (A \to (B \to C)) \to (B \to (A \to C))$
$(\mathcal{F}4)$ $\quad (A \to B) \to (\neg B \to \neg A)$
$(\mathcal{F}5)$ $\quad \neg\neg A \to A$
$(\mathcal{F}6)$ $\quad A \to \neg\neg A$

Further relevant axiom systems in $\mathcal{H}$ are those of D. Hilbert (cf. Hilbert & Ackermann, 1928) and in particular, because of the standard proof of completeness for this class, L. Kalmár's (Kalmár, 1935).

Proposition 11.14. *The following are the axiom schemata of Kalmár's axiom system:*

$(\mathcal{K}1)$ $\quad (A \to (B \to C)) \to ((A \to B) \to (A \to C))$
$(\mathcal{K}2)$ $\quad (A \to B) \to ((\neg A \to B) \to B)$
$(\mathcal{K}3)$ $\quad A \to \neg\neg A$
$(\mathcal{K}4)$ $\quad B \to (A \to B)$
$(\mathcal{K}5)$ $\quad \neg A \to (A \to B)$
$(\mathcal{K}6)$ $\quad A \to (\neg B \to \neg(A \to B))$

Kalmár provided further axioms for the remaining connectives of the set O_L, but he remarked that the six schemata above suffice for the adequateness of this axiom system.

Proposition 11.15. *The following are the axiom schemata of Hilbert's axiom system:*[8]

$(H1)$ $\quad A \to (B \to A)$
$(H2)$ $\quad (A \to (B \to C)) \to (B \to (A \to C))$
$(H3)$ $\quad (B \to C) \to ((A \to B) \to (A \to C))$
$(H4)$ $\quad A \to (\neg A \to B)$
$(H5)$ $\quad (A \to B) \to ((\neg A \to B) \to B)$

▶ *Do Exercises 11.13-14.*

---
[8] We write H (instead of $\mathcal{H}$) to distinguish these axiom schemata from those of the H propositional axiom system (cf. Prop. 11.6). Cf. Hilbert & Ackermann (1928).

11. Hilbert-style systems

11.2.2. Other systems

If we consider $\bot$ as a 0-ary connective, then the subset $O'_L = \{\bot, \to\}$ is functionally complete (cf. Exercise 5.5). The following two propositional axiom systems have their axiom schemata over this subset. MP is a rule of inference in both systems, and A. Church also considers explicitly SUB as a rule.

Proposition 11.16. *The following are the axiom schemata in Church's axiom system:*

$$
\begin{array}{ll}
(\mathcal{C}1) & A \to (B \to A) \\
(\mathcal{C}2) & (A \to (B \to C)) \to ((A \to B) \to (A \to C)) \\
(\mathcal{C}3) & ((A \to \bot) \to \bot) \to A
\end{array}
$$

This propositional axiom system, which A. Church called P_1, is elaborated on in Church (1956).

The following axiom system, named after Tarski, Bernays, and Wajsberg, is not a joint creation of these authors, but rather a more or less independent conception joining together axioms from different sources (e.g., Łukasiewicz & Tarski, 1930; Wajsberg, 1937). Wajsberg can be credited with adding the axiom schema involving $\bot$ to the three schemata involving $\to$ alone, which are, in turn, known as *Tarski-Bernays axiom system*. Note the following definition:

$$
(\bot_{def}) \qquad \neg(A \to A).
$$

Proposition 11.17. *The following are the axiom schemata in the Tarski-Bernays-Wajsberg axiom system:*

$$
\begin{array}{ll}
(\mathcal{T}1) & (A \to B) \to ((B \to C) \to (A \to C)) \\
(\mathcal{T}2) & A \to (B \to A) \\
(\mathcal{T}3) & ((A \to B) \to A) \to A \\
(\mathcal{T}4) & \bot \to A
\end{array}
$$

Nicod's axiom system has a single axiom schema and a single rule of inference (cf. Nicod, 1917).

Proposition 11.18. *Nicod's axiom is*

$$
(\mathcal{N}1) \qquad (A|(B|C))|(((D|(D|D))|((E|B)|((A|E)|(A|E))))
$$

11.2. Further Hilbert-style systems

and the single rule of inference is

$$(MP_N) \qquad A, A|(B|C) \vdash C.$$

▶ Do Exercises 11.14-15.

Exercises

Exercise 11.1. Prove $\mathcal{H}$5-9 (cf. Prop. 11.6) in $\mathcal{L}p$.

Exercise 11.2. Prove in $\mathcal{L}p$ the classical *validity* of the following formulas, derived axiom schemata, or arguments:

1. $P \to P$
2. $\neg P \to (P \to Q)$
3. $\{P \to Q, \neg P \to Q, Q\}$
4. $\{P \to R, R \to Q, \neg(\neg Q \land P)\}$
5. $\neg(P \land R) \to (R \to \neg P)$
6. $\{\neg P \to \neg Q, Q \to P\}$
7. $\{P \to R, (Q \land P) \to (Q \land R)\}$
8. $(\neg P \to P) \to P$
9. $\neg P \to (\neg R \to \neg(P \lor R))$
10. $\neg\neg\neg P \to \neg P$

Exercise 11.3. With respect to the argument of Example 11.4:

1. Prove its correctness in $\mathcal{L}p$ by applying at least one of the axioms $\mathcal{L}$1-3.

2. Show by means of the axiom system $\mathcal{L}p$ that, for the formula S,

$$S = Cn(X) = \bigcup \{Cn(X') \mid X' \subseteq X\}$$

for all the finite $X' \subseteq X$.

3. What have you proved with respect to $\mathcal{L}p$ in the Exercise immediately above?

11. Hilbert-style systems

Exercise 11.4. Give a proof of $DT_\vdash$ in $\mathcal{L}p$. (Hint: Consider the proof of $X, \phi \vdash_{\mathcal{L}p} \psi$ to be the sequence $\psi_1, ..., \psi_n$ where $\psi_n = \psi$. Apply induction on i for $1 \leq i \leq n$.)

Exercise 11.5. Prove, or complete the proof of, the following statements:

1. Proposition 11.3.

2. Theorem 11.5. (Consider also the strong version.)

3. Proposition 11.6.

4. Lemma 11.9.

Exercise 11.6. Our proof of Theorem 11.10 is called a *constructive proof*. Explain why.

Exercise 11.7. Prove the strong completeness of $\mathcal{H}p$.

Exercise 11.8. State the Corollary to Theorem 11.10. (Hint: Consider O_L.)

Exercise 11.9. Prove the following theorems in $\mathcal{L}q$:

1. $\neg \forall x (A \to C) \lor (\exists x A \to \exists x C)$

2. $\neg \forall x (A) \to \exists x (\neg A)$

3. $\forall x (A \to C) \to (\forall x A \to \forall x C)$

4. $\forall x A \to \forall x (A \lor C)$

5. $(A \to \forall x C) \leftrightarrow \forall x (A \to C)$

Exercise 11.10. Prove the syllogisms of Exercise 5.9 in the axiom system $\mathcal{L}q$.

Exercise 11.11. Prove the soundness of $\mathcal{L}q$.

Exercise 11.12. Adapt the proof given for Theorem 8.2 to the FO axiom system $\mathcal{L}$.

Exercise 11.13. With respect to the original Frege system (cf. Prop. 11.13):

1. Lukasiewicz showed that axiom $\mathcal{F}3$ could be derived from $\mathcal{F}1$-2 ($=$ $\mathcal{L}1$-2). Give the proof(s).

11.2. Further Hilbert-style systems

2. Lukasiewicz further substituted $\mathcal{F}$4-6 by $\mathcal{L}$3. How do you account for this substitution?

Exercise 11.14. Prove the items of Exercises 11.1-2 in the axiom systems of Section 11.2.

Exercise 11.15. In Lukasiewicz (1929), the following propositional axiom system is proposed:

$$
\begin{aligned}
&(\mathcal{L}'1) \quad (A \to B) \to ((B \to C) \to (A \to C)) \\
&(\mathcal{L}'2) \quad (\neg A \to A) \to A \\
&(\mathcal{L}'3) \quad A \to (\neg A \to B)
\end{aligned}
$$

Show that this axiom system is equivalent to Church's P_1 system. (Hint: Make $\neg A = A \to \bot$.)

Exercise 11.16. With respect to Church's P_1 system, check it for independence, consistency, and sufficiency.

12. Gentzen systems

We use here the label "Gentzen systems" for both the natural calculus $\mathcal{NK}$ and the sequent calculus $\mathcal{LK}$. Although $\mathcal{NK}$ was also independently conceived at around 1934 by S. Jaśkowski, it is often spoken of as being a Gentzen system, and we keep to this usage that lends unity to the two above-mentioned systems–a unity that is also in kind, as both $\mathcal{NK}$ and $\mathcal{LK}$ are *rule-based proof systems* (vs. axiomatic systems proper; cf. Chapter 11). As a matter of fact, they are equivalent proof systems–a result whose proof we leave as an exercise.

12.1. The natural deduction calculus $\mathcal{NK}$

Around 1930, there was in Europe a concern with natural ways of reasoning in mathematical proofs. More specifically, axiom systems, or Hilbert(-like) systems, were seen as quite remote from the reasoning instances typically carried out by mathematicians, and other proof methods closer to these were actively searched for. In this quest, *natural deduction*, denoted here by $\mathcal{NK}$ (abbreviating the German expression **n**atürliches **K**alkül), was invented independently by Gentzen (1934-5) and Jaśkowski (1934) as a proof system for CFOL, and it has passed the test of time, being today still a central calculus for direct proofs–even if its role in the field of automated deduction is negligible.

We shall consider this calculus as the triple $\mathcal{NK} = (F_L, RI, \emptyset)$, i.e. as a calculus over L constituted by a (small) set of inference rules and an empty set of axioms. We give here the basic inference rules as they appear in Prawitz (1965), to which we refer the reader for further contents on this calculus. For ease of exposition, we discuss the $\mathcal{NK}$ inference rules for the connectives and for the quantifiers in separate Sections; by and large, this segregation contributes to our distinguishing between a propositional and a FO $\mathcal{NK}$ calculi.

The inference rules of the calculus $\mathcal{NK}$ are often also called *formation rules*, as they are actually rules of formation of formulas: Given some formula ϕ, by means of the application of one of these rules we obtain a (well-formed) formula ϕ'. In effect, the inference rules of natural deduction are essentially rules for *introduction* (denoted by I) and *elimination* (E) of connectives and quantifiers, a feature that makes of this an ideal calculus for the analysis of the classical behavior of the individual logical

12. Gentzen systems

connectives and quantifiers.

12.1.1. The propositional calculus $\mathcal{NK}$

In what follows, a sequence of the form

$$\begin{array}{c}[\![\phi]\!]\\ \vdots\\ \psi\end{array}$$

in an inference rule denotes that the assumption ϕ is discharged when ψ is obtained (see below). We leave the proof of the following proposition as an exercise, providing the hint that all the inference rules of $\mathcal{NK}$ for the connectives of O_L satisfy the classicality conditions for the syntactical $\heartsuit$-consequence relation.

Proposition 12.1. *The following are the inference (or formation) rules of $\mathcal{NK}$ for the logical connectives of O_L:*

1. $(\wedge I) \quad \dfrac{A \quad B}{A \wedge B}$

2. $(\wedge E) \quad \dfrac{A \wedge B}{A} \ ; \ \dfrac{A \wedge B}{B}$

3. $(\vee I) \quad \dfrac{A}{A \vee B} \ ; \ \dfrac{B}{A \vee B}$

4. $(\vee E) \quad \dfrac{A \vee B \quad \begin{array}{c}[\![A]\!]\\ \vdots\\ C\end{array} \quad \begin{array}{c}[\![B]\!]\\ \vdots\\ C\end{array}}{C}$

5. $(\to I) \quad \dfrac{\begin{array}{c}[\![A]\!]\\ \vdots\\ B\end{array}}{A \to B}$

6. $(\to E) \quad \dfrac{A \quad A \to B}{B}$

12.1. The natural deduction calculus $\mathcal{NK}$

7.
$$(\neg I) \quad \frac{\begin{array}{c}[\![A]\!]\\ \vdots \\ \bot\end{array}}{\neg A}$$

8.
$$(\neg E) \quad \frac{A \quad \neg A}{\bot}$$

Remark 12.2. Concerning negation, rules 7 and 8 are considered by Gentzen, but Prawitz (1965), who considers instead the rule

$$(\bot) \quad \frac{\begin{array}{c}[\![\neg A]\!]\\ \vdots \\ \bot\end{array}}{A}$$

where (*Restrictions:*) A should be different from $\bot$ and it should not have the form $B \to \bot$, sees them as special cases of rules 5 and 6, respectively.

Recall the notion of proof or derivation in some proof system $\mathcal{P}$, and, more specifically, the definition of a direct proof (cf. Def.s 4.16 and 10.5, respectively). We shall now represent the sequence $\psi_1, \psi_2, ..., \psi_n$ constituting a proof of $\psi = \psi_n$ in the form

$$\begin{array}{rl} 1. & \psi_1 \\ 2. & \psi_2 \\ \vdots & \vdots \\ n. & \vdash \psi_n \end{array}$$

Furthermore, we shall indicate on the right of each ψ_i whether this is a premise or an assumption $\psi_{i<n}$, or the result of the application of a rule of inference $\psi_m = \mathbf{r}(\psi_i, ..., \psi_j)$ on one or more premises/assumptions such that $i, j < m$.

Example 12.3. Figure 12.1.1 shows a proof of the derivation

$$\{P \leftrightarrow (Q \vee R), R\} \vdash P$$

in $\mathcal{NK}$. Note in this Example the inference rule in $\mathcal{NK}$ for the elimination of the connective $\leftrightarrow$,

$$\leftrightarrow E \qquad \frac{A \leftrightarrow B}{A \to B} ; \frac{A \leftrightarrow B}{B \to A}$$

where, just as in the case of the rules in Proposition 12.1, the punctuation symbol ";" denotes "either ... or." This is clearly a secondary rule of inference, as $\leftrightarrow$ is not considered a primary connective in $\mathcal{NK}$. We remark the omission of the symbol $\vdash$ in steps 3-5: indeed, this is a symbol of the metalanguage of $\mathcal{NK}$, whereas the proof at hand is a construct of the object language L by means of $\mathcal{NK} = (F_L, RI)$.

1.	$P \leftrightarrow (Q \vee R)$	Premise
2.	R	Premise
3.	$(Q \vee R) \to P$	$\leftrightarrow E$ (1)
4.	$Q \vee R$	$\vee I$ (2)
5.	P	$\to E$ (3, 4)

Figure 12.1.1.: A proof of a propositional derivation in $\mathcal{NK}$.

Discharge (of assumptions) is a central feature of the $\mathcal{NK}$ calculus. It is present in the rules $\vee E$, $\to I$, $\neg I$, and $\bot$. In these, this feature is abbreviated by means of the square parentheses around some formula ϕ (i.e. $[\![\phi]\!]$) and the vertical ellipsis such that we have a sequence of the form

$$\begin{array}{c} [\![\phi]\!] \\ \vdots \\ \psi \end{array}$$

and we now elaborate on this important property of $\mathcal{NK}$ proofs. In this elaboration, we shall retake the notion of assumption or hypothesis, first presented in Definition 10.6, in order to specify it in the context of discharge.

Definition 12.4. Let $\mathcal{D}$ be a $\mathcal{NK}$ proof or derivation. We say that $\mathcal{D}' \subset \mathcal{D}$ is a *sub-proof*, or *sub-derivation*, of $\mathcal{D}$ if there is a sequence of formulas $\psi'_1, \psi'_2, ..., \psi'_k \in \mathcal{D}'$ such that, given the sequence $\psi_1, \psi_2, ..., \psi_n \in \mathcal{D}$, we have

$$\underbrace{\psi_1, \psi_2, ..., \psi_i, \underbrace{\psi'_1, \psi'_2, ..., \psi'_k}_{\mathcal{D}'}, \psi_{(i+k+1)=l}, ..., \psi_n}_{\mathcal{D}}$$

12.1. The natural deduction calculus $\mathcal{NK}$

such that we have the sequence

$$
\begin{array}{ll}
1. & \psi_1 \\
2. & \psi_2 \\
\vdots & \vdots \\
i & \psi_i \\
(i+1). & \quad\quad \psi'_1 \\
(i+2). & \quad\quad \psi'_2 \\
& \quad\quad \vdots \\
(i+k). & \quad\quad \vdash \psi'_k \\
(i+k+1). & \vdash \psi_l \\
\vdots & \\
n. & \vdash \psi_n
\end{array}
$$

where $\psi_l = \mathbf{r}\left(\psi'_1 \text{-} \psi'_k\right)$ for some inference rule $\mathbf{r}$. In $\psi'_1, \psi'_2, ..., \psi'_k \in \mathcal{D}'$, $\psi'_1, \psi'_2, ..., \psi'_{k-j}$, $j < k$, are called the *assumptions* or *hypotheses*, and ψ'_k is the *conclusion* of the sub-derivation.

Note the indentation for the sequence $\mathcal{D}'$ with respect to $\mathcal{D}$ as a graphical means to indicate that $\mathcal{D}' \subset \mathcal{D}$.

We now require a specific notion of assumption or hypothesis:

Definition 12.5. Let $\mathcal{D}' = \psi'_1, \psi'_2, ..., \psi'_k$ be a sub-derivation or sub-proof of some proof $\mathcal{D}$. Then, the formula $\psi'_1 \in \mathcal{D}'$ is said to be an *assumption* or *hypothesis* in $\mathcal{D}$.

Obviously, there may be more than one single assumption or hypothesis in the same derivation $\mathcal{D}$, but then we have sub-derivations $\mathcal{D}', \mathcal{D}'', \mathcal{D}'''$, etc. such that for, say, $\mathcal{D}'' \subset \mathcal{D}' \subset \mathcal{D}$ we have the construction

$$
\begin{array}{l}
\psi_1 \\
\vdots \\
\quad \psi'_1 \\
\quad\quad \psi''_1 \\
\quad\quad \vdots \\
\quad\quad \vdash \psi''_h \\
\quad \vdash \psi'_k \\
\vdash \psi_l \\
\vdots \\
\vdash \psi_n
\end{array}
$$

12. Gentzen systems

Assumptions or hypotheses, as their name indicates, are not premises of the argument to be proven, but formulas that are invoked because they help us to obtain a formula $\psi_l = \mathbf{r}\left(\psi'_1\text{-}\psi'_k\right)$ that constitutes an important step in the main derivation $\mathcal{D}$. When we reach $\psi_l = \mathbf{r}\left(\psi'_1\text{-}\psi'_k\right)$, we close the indentation opened for $\mathcal{D}'$ and say that the assumption or hypothesis $\psi'_1 \in \mathcal{D}'$ has been *discharged*–in other words, it has served its purpose and is no longer necessary.

Importantly, the invocation of assumptions or hypotheses and their discharge actually follows specific rules.

Remark 12.6. With respect to the $\vee E$-rule, in order to eliminate the connective for disjunction in $\phi \vee \psi \vdash \chi$ we must show that, by assuming both ϕ and ψ, we independently obtain the same conclusion χ. Thus, what we in fact have is the derivation

$$\{\phi \vee \psi, \phi \to \chi, \psi \to \chi\} \vdash \chi.$$

Therefore, given the premise $(\phi \vee \psi) \in \mathcal{D}$ for a derivation $\mathcal{D}$, we must invoke ϕ and ψ as hypotheses or assumptions in two sub-derivations $\mathcal{D}'_1, \mathcal{D}'_2$. Graphically, we have the construction

$$
\begin{array}{c}
\vdots \\
\phi \vee \psi \\
\begin{array}{c|c} \phi & \psi \\ \vdots & \vdots \\ \vdash \chi & \vdash \chi \end{array} \\
\vdash \chi \\
\vdots
\end{array}
$$

Example 12.7. Figure 12.1.2 shows the application of the $\vee E$-rule in the proof of
$$\{P \wedge (Q \vee R)\} \vdash (P \wedge Q) \vee (P \wedge R).$$

Remark 12.8. Concerning the $\to I$-rule, the rationale behind it is that if one assumes some formula ϕ from which some other formula ψ is proven to follow, then one has proven $\vdash \phi \to \psi$. In the context of a

12.1. The natural deduction calculus $\mathcal{NK}$

1.	$P \wedge (Q \vee R)$		Premise
2.	P		$\wedge E\ (1)$
3.	$Q \vee R$		$\wedge E\ (1)$
4.	Q	R	Assumption
5.	$P \wedge Q$	$P \wedge R$	$\wedge I\ (2, 4)$
6.	$(P \wedge Q) \vee (P \wedge R)$	$(P \wedge Q) \vee (P \wedge R)$	$\vee I\ (5)$
7.	$(P \wedge Q) \vee (P \wedge R)$		$\vee E\ (4\text{-}6)$

Figure 12.1.2.: A proof in $\mathcal{NK}$ of the distributivity property for $\wedge$.

sub-derivation $\mathcal{D}' \subset \mathcal{D}$, and graphically, we have the construction

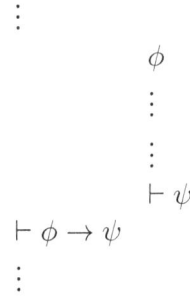

Example 12.9. Figure 12.1.3 shows a $\mathcal{NK}$ proof of the theorem

$$\vdash ((P \rightarrow Q) \wedge (P \rightarrow R)) \rightarrow (P \rightarrow (Q \wedge R)).$$

This is the proof of a conditional statement, so the best strategy is to apply $\rightarrow I$. However, this is a theorem, and for this reason we have no premises but solely assumptions or hypotheses. We begin by assuming the antecedent (step 1). This antecedent has itself two conditional propositions in which P is an antecedent, and we actually want to prove first that from this antecedent $(Q \wedge R)$ can be proved. We thus assume P (step 2) and enter a new level of the proof (a sub-proof). Steps 3-7 represent the application of the rules of inference of $\mathcal{NK}$ with respect to the former steps to obtain a proof that $P \rightarrow (Q \wedge R)$ (step 8). Note in 8 that the indentation started in 2 is closed; this represents graphically the fact that the assumption in 2 was discharged. Step 9 concludes the proof of the main conditional, and resumes the indentation in 1, i.e. it discharges the first assumption in 1. Note thus that a (sub-)formula is proved iff it is proved under certain assumptions, unless it is a theorem.

12. Gentzen systems

1.	$(P \to Q) \land (P \to R)$	Assumption
2.	$\quad P$	Assumption
3.	$\quad P \to Q$	$\land E1$ (1)
4.	$\quad P \to R$	$\land E2$ (1)
5.	$\quad Q$	$\to E$ (3, 2)
6.	$\quad R$	$\to E$ (4, 2)
7.	$\quad Q \land R$	$\land I$ (5, 6)
8.	$\quad P \to (Q \land R)$	$\to I$ (2, 7)
9.	$((P \to Q) \land (P \to R)) \to (P \to (Q \land R))$	$\to I$ (1, 8)

Figure 12.1.3.: Proof of $\vdash_{\mathcal{NK}} ((P \to Q) \land (P \to R)) \to (P \to (Q \land R))$.

Remark 12.10. As for the $\neg I$-rule, this is just a version of *reductio ad absurdum* (RA). In effect, if by assuming some formula $\phi \in \mathcal{D}'$ we end up reaching a contradiction $(\psi \land \neg \psi) \in \mathcal{D}'$, then it is safe to state that $\vdash \neg \phi$ in the main derivation $\mathcal{D}$. Graphically, we have the construction

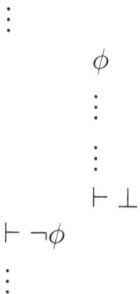

Example 12.11. In the proof of Figure 12.1.4 below there is to be seen the application of rule $\neg I$.

Remark 12.12. We leave the rationale of the inference rule for $\bot$ (cf. Remark 12.2) as an exercise.

Although the proofs in the Examples above were completed solely by means of the inference rules of $\mathcal{NK}$, this is actually not the common practice, especially in the case of complex arguments. Clearly, any classical theorem can be proven from these and only these rules (otherwise, it would not be a classical theorem), but restricting the proofs to these rules can be cumbersome and impractical. In order to avoid this we may apply any *derived rule*, i.e. any rule that has been derived from these main rules, and thus any known theorem, as well as *inder-definitions*, *equivalences*, and *properties* (e.g., distributivity of $\land$ and $\lor$) can be used

12.1. The natural deduction calculus $\mathcal{NK}$

in proving theoremhood in the propositional calculus $\mathcal{NK}$. This holds also for any argument form of Figure 2.3.1, as well as for all the classical $\heartsuit$-consequences of Chapter 6. If one so wishes, this calculus can be referred to as *extended* $\mathcal{NK}$.

Example 12.13. Figure 12.1.4 shows a proof in (extended) $\mathcal{NK}$ of the argument

$$\{(P \vee R) \rightarrow \neg(Q \wedge T), \neg S \rightarrow Q, P \wedge T\} \vdash S$$

In this proof, besides the $\mathcal{NK}$ rules proper, De Morgan's law for conjunction (DeM$_\wedge$; cf. Prop. 7.1, $\models$ 10), the argument form known as *modus tollendo ponens* (TP; cf. Fig. 2.3.1), and the double-negation law (DN; cf. Prop. 7.1, $\models$ 4) were employed.

1.	$(P \vee R) \rightarrow \neg(Q \wedge T)$	Premise
2.	$\neg S \rightarrow Q$	Premise
3.	$P \wedge T$	Premise
4.	P	$\wedge E$ (3)
5.	T	$\wedge E$ (3)
6.	$P \vee R$	$\vee I$ (4)
7.	$\neg(Q \wedge T)$	$\rightarrow E$ (6, 1)
8.	$\neg Q \vee \neg T$	DeM$_\wedge$ (7)
9.	$\quad \neg S$	Assumption
10.	$\quad Q$	$\rightarrow E$ (9, 2)
11.	$\quad \neg T$	TP (8, 10)
12.	$\quad T \wedge \neg T$	$\wedge I$ (5, 11)
13.	$\neg \neg S$	$\neg I$ (9, 12)
14.	S	DN (13)

Figure 12.1.4.: Proof of an argument in (extended) $\mathcal{NK}$.

Finally, (extended) $\mathcal{NK}$ can also be used as a hybrid with an indirect proof system: if one assumes the negation of the formula ϕ (the conclusion of the argument α) to be proven and derives a contradiction, this is a *reductio ad absurdum* proof, or simply a proof by contradiction (cf. Def. 1.6), of ϕ (α, respectively).

As said above, the rules for the elimination and introduction of the connectives suffice for proving correct reasoning instances in propositional $\mathcal{NK}$. Actually, they also suffice for proving their *validity*, as $\mathcal{NK}$ is an adequate proof system, i.e. it is both sound and complete. We give

12. Gentzen systems

here summary proofs for both soundness and completeness of propositional $\mathcal{NK}$.

Theorem 12.14. *(Soundness of the propositional $\mathcal{NK}$ calculus) Let X be a (possibly empty) set of propositional formulas. Then, for some propositional formula ψ,*

$$\text{if } X \vdash_{\mathcal{NK}} \psi, \text{ then } X \models \psi.$$

Proof: (Sketch) The proof is by structural induction on $\ell(\mathcal{D})$ where $\mathcal{D}$ is some derivation of the form $\chi_1, ..., \chi_n \vdash_{\mathcal{NK}} \psi$ and ℓ denotes length in number of steps. For the basis, we have $\ell(\mathcal{D}) = 1$; thus, ψ must actually be one of the premises $\chi_i \in X$. (This is possible even if $X = \emptyset$ by applying the reflexivity condition of logical consequence, i.e. if $\vdash_{\mathcal{NK}} \phi$, then $\phi \vdash_{\mathcal{NK}} \phi$, as clearly we have $\vdash_{\mathcal{NK}} \phi \to \phi$.) Let now there be given an interpretation $\mathcal{I}$ such that there is a valuation $val_\mathcal{I}$ (abbreviated val) such that $val(\chi_i) = \mathbf{t}$ for every $\chi_i \in X$. Then, $val(\psi) = \mathbf{t}$ and we have $X \models \psi$. For the induction step, we consider the different cases for the last rule applied. For instance, let $\wedge I$ be the last rule applied in $\mathcal{D}$. Then, $\psi = \psi_1 \wedge \psi_2$, and we have derivations $\mathcal{D}_1 = X \vdash_{\mathcal{NK}} \psi_1$ and $\mathcal{D}_2 = X \vdash_{\mathcal{NK}} \psi_2$. From the induction hypothesis, we have $X \models \psi_1$ and $X \models \psi_2$. We assume again that $val(\chi_i) = \mathbf{t}$ for every $\chi_i \in X$. Then, we have $val(\psi_1) = \mathbf{t}$ and $val(\psi_2) = \mathbf{t}$, which allows us to conclude $val(\psi) = \mathbf{t}$. Hence, we have $X \models \psi$. We leave the remaining rules as an exercise. **QED**

As for the proof of the completeness of the propositional calculus $\mathcal{NK}$, more convoluted than that of the soundness of this calculus, we shall require the (generalization) of Theorems 4.25 and 4.42, which for convenience we restate as a lemma (Lemma 12.15), as well as lemmas 12.16-18.

Lemma 12.15. *Let $X = \{\chi_1, ..., \chi_n\}$ be a set of propositional formulas and let ψ be a propositional formula. Then, we have*

1. $\chi_1, ..., \chi_n \models \psi$ *iff* $\models \chi_1 \to (\chi_2 \to (\chi_3 \to (\cdots \to (\chi_n \to \psi))))$

2. $\chi_1, ..., \chi_n \vdash_{\mathcal{NK}} \psi$ *iff* $\vdash_{\mathcal{NK}} \chi_1 \to (\chi_2 \to (\chi_3 \to (\cdots \to (\chi_n \to \psi))))$

Proof: (Sketch) We let

$$\chi_1 \to (\chi_2 \to (\chi_3 \to (\cdots \to (\chi_n \to \psi)))) = \varphi.$$

With respect to 1: ($\Rightarrow$) any truth-value assignment to $\{\chi_1, ..., \chi_n\}$ such that we have $val(\chi_1, ..., \chi_n) = \mathbf{t}$ will also give us $val(\varphi) = \mathbf{t}$, and hence

12.1. The natural deduction calculus $\mathcal{NK}$

$\models \varphi$, and ($\Leftarrow$) if $\models \varphi$, then by applying the $\Leftarrow$-direction of DDT$_\models$ to every implication in φ one at a time we arrive at $\chi_1, ..., \chi_n \models \psi$. As for 2: ($\Rightarrow$) starting from the assumption that $\chi_1, ..., \chi_n \vdash_{\mathcal{NK}} \psi$ we obtain $\vdash_{\mathcal{NK}} \varphi$ by n applications of rule $\to I$, and ($\Leftarrow$) assuming that $\vdash_{\mathcal{NK}} \varphi$ we obtain $\chi_1, ..., \chi_n \vdash_{\mathcal{NK}} \psi$ by n applications of rule $\to E$. **QED**

Lemma 12.16. *For propositional formulas ϕ, χ, ψ we have:*

1. $\vdash_{\mathcal{NK}} \phi \leftrightarrow \neg\neg\phi$
2. $\vdash_{\mathcal{NK}} \phi \vee \neg\phi$
3. $\vdash_{\mathcal{NK}} \neg\phi \wedge \neg\psi \leftrightarrow \neg(\phi \vee \psi)$
4. $\vdash_{\mathcal{NK}} \neg\phi \vee \neg\psi \leftrightarrow \neg(\phi \wedge \psi)$
5. $\vdash_{\mathcal{NK}} (\phi \to \psi) \leftrightarrow \neg\phi \vee \psi$
6. $\vdash_{\mathcal{NK}} \neg(\phi \to \psi) \leftrightarrow \phi \wedge \neg\psi$

Proof: (Sketch) We have to prove each of the above theorems 1-6 in $\mathcal{NK}$. We show the proof for theorem 1 (cf. Fig. 12.1.5) and leave the proofs of the remaining theorems as an exercise. In the case of this particular theorem, we give a single proof, but it might be more convenient for the remaining theorems above involving the connective $\leftrightarrow$ to segregate both directions of the implication in terms of proof. **QED**

1.	ϕ	Assumption
2.	$\neg\phi$	Assumption
3.	$\phi \wedge \neg\phi$	$\wedge I$ (1, 2)
4.	$\neg\neg\phi$	$\neg I$ (2-3)
5.	$\phi \to \neg\neg\phi$	$\to I$ (1, 4)
6.	$\neg\neg\phi$	Assumption
7.	ϕ	DN (6)
8.	$\neg\neg\phi \to \phi$	$\to I$ (6-7)
9.	$\phi \leftrightarrow \neg\neg\phi$	$\leftrightarrow I$ (5, 8)

Figure 12.1.5.: Proof of $\vdash_{\mathcal{NK}} \phi \leftrightarrow \neg\neg\phi$.

Lemma 12.17. *For a formula $\phi(p_1, ..., p_n)$ and a valuation $val : F_{L0} \longrightarrow W$, we define*

$$q_i := \begin{cases} p_i & \text{if } val(p_i) = \mathtt{t} \\ \neg p_i & \text{if } val(p_i) = \mathtt{f} \end{cases}.$$

12. Gentzen systems

Then, we have
$$q_1, \ldots, q_n \vdash_{\mathcal{NK}} \phi \text{ if } val(\phi) = \mathtt{t}$$

and
$$q_1, \ldots, q_n \vdash_{\mathcal{NK}} \neg\phi \text{ if } val(\phi) = \mathtt{f}.$$

Proof: (Sketch) The proof is by structural induction on ϕ. For the basis, let $\phi = p$; then, we have $p \vdash_{\mathcal{NK}} p$ and $\neg p \vdash_{\mathcal{NK}} \neg p$, whose proofs are trivial. If we let $\phi = \bot$, then we have $val(\phi) = \mathtt{f}$ and

$$\frac{[\![\bot]\!]}{\neg \bot}$$

by an application of rule $\neg I$. For the induction steps, we prove the case of negation, and leave the remaining connectives as an exercise. Let $\phi = \neg\phi'$. Then, ϕ' is a smaller formula, as it has one connective fewer than ϕ, namely $\neg$. We have two cases: 1. If $val(\phi) = \mathtt{t}$, then $val(\phi') = \mathtt{f}$, and by the induction hypothesis there is a derivation $q_1, \ldots, q_n \vdash_{\mathcal{NK}} \neg\phi'$. Because $\neg\phi' = \phi$, we are done, i.e. we have $q_1, \ldots, q_n \vdash_{\mathcal{NK}} \phi$. 2. If $val(\phi') = \mathtt{t}$, then $val(\phi) = \mathtt{f}$, and by the induction hypothesis there is a derivation $q_1, \ldots, q_n \vdash_{\mathcal{NK}} \phi'$. By assuming ϕ', we obtain $\phi' \to \neg\neg\phi'$ (cf. proof of Lemma 12.16.1, Steps 1-5), and by an application of rule $\to E$, we have $\neg\neg\phi'$, and hence have the derivation $q_1, \ldots, q_n \vdash_{\mathcal{NK}} \neg\phi$. **QED**

Lemma 12.18. *Let $\chi_1, \ldots, \chi_n, \phi, \psi$ be propositional formulas. If it is the case that*
$$\chi_1, \ldots, \chi_n, \phi \vdash_{\mathcal{NK}} \psi$$

and
$$\chi_1, \ldots, \chi_n, \neg\phi \vdash_{\mathcal{NK}} \psi,$$

then
$$\chi_1, \ldots, \chi_n \vdash_{\mathcal{NK}} \psi.$$

Proof: We first obtain $\vdash_{\mathcal{NK}} \phi \vee \neg\phi$ (cf. Lemma 12.16.2); by assuming both ϕ and $\neg\phi$, we independently obtain ψ. We then apply $\vee E$. **QED**

We are now ready to prove the completeness of the propositional $\mathcal{NK}$ calculus, but leave the proof as an exercise.

Theorem 12.19. *(Completeness of the propositional $\mathcal{NK}$ calculus) Let X be a (possibly empty) set of propositional formulas. Then, for some propositional formula ψ,*

$$\text{if } X \models \psi, \text{ then } X \vdash_{\mathcal{NK}} \psi.$$

Proof: (Idea) We let

$$\varphi = \chi_1 \to (\chi_2 \to (\chi_3 \to (\cdots \to (\chi_n \to \psi)))).$$

By Lemma 12.15.1, we have $\models \varphi$. We next must obtain $\vdash_{\mathcal{NK}} \varphi$, for which proof we invoke Lemmas 12.16-18. Finally, by Lemma 12.15.2, we obtain $X \vdash_{\mathcal{NK}} \psi$. **QED**

▶ *Do Exercises 12.1-7.*

12.1.2. The FO predicate calculus $\mathcal{NK}$

The extension of $\mathcal{NK}$ to and over L1 comprises four additional inference rules for the quantifiers of $Q_{\mathsf{L1}} = \{\forall, \exists\}$. These rules implicitly apply substitutions such as ϕ_x^a (ϕ_t^x), denoting the substitution in ϕ of the variable x for the constant a (the term t free for the variable x, respectively).[1] We indicate in an abbreviated way the discharge of assumptions by means of square parentheses and a vertical ellipsis (cf. Prop. 12.1). We leave the proof of the next proposition as an exercise.

Proposition 12.20. *The following are the inference (or formation) rules of $\mathcal{NK}$ for the quantifiers of Q_{L1}:*

1.
$$(\forall I) \quad \frac{A(a)}{\forall x A(x)}$$

   Restriction: a must not occur in any assumption on which $A(a)$ depends.

2.
$$(\forall E) \quad \frac{\forall x A(x)}{A(t)}$$

3.
$$(\exists I) \quad \frac{A(t)}{\exists x A(x)}$$

4.
$$(\exists E) \quad \frac{\exists x A(x) \quad \begin{array}{c} [A(a)] \\ \vdots \\ B \end{array}}{B}$$

---

[1] More strictly, a is a *parameter*, i.e. a free variable. But as $[\forall x \phi(x)]_a^x = \phi(x)$, we can apply "directly" $\phi_a^x = \phi(a)$ where a is a constant, namely for an interpretation $\mathcal{I}_a^x$. Equivalently but more formally, take a parameter to be a new constant symbol $c \in \mathit{Cons}^* \subset \mathsf{L1}^*$, where $\mathsf{L1}^*$ extends $\mathsf{L1}$ (cf. Chapter 8).

Restriction: a must not occur in either $\exists x A(x)$ or B, or in any assumption on which the upper occurrence of B depends other than $A(a)$.

Remark 12.21. Rule $\forall I$ is also called *universal generalization*, a label that has to do with the restriction indicated above in Proposition 12.20.1: by not allowing a to occur in any assumption on which A depends, we make sure that no assumptions are made that might distinguish a from any other individual; this gives to a the required generality in the proof at hand.

Example 12.22. We show (in Figure 12.1.6) a simple proof in which rule $\forall I$ is correctly applied. The proof is for the reasoning instance

$$\{\forall x\,(P(x) \wedge Q(x))\} \vdash_{\mathcal{NK}} \forall x\,(P(x)) \wedge \forall x\,(Q(x)).$$

1.	$\forall x\,(P(x) \wedge Q(x))$	Premise
2.	$P(a) \wedge Q(a)$	$\forall E\,(1)$
3.	$P(a)$	$\wedge E\,(2)$
4.	$Q(a)$	$\wedge E\,(2)$
5.	$\forall x P(x)$	$\forall I\,(3)$
6.	$\forall x Q(x)$	$\forall I\,(4)$
7.	$\forall x P(x) \wedge \forall x Q(x)$	$\wedge I\,(5,\,6)$

Figure 12.1.6.: A proof with universal generalization.

Remark 12.23. Rule $\forall E$ is also called *universal instantiation*, because whatever is true of every unspecified individual x of a domain is obviously true of some specific individual (i.e. an instance) t of that domain. This is the rule applied in the proof of the famous argument involving Socrates and his mortality.

Example 12.24. We show (in Figure 12.1.7) that if all humans are mortal and Socrates is a human, then Socrates is mortal. We let $H(x)$ and $M(s)$ stand for "x is a human" and "x is mortal," respectively, and we further isolate Socrates, denoted by s, from the domain of all humans. Formalized, the argument is as follows:

$$\{\forall x\,(H(x) \to M(x)), H(s)\} \vdash_{\mathcal{NK}} M(s).$$

12.1. The natural deduction calculus $\mathcal{NK}$

```
1.  ∀x (H(x) → M(x))        Premise
2.  H(s)                     Premise
3.  H(s) → M(s)              ∀E (1)
4.  M(s)                     → E (2, 3)
```

Figure 12.1.7.: An example of universal instantiation.

Remark 12.25. The rationale behind rule $\exists I$ is very straightforward: If, say, P holds for some individual t in a specified domain, i.e. $P(t)$, then there is some unspecified individual x such that $P(x)$ holds. Thus, given some formula ϕ, we may replace one or more occurrences of t in ϕ by some variable x not occurring in ϕ. This rule is also often called *existential generalization*. Straightforward, or intuitive, as this rule might be, one must beware of non-existing entities in the following sense: Let $F(x)$ denote "x is a factitious being," and let the Yeti be one such being; then, from $F(yeti)$ one cannot deduce $\exists x F(x)$, because factitious beings do not actually exist.

Example 12.26. Let anyone who wins a million dollars be considered a millionaire. We have the domain of all people and the unary predicate symbols W and M standing for the properties of winning a million dollars and being a millionaire, respectively. Let us suppose that John has won a million dollars. Then, John is a millionaire. Hence, some person is a millionaire. This valid FO $\mathcal{NK}$ deduction employing rule $\exists I$ is shown in Figure 12.1.8.

```
1.  ∀x (W(x) → M(x))           Premise
2.  W(john)                     Premise
3.  W(john) → M(john)           ∀E (1)
4.  M(john)                     → E (2, 3)
5.  ∃x M(x)                     ∃I (4)
```

Figure 12.1.8.: An example of existential generalization.

Remark 12.27. Rule $\exists E$, also called *existential instantiation*, is the most complex rule of Proposition 12.20. In effect, in this rule we are required to make assumptions and discharge them in a very specific way. The rationale is as follows: By the formula $\exists x (\phi(x))$, we state that there is

267

12. Gentzen systems

some unspecified individual of a domain that has property ϕ; by making the assumption $\phi(a)$ for some individual a we hypothesize in fact that a has property ϕ, and if from this assumption we prove that some fact ψ follows, then we can deduce ψ in the main deduction. Graphically, we have

$$
\begin{array}{c}
\vdots \\
\exists x \phi(x) \\
\quad\quad\quad \phi(a) \\
\quad\quad\quad \vdots \\
\quad\quad\quad \vdash \psi \\
\vdash \phi(a) \to \psi \\
\vdash \psi \\
\vdots
\end{array}
$$

As a matter of fact, we may just *reassert* ψ in the (sub-)deduction, skipping the line for $\vdash \phi(a) \to \psi$. Importantly, we must have no additional information about the individual a, which accounts for the restriction stated in Proposition 12.20.4 above; so to say, a is a mere "tag." For instance, suppose one suspects there is some bacterium causing a particular disease; then, one may tag it as "nasty_bacterium," and conclude that there is indeed some bacterium causing the disease, though the information we have on it may be all but negligible.

Example 12.28. Figure 12.1.9 shows a correct application of rule $\exists E$ in the proof of

$$\{\exists x \forall y \, (P(x, y))\} \vdash_{\mathcal{NK}} \forall x \exists y \, (P(y, x)).$$

1.	$\exists x \forall y \, (P(x,y))$	Premise
2.	$\forall y \, (P(a,y))$	Assumption
3.	$P(a,b)$	$\forall E$ (2)
4.	$\exists y \, (P(y,b))$	$\exists I$ (3)
5.	$\forall x \exists y \, (P(y,x))$	$\forall I$ (4)
6.	$\forall x \exists y \, (P(y,x))$	$\exists E$ (1, 2-5)

Figure 12.1.9.: An example of existential instantiation.

Just as in the case of the propositional $\mathcal{NK}$ calculus, we allow other,

12.1. The natural deduction calculus $\mathcal{NK}$

additional, rules for the quantifiers than those in Proposition 12.20. In particular, we may employ in a FO $\mathcal{NK}$ proof the properties of quantifier reversal (cf. Prop. 2.18.1) and of quantifier duality or negation distribution (cf. Prop. 2.32.4 and Prop. 2.18.2, respectively), the equivalences 3 and 4 of Exercise 5.7, the equivalences in Proposition 3.32, and the derived rules in Exercise 12.9.

Example 12.29. Figure 12.1.10 shows a proof of the FO argument

$$\frac{\forall x\,(R\,(x)) \vee \forall y\,(\neg P\,(y))}{\neg \exists z\,(P\,(z) \wedge R\,(z))}$$
$$\forall y\,(\neg P\,(y))$$

in the extended FO $\mathcal{NK}$ calculus.

1.	$\forall x\,(R\,(x)) \vee \forall y\,(\neg P\,(y))$	Premise
2.	$\neg \exists z\,(P\,(z) \wedge R\,(z))$	Premise
3.	$\forall z \neg (P\,(z) \wedge R\,(z))$	QN$_\exists$ (2)
4.	$\forall z\,(\neg P\,(z) \vee \neg R\,(z))$	DeM$_\wedge$ (3)
5.	$\quad y_1$	
6.	$\quad \neg P\,(y_1) \vee \neg R\,(y_1)$	$\forall E$ (4)
7.	$\quad\quad \forall x\,(R\,(x))$	Assumption
8.	$\quad\quad R\,(y_1)$	$\forall E$ (7)
9.	$\quad\quad \neg\neg R\,(y_1)$	DN (8)
10.	$\quad\quad P\,(y_1)$	TP (6, 9)
11.	$\quad\quad \forall y\,(\neg P\,(y_1))$	Assumption
12.	$\quad\quad \neg P\,(y_1)$	$\forall E$ (11)
13.	$\quad \neg P\,(y_1)$	$\vee E$ (1, (7-12))
14.	$\forall y\,(\neg P\,(y))$	$\forall I$ (5, 13)

Figure 12.1.10.: A FO $\mathcal{NK}$ proof.

We now turn to the soundness and completeness properties of the FO $\mathcal{NK}$ calculus.

Theorem 12.30. *(Soundness of the FO $\mathcal{NK}$ calculus) Let X be a (possibly empty) set of FO formulas and ψ a FO formula. Then,*

$$\text{if } X \vdash_{\mathcal{NK}} \psi, \text{ then } X \models \psi.$$

Proof: (Sketch) The proof is by induction on $\mathcal{D} = \chi_1, ..., \chi_n \vdash_{\mathcal{NK}} \psi$. The basis is just as for the propositional case (cf. Theorem 12.14) and we omit it. For the induction step, there are four additional cases in FO predicate logic, corresponding to the four inference rules of $\mathcal{NK}$ for the quantifiers. We give a sketch of the proof for the $\forall E$-rule and leave the remaining cases as an exercise. This rule states that if we have $\forall x \phi(x)$, where x is a bound variable, then we can simply assert $\phi(t)$ for any term t of a particular domain. We omit the case where t is a variable as a simple change of variable ϕ_y^x (but see Exercise 12.9.4) and we focus on the change ϕ_a^x for some constant a of the given domain. Recall from Proposition 3.27.1 that, for a FO formula $\forall x \phi$ and some interpretation $\mathcal{I} = (\mathcal{D}, \Theta, \varpi)$, we have $val_\mathcal{I}(\forall x \phi) = \mathbf{t}$ if $val_{\mathcal{I}_a^x}(\phi) = \mathbf{t}$ for all $a \in \mathcal{D}$. Let $\psi = \psi'\binom{x}{a}$, where in brackets we indicate that a substitutes for x, so that we have the derivation $\chi_1, ..., \chi_n \vdash_{\mathcal{NK}} \forall x (\psi')$. We assume that there is a model $\mathcal{M}$ such that $\models_\mathcal{M} \chi_i \binom{x}{a}$ for every $\chi_i \in X$. By the induction hypothesis, we have $\models_\mathcal{M} \forall x (\psi')$ and thus $\models_\mathcal{M} \psi' \binom{x}{a}$. **QED**

As for the completeness of the FO $\mathcal{NK}$ calculus, Henkin's proof of the completeness of CFOL (cf. Chapter 8) can be easily adapted to prove it. This is left as an exercise.

▶ *Do Exercises 12.8-14.*

12.1.3. The extension $\mathcal{NK}^=$ for CL$^=$

Finally, we can extend the FO $\mathcal{NK}$ calculus to $\mathcal{NK}^=$, i.e. the *FO $\mathcal{NK}$ calculus with equality*, a useful calculus in the context of CL$^=$. In order to obtain this extension it suffices to add the following two rules to the FO $\mathcal{NK}$ calculus:

Proposition 12.31. *The following are the inference (or formation) rules for the FO $\mathcal{NK}$ calculus with equality, denoted by $\mathcal{NK}^=$:*

1.
$$(= I) \qquad \overline{t = t}$$

2.
$$(= E) \qquad \frac{A(t) \quad (t = s)}{A(s)}$$

Remark 12.32. The $=$ I-rule simply states that at any line of a proof we may assert $t = t$ for any term t. The reason for this atypical natural deduction rule is that every object is simply identical to itself. This is actually the SubP-law (the substitution principle) of the theory of logical identity Θ_{Id}, or, more simply, of **CL**$^=$ (cf. Section 7.2). As a matter of fact, we may consider $\forall x (x = x)$ as an axiom (cf. $\mathcal{E}1$ in Prop. 7.5), and thus dispense with the previous introduction of $t = t$.

Example 12.33. To prove the theorem $\vdash_{\mathcal{NK}^=} \forall x \, (x = x)$, it suffices to apply the $=I$-rule (e.g., $a = a$).

Remark 12.34. With respect to the $=E$-rule, it simply states that we may replace one or more occurrences of t in $A(t)$ (either $t = s$ or $s = t$) by s. Given the formula, say, $P(a, a)$ and $a = b$, we may have $P(b, a)$, $P(b, b)$, or $P(a, b)$. Actually, this inference rule is the IdI-law, called identity of indiscernibles, in the context of the theory of logical identity Θ_{Id}, or, more simply, $\mathbf{CL}^=$ (cf. Section 7.2).

Example 12.35. In Figure 12.1.11, we show a proof with the $=E$-rule of the argument

$$\{P(a), \forall x \, (x = f(b))\} \vdash_{\mathcal{NK}^=} P(f(b)).$$

1.	$P(a)$	Premise
2.	$\forall x \, (x = f(b))$	Premise
3.	$(a = f(b))$	$\forall E \, (2)$
4.	$P(f(b))$	$= E \, (1, 3)$

Figure 12.1.11.: A proof in $\mathcal{NK}^=$.

▶ *Do Exercises 12.15-16.*

12.2. The sequent calculus $\mathcal{LK}$

Sequent calculi are rather more interesting for the analysis of proofs than for theorem proving, for which natural deduction is more adequate. For this reason, we give here only a short elaboration on Gentzen's sequent calculus $\mathcal{LK}$. This shortness, however, is sufficient for the recognized interest of $\mathcal{LK}$ to the study of the analytic tableaux calculus (cf. Chapter 15).

The sequent calculus for classical logic $\mathcal{LK}$ (abbreviating the German expression **k**lassische Prädikaten**l**ogik, "classical predicate logic" in English) was firstly conceived by Gentzen (1934-5). As the name of this calculus indicates, sequents are the central object of $\mathcal{LK}$, and we begin our short exposition of this calculus by providing the corresponding definition.

12. Gentzen systems

Definition 12.36. A *sequent s* is an expression of the form

$$(s) \quad A_1, ..., A_n \Rightarrow B_1, ..., B_k$$

where A_i, $0 \leq i \leq n$, and B_j, $0 \leq j \leq k$, are formulas, and $\Rightarrow$ is a new symbol denoting that the disjunction of the B_j *follows from* the conjunction of the A_i, i.e.

$$(s) \quad \left(\bigwedge_{i=0}^{n} A_i\right) \Rightarrow \left(\bigvee_{j=0}^{k} B_j\right)$$

which, in turn, is equivalent to

$$(s') \quad \Rightarrow \left(\bigwedge_{i=0}^{n} A_i\right) \rightarrow \left(\bigvee_{j=0}^{k} B_j\right)$$

it being the case that a sequent of the form "$\Rightarrow A$" denotes a theorem. This is the same as saying that a sequent is actually a structure asserting that whenever all the A_i are true at least one of the B_j will also be true, and we can replace $\Rightarrow$ by $\vdash$ (see Exercise 12.18). We call $\bigwedge_{i=0}^{n} A_i$ the *antecedent* and $\bigvee_{j=0}^{k} B_j$ the *succedent* (or *consequent*).

Definition 12.37. A *context* (or *side formula*), denoted by $\Gamma, \Lambda, \Sigma, \Pi$, is a finite, possibly empty, sequence of formulas. In the conclusion of each rule, the formula not in the context is the *principal* (or *main*) formula.

Definition 12.38. A *sequent rule s* is of the general form

$$(\mathbf{s}) \quad \frac{s_n}{s_{n-1}}$$

and a *sequent axiom* $\mathbf{a}_s$ has the form

$$(\mathbf{a}_s) \quad \frac{}{s_n}$$

Note the "upward direction" of the sequent rules and axioms.

Definition 12.39. A *proof* in a sequent calculus is a (branching) sequence

$$s_0, s_{11}, ..., s_{ij}..., s_{nm}$$

where s_0 is the proved sequent / formula, and each of the sequents s_{ij} in the proof is inferred (or derived) from earlier sequents in the sequence by means of rules of inference (or derivation). This sequence is an (upwards-growing) *ordered finite tree*, with the root s_0, the sequents s_{ij}, $0 < i < n$

are the i-th *node* in the j-th *branch*, $0 < j \leq m$, and s_{nj} is the *leaf* of the j-th branch.[2]

Importantly, this proof structure without rule CUT (see below) guarantees the following property of the sequent calculus known as *sub-formula property*:

Proposition 12.40. *All formulas occurring in a proof of a sequent s_0 are sub-formulas of the formulas in s_0.*

Proof: Trivial. **QED**

In effect, the possibility of applying rules along each j-th branch stops when we have obtained atomic sub-formulas, the above s_{nj} sequents. The semantical explanation for this proof structure is as follows for material implication (for example): if all branches terminate in sequents of the form $\Gamma', L \vdash L, \Lambda'$, then there is *no* interpretation $\mathcal{I}$ for which $val_\mathcal{I}(\Gamma) = \mathtt{t}$ and $val_\mathcal{I}(\Lambda) = \mathtt{f}$.

Definition 12.41. *A sequent calculus for CL over L is a triple* $\mathcal{S} = (F_\mathsf{L}, RI_s, AX_s)$ *where* RI_s, AX_s *denote the sets of rules of inference and axioms for sequents.*

We focus now on the sequent calculus $\mathcal{LK}$ over L1. $\mathcal{LK}$ has several rules of inference (see below) and a single axiom:

Definition 12.42. *Axiom of identity:*

$$(\text{Ax}) \quad \frac{}{A \vdash A}$$

The inference rules are of two kinds, *structural* and *logical* (or *operational*), the latter being rules for the use of the logical operators (connectives, as well as quantifiers). Each rule, structural or logical, has a right and left version, denoted by R and L, respectively. We begin by the structural rules.

Definition 12.43. *The following are the structural rules of* $\mathcal{LK}$:

1. *Weakening* (W):

$$(\text{WL}) \quad \frac{\Gamma \vdash \Lambda}{\Gamma, A \vdash \Lambda} \qquad (\text{WR}) \quad \frac{\Gamma \vdash \Lambda}{\Gamma \vdash A, \Lambda}$$

---

[2] There are variations to this structure; see, e.g., Troelstra & Schwichtenberg (2000).

12. Gentzen systems

2. *Contraction* (C):

$$(\text{CL}) \quad \frac{\Gamma, A, A \vdash \Lambda}{\Gamma, A \vdash \Lambda} \qquad (\text{CR}) \quad \frac{\Gamma \vdash A, A, \Lambda}{\Gamma \vdash A, \Lambda}$$

3. *Permutation* (P):

$$(\text{PL}) \quad \frac{\Gamma, A, B, \Sigma \vdash \Lambda}{\Gamma, B, A, \Sigma \vdash \Lambda} \qquad (\text{PR}) \quad \frac{\Gamma \vdash \Lambda, A, B, \Pi}{\Gamma \vdash \Lambda, B, A, \Pi}$$

These three rules can be entirely or selectively rejected, giving rise to the so called *substructural logics*, which constitute an interesting contrast with respect to classical consequence (cf. Augusto, 2017, Section 4.6). However, our interest falls more immediately on the logical rules of $\mathcal{LK}$.

Definition 12.44. The following are the logical rules of $\mathcal{LK}$ for the connectives of O_L:

1. $\wedge$:

$$(\wedge \text{L}_1) \quad \frac{\Gamma, A \vdash \Lambda}{\Gamma, A \wedge B \vdash \Lambda} \qquad (\wedge \text{L}_2) \quad \frac{\Gamma, B \vdash \Lambda}{\Gamma, A \wedge B \vdash \Lambda}$$

$$(\wedge \text{R}) \quad \frac{\Gamma \vdash A, \Lambda \quad \Sigma \vdash B, \Pi}{\Gamma, \Sigma \vdash A \wedge B, \Lambda, \Pi}$$

2. $\vee$:

$$(\vee \text{L}) \quad \frac{\Gamma, A \vdash \Lambda \quad \Sigma, B \vdash \Pi}{\Gamma, \Sigma, A \vee B \vdash \Lambda, \Pi}$$

$$(\vee \text{R}_1) \quad \frac{\Gamma \vdash A, \Lambda}{\Gamma \vdash A \vee B, \Lambda} \qquad (\vee \text{R}_2) \quad \frac{\Gamma \vdash B, \Lambda}{\Gamma \vdash A \vee B, \Lambda}$$

3. $\rightarrow$:

$$(\rightarrow \text{L}) \quad \frac{\Gamma \vdash A, \Lambda \quad \Sigma, B \vdash \Pi}{\Gamma, \Sigma, A \rightarrow B \vdash \Lambda, \Pi}$$

$$(\rightarrow \text{R}) \quad \frac{\Gamma, A \vdash B, \Lambda}{\Gamma \vdash A \rightarrow B, \Lambda}$$

4. $\neg$:

$$(\neg \text{L}) \quad \frac{\Gamma \vdash A, \Lambda}{\Gamma, \neg A \vdash \Lambda}$$

$$(\neg \text{R}) \quad \frac{\Gamma, A \vdash \Lambda}{\Gamma \vdash \neg A, \Lambda}$$

12.2. The sequent calculus $\mathcal{LK}$

Note that, contrarily to natural deduction calculi, in which there are rules for the introduction and the elimination of the connectives, in the sequent calculi there are only rules for the introduction of the connectives in the antecedent or in the succedent of a sequent; the eliminations in the latter calculi take the form of introductions in the antecedent.

Example 12.45. Figure 12.2.1 shows a proof in $\mathcal{LK}$ of axiom $\mathscr{L}2$.

The sequent calculus $\mathcal{LK}$ can be extended to classical FOL, sufficing to that end to add two more logical rules for the quantifiers.

Definition 12.46. Let t be a term free for x and let a be a constant (more strictly: a parameter). The following are the logical rules of $\mathcal{LK}$ for the quantifiers:

1. ($\forall$):

$$(\forall L) \quad \frac{\Gamma, A(t) \vdash \Lambda}{\Gamma, \forall x A(x) \vdash \Lambda}$$

$$(\forall R) \quad \frac{\Gamma \vdash A(a), \Lambda}{\Gamma \vdash \forall x A(x), \Lambda}$$

2. ($\exists$):

$$(\exists L) \quad \frac{\Gamma, A(a) \vdash \Lambda}{\Gamma, \exists x A(x) \vdash \Lambda}$$

$$(\exists R) \quad \frac{\Gamma \vdash A(t), \Lambda}{\Gamma \vdash \exists x A(x), \Lambda}$$

Note that restrictions apply to rules $\forall R$ and $\exists L$: a must not occur within Γ and Λ, *or* it must not appear anywhere in the respective lower sequents.

Example 12.47. Figure 12.2.2 shows a proof in $\mathcal{LK}$ of the CFOL theorem $\vdash \forall x (A(x) \to B) \to \exists x (A(x) \to B)$. We make $t = a$.

12. Gentzen systems

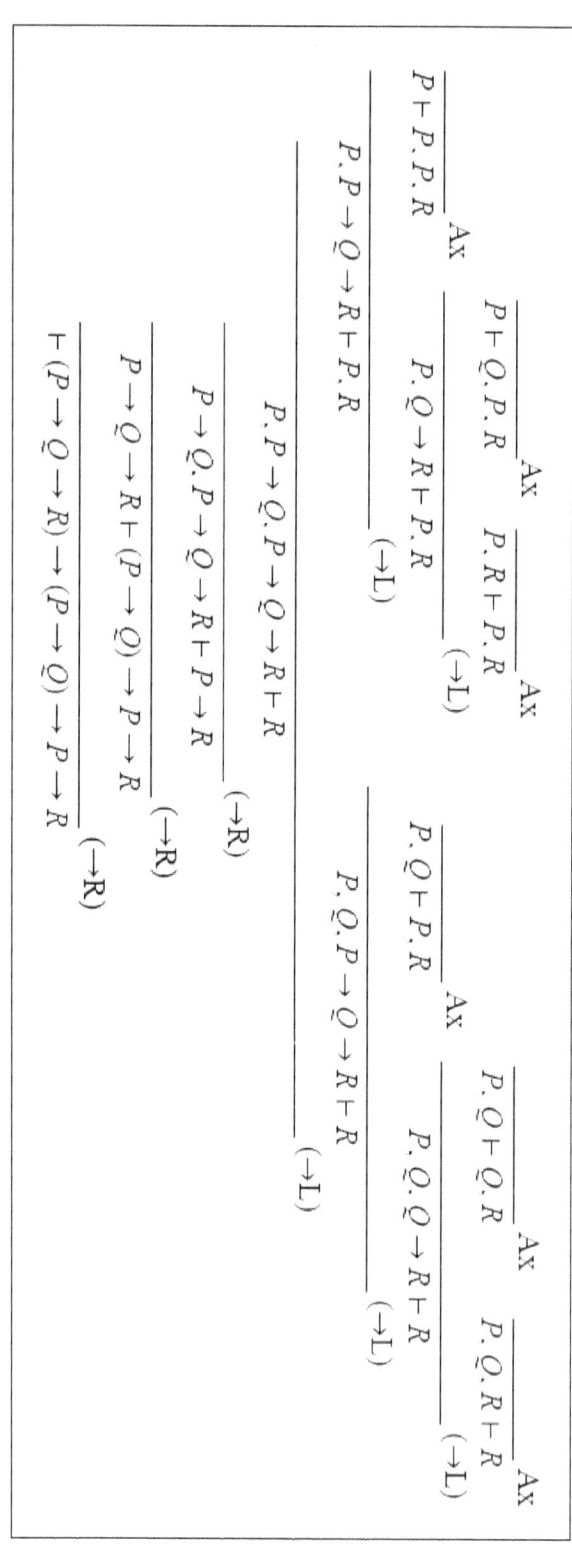

Figure 12.2.1.: Proof in $\mathcal{LK}$ of axiom $\mathcal{L}2$ of the axiom system $\mathcal{L}$.

12.2. The sequent calculus $\mathcal{LK}$

$$\dfrac{\dfrac{\dfrac{\dfrac{\dfrac{\dfrac{\dfrac{A(a) \vdash A(a), B}{\vdash A(a), A(a) \to B}(\to R)}{\vdash A(a), \exists x A(x) \to B}(\exists R) \quad \dfrac{\dfrac{\dfrac{A(a), B \vdash B}{B \vdash A(a) \to B}(\to R)}{B \vdash \exists x A(x) \to B}(\exists R)}{A(a) \to B \vdash \exists x A(x) \to B}(\to L)}{\forall x A(x) \to B \vdash \exists x A(x) \to B}(\forall L)}{\vdash \forall x (A(x) \to B) \to \exists x (A(x) \to B)}(\to R)$$

Figure 12.2.2.: Proof in $\mathcal{LK}$ of a FO theorem.

Finally, an additional structural rule of $\mathcal{LK}$ is the following:

Definition 12.48. *Cut*:

$$\text{(CUT)} \quad \dfrac{\Gamma \vdash \Lambda, A \quad A, \Sigma \vdash \Pi}{\Gamma, \Sigma \vdash \Lambda, \Pi}$$

Rule CUT has motivated much work–if not furore–in modern logic, in particular in the field of automated deduction. For its "special" character and status in the sequent calculus above, see, e.g., Fitting (1999).

▶ *Do Exercises 12.17-19.*

Exercises

Exercise 12.1. Complete the following proofs in the extended $\mathcal{NK}$ calculus (all the premises are given):

1.
   1. $\neg P \to P$ Premise
   2. $\neg\neg P \vee P$ ─────
   3. ───────────   DN (2)
   4. P ─────

2.
   1. $(\neg P \vee Q) \vee R$ Premise
   2. $(Q \vee R) \to S$ Premise
   3. $\neg P \vee (Q \vee R)$ ─────
   4. ─────────── $\to_{def}$ (3)
   5. $P \to S$ ─────

12. Gentzen systems

3.
1. $(P \vee Q) \wedge (P \vee R)$ Premise
2. $P \vee (Q \wedge R)$ ─────────
3. ────────────────   DN (2)
4. $\neg P \to (Q \wedge R)$ ─────────

4.
1. $(P \wedge Q) \to (R \wedge S)$ Premise
2. $\neg\neg P$ Premise
3. Q Premise
4. P ─────────
5. ────────────────   ─────────
6. $R \wedge S$ $\to E$ (1,5)
7. S ─────────

5.
1. $(P \vee Q) \wedge (P \vee R)$ Premise
2. $P \to S$ Premise
3. $Q \to S$ Premise
4. $P \to T$ Premise
5. $R \to T$ Premise
6. $P \vee Q$ $\wedge E$ (1)
7. $\neg P \vee S$ ─────────
8. S ─────────
9. $P \vee R$ $\wedge E$ (1)
10. ────────────────   ─────────
11. T ─────────
12. $S \wedge T$ ─────────

6.
1. $(P \vee Q) \to \neg R$ Premise
2. ────────────────   Assumption
3. $\neg(P \vee Q)$ ─────────
4. $\neg\neg R \to \neg(P \vee Q)$ ─────────

7.
1. $P \to Q$ Premise
2. $\neg Q$ Premise
3. _____ Assumption
4. Q $\to E\ (1,\ 3)$
5. $Q \land \neg Q$ _____
6. $\neg P$ _____

8.
1. $(P \land Q) \to R$ Premise
2. P _____
3. _____ Assumption
4. $P \land Q$ _____
5. R $\to E\ (1,\ 4)$
6. $Q \to R$ _____
7. $P \to (Q \to R)$ _____

9.
1. R Premise
2. $(R \land Q) \to \neg S$ Premise
3. $\neg S \to \neg P$ Premise
4. _____ Assumption
5. $R \land Q$ _____
6. $\neg S$ $\to E\ (2,\ 5)$
7. _____ $\to E\ (3,\ 6)$
8. $Q \to \neg P$ _____

Exercise 12.2. With respect to the inference rule for $\bot$ (cf. Remark 12.2) in $\mathcal{NK}$:

1. Comment on the rationale behind it.

2. Give an example of its application.

3. Comment on the possible reason(s) why Prawitz considers it instead of rules 7 and 8 of Proposition 12.1.

Exercise 12.3. Formalize the $\mathcal{NK}$ rule for the introduction of the connective $\leftrightarrow$, i.e. $\leftrightarrow I$.

Exercise 12.4. Prove in the propositional calculus $\mathcal{NK}$ that all the sentences of Proposition 7.1 are theorems.

12. Gentzen systems

Exercise 12.5. Complete the proof of Theorem 12.14. Hints for the rules with discharge:

- For $\to I$: Assume $\psi = \psi_1 \to \psi_2$ so that there is the derivation $\mathcal{D} = \chi_1, ..., \chi_n, \psi_1 \vdash_{\mathcal{NK}} \psi_2$.

- For $\neg I$: Assume $\psi = \neg \psi'$ so that there is the derivation $\mathcal{D} = \chi_1, ..., \chi_n, \psi' \vdash_{\mathcal{NK}} \bot$.

- For $\vee E$: Assume now the derivations $\mathcal{D} = \chi_1, ..., \chi_n \vdash_{\mathcal{NK}} \psi_1 \vee \psi_2$, $\mathcal{D}_1 = \chi_1, ..., \chi_n, \psi_1 \vdash_{\mathcal{NK}} \psi$, and $\mathcal{D}_2 = \chi_1, ..., \chi_n, \psi_2 \vdash_{\mathcal{NK}} \psi$.

Exercise 12.6. Prove theorems 2-6 in Lemma 12.16.

Exercise 12.7. Prove, or complete the proof of, the following statements:

1. Lemma 12.15.

2. Lemma 12.17.

3. Theorem 12.19.

Exercise 12.8. Determine whether the following applications of the $\mathcal{NK}$ rules for the quantifiers are correct or incorrect. If the latter, explain why and make the necessary corrections.

1.
1.	$\forall x P(x, x)$	Premise
2.	$P(a, a)$	$\forall E$ (1)
3.	$\forall y P(y, a)$	$\forall I$ (2)

2.
1.	$\neg \forall x P(x)$	Premise
2.	$\neg P(a)$	$\forall E$ (1)

3.
1.	$\forall x \exists y (P(y, x))$	Premise
2.	$\exists y P(y, a)$	$\forall E$ (1)
3.	$P(a, a)$	Assumption
4.	$\exists x P(x, x)$	$\exists I$ (3)
5.	$\exists x P(x, x)$	$\exists E$ (2, 3-4)

280

12.2. The sequent calculus $\mathcal{LK}$

4.
1.	$P(a) \to Q(a)$	Premise
2.	$\exists x P(x) \to Q(a)$	$\exists I$ (1)

5.
1.	$\exists x Q(x)$	Premise
2.	$P(a)$	Premise
3.	$\quad Q(a)$	Assumption
4.	$\quad P(a) \wedge Q(a)$	$\wedge I$ (2, 3)
5.	$\quad \exists x (P(x) \wedge Q(x))$	$\exists I$ (4)
6.	$\exists x (P(x) \wedge Q(x))$	$\exists E$ (1, 3-5)

6.
1.	$P(a)$	Premise
2.	$\forall x P(x)$	$\forall I$ (1)

7.
1.	$\forall x P(x) \to \forall x Q(x)$	Premise
2.	$P(x) \to Q(x)$	$\forall E$ (1)

8.
1.	$P(a) \wedge Q(a)$	Premise
2.	$\exists x P(x) \wedge Q(a)$	$\exists I$ (1)

9.
1.	$P(y) \to Q(y)$	Premise
2.	$\forall x (P(x) \to Q(y))$	$\forall I$ (1)

10.
1.	$P(a) \to Q(a)$	Premise
2.	$\forall x (P(x) \to Q(x))$	$\forall I$ (1)

11.
1.	$\exists x Q(x, x)$	Premise
2.	$\quad Q(a, a)$	Assumption
3.	$\quad \exists x Q(a, x)$	$\exists I$ (2)
4.	$\exists x Q(a, x)$	$\exists E$ (1, 2-3)

12. Gentzen systems

12.
1. $R(b)$ — Premise
2. $\exists y\, (R(y)) \rightarrow T(a)$ — Premise
3. $\exists y R(y)$ — $\exists I$ (1)
4. $T(a)$ — $\rightarrow E$ (2, 3)

13.
1. $P(zeus)$ ($Zeus$ is here the Greek god) — Premise
2. $\exists x P(x)$ — $\exists I$ (1)

14.
1. $\exists x Q(x)$ — Premise
2. $\quad P(a)$ — Assumption
3. $\quad\quad Q(a)$ — Assumption
4. $\quad\quad P(a) \wedge Q(a)$ — $\wedge I$ (2, 3)
5. $\quad\quad \exists x\, (P(x) \wedge Q(x))$ — $\exists I$ (4)
6. $\quad \exists x\, (P(x) \wedge Q(x))$ — $\exists E$ (1, 3-5)
7. $P(a) \rightarrow \exists x\, (P(x) \wedge Q(x))$ — $\rightarrow I$ (2-6)

15.
1. $\forall x\, (P(x) \rightarrow Q(x))$ — Premise
2. $P(a) \rightarrow Q(a)$ — $\forall E$ (1)
3. $\quad P(a)$ — Assumption
4. $\quad Q(a)$ — $\rightarrow E$ (2, 3)
5. $\quad \forall x Q(x)$ — $\forall I$ (4)
6. $P(a) \rightarrow \forall x Q(x)$ — $\rightarrow I$ (3, 5)

16.
1. $P(a) \wedge Q(a)$ — Premise
2. $\exists x\, (P(a) \wedge Q(x))$ — $\exists I$ (1)

17.
1. $\forall x P(a, x)$ — Premise
2. $\forall x \forall y (P(x, y)) \to Q(y, x)$ — Premise
3. $P(a, b)$ — $\forall E$ (1)
4. $\forall y P(a, y) \to Q(y, a)$ — $\forall E$ (2)
5. $P(a, b) \to Q(b, a)$ — $\forall E$ (4)
6. $Q(b, a)$ — $\to E$ (3, 5)
7. $\forall x Q(x, a)$ — $\forall I$ (6)

Exercise 12.9. Prove in the FO $\mathcal{NK}$ calculus:

1. $\forall x \phi(x) \vdash \exists x (\phi(x))$
2. $\neg \forall x \phi(x) \vdash \neg \phi(a)$
3. $\neg \exists x \phi(x) \vdash \neg \phi(t)$
4. $\forall x \phi(x) \dashv\vdash \forall y \phi(y)$
5. $\exists x \phi(x) \dashv\vdash \exists y \phi(y)$

Exercise 12.10. Prove in the FO $\mathcal{NK}$ calculus the equivalences in Proposition 3.32.

Exercise 12.11. Prove in the FO $\mathcal{NK}$ calculus:

1. $Q(a, a) \vdash \exists x \exists y (Q(x, y))$
2. $\{\forall x (S(x) \to T(x)), \exists x S(x)\} \vdash \exists x T(x)$
3. $\exists x \forall y (T(x, y)) \vdash \forall x \exists y (T(y, x))$
4. $\{\forall x \forall y \forall z ((P(x, y) \land P(y, z)) \to P(x, z)), P(a, b)\}$
$\vdash P(b, c) \to P(a, c)$
5. $\{\forall x \forall y (P(x, y) \to Q(x)), \forall x P(x, x)\} \vdash \forall x Q(x)$
6. $\forall x (\neg P(x)) \to \forall x (\neg Q(x)) \vdash \exists x Q(x) \to \exists x P(x)$

Exercise 12.12. In the proof in Figure 12.1.10, what should be written on the right side of y_1 (Step 5)?

Exercise 12.13. Complete the proof of Theorem 12.30.

Exercise 12.14. Consider Henkin's proof of the completeness of CFOL (cf. Chapter 8). How can it be adapted to prove the completeness of the FO $\mathcal{NK}$ calculus?

Exercise 12.15. Prove in the FO $\mathcal{NK}^=$ calculus:

12. Gentzen systems

1. $\{P(a), \neg P(b)\} \vdash (a \neq b)$ (Note: $a \neq b$ just is an alternative way to write $\neg(a = b)$.)

2. $(a \neq a) \vdash P$

3. $\vdash P \vee (a = a)$

Exercise 12.16. Comment on the additional proofs required to prove the soundness and completeness of the calculus $\mathcal{NK}^=$.

Exercise 12.17. Prove in $\mathcal{LK}$ all the items in the following Exercises:

1. Exercise 11.2

2. Exercise 11.9.

3. Exercise 12.9.

4. Exercise 12.11.

Exercise 12.18. Prove the following theorem:

Theorem 12.49. *For $\mathcal{LK}$ the sequent calculus with the CUT-rule and the natural calculus $\mathcal{NK}$ we have*

$$X \vdash_{\mathcal{LK}} \phi \quad \text{iff} \quad X \vdash_{\mathcal{NK}} \phi$$

for a (possibly empty) set of formulas X and a formula ϕ.

Part V.

Classical Proofs II: Indirect Proofs

13. The satisfiability problem, or *SAT*

As seen above in Chapter 10, testing directly for validity in CFOL is not only cumbersome; it lacks the algorithmic character required of a decision procedure. In this Chapter, we show that testing for (un)satisfiability fares much better, especially in the context of automated deduction.

Together with logic programming and logic circuits, *automated theorem proving* (ATP), also spoken of as *automated deduction*, is a major field of application of CFOL, with top applications in software verification and knowledge representation. Born in the late 1950s out of the desire to automate the ways humans produce mathematical proofs, ATP soon became an indispensable component of contemporary applications of logic. Although mathematical proof remains a central concern, nowadays automated deduction has many more applications, namely in technology and industry. Indeed, this field has extended to automating deduction with non-classical logics, an extension that was motivated by new key technologies for which classical logic is not adequate.[1] Nevertheless, automated deduction with classical logic remains a fruitful area of research, as many technological applications still require the employment of CFOL.

In order to develop a fully automated prover it is required that there be an algorithm for this end. In other words, the proof calculi of interest for ATP must be algorithmic in nature. Moreover, the rules or axioms should not be such that the simple or selective removal of one or more of them determines a logic other than CL. The classical proof systems of Part IV either lack an algorithmic nature (e.g., the Hilbert-style systems and the calculus $\mathcal{NK}$), or have "unstable" rules or axioms (e.g., the structural rules in the sequent calculus $\mathcal{LK}$).

Recall that these proof calculi are employed to test for validity, being thus called direct proof calculi. Contrarily to these, the refutation, or indirect proof, calculi are essentially algorithmic, and thus provide the best terrain for automation. We speak now of *indirect proof calculi*

---

[1] See Augusto (2020c) for automated deduction with many-valued logics, as well as for bibliography on other non-classical logics.

13. The satisfiability problem, or SAT

or *refutation calculi*, because the aim is *not* to prove directly that a formula (an argument) is valid, but to prove indirectly that it is so by proving instead that its negation (the negation of its conclusion together with the original premises, respectively) is unsatisfiable. This method is sanctioned by Proposition 4.39.1, and the problem of testing for the satisfiability of an argument or of a formula is the *satisfiability problem*, or, abbreviated, *SAT* (cf. Def. 4.63.2).[2]

Although the refutation calculi target unsatisfiability rather than satisfiability, the testing for unsatisfiability is so in the general context of the SAT, and we give the essentials of this problem in this Chapter. Two refutation calculi have been proven well suited to automation, to wit, resolution and analytic tableaux, and we elaborate comprehensively on these in Chapters 14 and 15, respectively. Importantly, the small sets of the basic inference rules of these two calculi give us the guarantee that these are "stable" calculi for CFOL.

13.1. *SAT* and refutation proofs

13.1.1. The different forms of *SAT*

Broadly conceived, the SAT is actually not a single decision problem, but rather a cluster of decision problems that all involve satisfiability and Boolean expressions. We give now the different forms the language *SAT* can take.

Definition 13.1. The language

$$SAT = \{\langle \phi \rangle \mid \phi \text{ is satisfiable}\}$$

for a given formula ϕ and $\langle \phi \rangle$ an adequate encoding thereof, is called the *satisfiability problem*, or, abbreviated, *SAT*.

We give the adequate encoding. Recall Definitions 3.5-6.

Definition 13.2. An instance of *SAT* is a formula $\phi = \phi(x_1, ..., x_n)$ composed of

1. $x_1, ..., x_n$ Boolean variables;

---

[2] As for the validity problem (see Chapter 10), we write "*SAT*" for the language, and "the SAT" for the broad framework of the (sub-)problem(s) and solutions proposed. Unless otherwise stated, we shall proceed in the same way for the various forms of *SAT* (e.g., *2-SAT*, *QBF-SAT*).

13.1. SAT and refutation proofs

2. k Boolean connectives, i.e. connectives in the set $O'_L = \{\neg, \wedge, \vee\}$.

In other words, ϕ is a *Boolean formula*. We can actually make the definition of SAT more specific in the following way:

Definition 13.3. The language

$$SAT = \{\phi | \phi \text{ is a satisfiable Boolean formula}\}$$

is called the *Boolean satisfiability problem*, or, abbreviated, SAT.

From now on, when we write "SAT" we mean the Boolean satisfiability problem, even if we do not write "Boolean." We can reformulate this problem in a more formal way, namely in model-theoretical terms, as follows:

Definition 13.4. The language

$$SAT = \{\phi | \models_{\mathcal{M}} \phi\}$$

for ϕ a Boolean formula and some model $\mathcal{M}$ is the satisfiability problem, or SAT.

We can now give a strictly logical definition of the satisfiability problem:

Definition 13.5. Given a (propositional) formula $\phi(x_1, ..., x_n)$ with connectives $\heartsuit_j \in \{\neg, \wedge, \vee\}$, $j = 0, 1, ..., m$, it is asked if ϕ can be evaluated to $\mathbf{t}$ by some assignment of the truth-value set $W = \{\mathbf{f}, \mathbf{t}\}$ to the x_i, $1 \leq i \leq n$. We call this question the (Boolean) satisfiability problem, or SAT.

This entails that a Boolean formula $\phi = \phi(x_1, ..., x_n)$ belongs to SAT iff there is a truth-value assignment to each and every of its n variables that makes ϕ **true**. Thus, the SAT reduces to finding a model for a Boolean formula ϕ. Moreover, because this problem is formulated in Boolean terms, any formula or theory can be easily and adequately encoded in the binary notation using the alphabet $\Sigma = \{0, 1\}$. We need not here carry out a full encoding, being satisfied with Boolean formulas and their corresponding Boolean truth-value assignments, as illustrated in Example 13.6.

Example 13.6. Let there be given the formula

$$\phi = (x_1 \vee x_2) \wedge ((\neg x_1 \wedge x_2) \vee x_3)$$

13. The satisfiability problem, or SAT

The truth-value assignment $x_1 = 1$, $x_2 = 0$, $x_3 = 1$ produces a model for ϕ. In effect, applying the laws of Boolean algebra (cf. Def. 3.4) we have
$$(1 \vee 0) \wedge ((0 \wedge 0) \vee 1) =$$
$$= 1 \wedge (0 \vee 1) =$$
$$= 1 \wedge 1 = 1$$

We conclude $\phi \in SAT$.

The formula in Example 13.6 is actually given above (Example 3.34) in a propositional form, along with a truth table for it in which all the models of this formula can be verified. In effect, the variables in Example 13.6 are to be considered as propositional variables, and we shall write "SAT" only in the case of CPL. Below we specify the satisfiability problem for quantified Boolean formulas, but we anticipate the information that, by means of the rewritings known as normal forms (cf. Section 2.4) together with Herbrand semantics, classical FO formulas can be processed as classical propositional formulas, with the advantages associated to these, namely decidability. The reader can benefit here from a (re)reading of Sections 4.2.2.1 and 9.2 above, and Section 13.3 below.

Although SAT is generally formulated in terms of the set of connectives $O'_L = \{\neg, \wedge, \vee\}$ (cf. Def. 13.5), it is in fact more often than not formulated for formulas in CNF, and this to such a point that SAT is often synonymous with $CNF\text{-}SAT$. However, we can further specify instances of this problem according to the number of literals per clause, their kind (positive or negative), and/or the order of the formulas. Far from being a mere specification, this actually involves different complexity classes (see Exercise 13.10), so that if we can formulate or even convert an instance of SAT into another we may end up with significant computational gains.

Definition 13.7. The language

$$2\text{-}SAT = \{\phi|\models_\mathcal{M} \phi, \phi \text{ is a 2-CNF formula}\}$$

where "2-CNF formula" means a CNF formula whose clauses have at most two literals, is called the *satisfiability problem for 2-CNF formulas*, or, abbreviated, *2-SAT*.

Adding one single literal to a clause in a 2-CNF formula has a significant computational impact, so that this actually constitutes a separate instance of SAT.

13.1. SAT and refutation proofs

Definition 13.8. The language

$$3\text{-}SAT = \{\phi | \models_{\mathcal{M}} \phi, \phi \text{ is a 3-CNF formula}\}$$

where "3-CNF formula" means a CNF formula whose clauses have at most three literals, is called the *satisfiability problem for 3-CNF formulas*, or, abbreviated, *3-SAT*.

For clauses with more than three literals we write "k-CNF formulas" and speak of *k-SAT*.

The following instance of SAT is of import for logic programming (see Section 14.2.2 below):

Definition 13.9. The language

$$HORN\text{-}SAT = \{\phi | \models_{\mathcal{M}} \phi, \phi \text{ is a Horn formula}\}$$

for a Horn formula a CNF formula whose clauses have at most one positive literal, is called the *satisfiability problem for Horn formulas*, or, abbreviated, *HORN-SAT*.

Although the Boolean propositional satisfiability problem is typically formulated in terms of formulas in CNF, it cannot be ignored that it can also be formulated for formulas in DNF. In effect, this is sanctioned by the following theorem:

Theorem 13.10. *A formula in CNF is a tautology iff each of its clauses is a tautology. Dually, a formula in DNF is unsatisfiable iff each of its conjunctions of literals is unsatisfiable.*

Proof: (Sketch) Let us abbreviate

$$A = \bigwedge_{i=1}^{n} \left(\bigvee_{j=1}^{m_i} L_{i,j} \right)$$

as

$$A = \bigwedge_{i=1}^{n} \mathcal{C}_i.$$

Clearly, if each $\mathcal{C}_i$, $1 \leq i \leq n$, is a tautology, then A is a tautology, and if some $\mathcal{C}_i$ is not a tautology, then A is not a tautology. Apply now the duality principle to prove the dual result for a formula in DNF. **QED**

Nevertheless, a formulation of SAT in terms of DNFs is far less usual, and we shall be contented with a very general definition.

13. The satisfiability problem, or SAT

Definition 13.11. The language

$$DNF\text{-}SAT = \{\phi|\models_{\mathcal{M}} \phi, \phi \text{ is a DNF formula}\}$$

is called the *satisfiability problem for formulas in DNF*, or, abbreviated, *DNF-SAT*.

Last but not least, let us now consider FO formulas, which in the framework of Boolean logic we call *quantified Boolean formulas (QBFs)*:

Definition 13.12. The language

$$QBF\text{-}SAT = \{\phi|\models_{\mathcal{M}} \phi, \phi \text{ is a QBF}\}$$

is called the *satisfiability problem for QBFs*, or, abbreviated, *QBF-SAT*.

▶ *Do Exercises 13.1-2*

13.1.2. Indirect proofs

Were it not for the fact that truth tables are inefficient SAT solvers for more than a few atomic formulas, we would happily resort to them; but they are not very efficient for CPL–at all, for FO formulas. In effect, there is no decision procedure for CFOL, which means that SAT is basically undecidable for Boolean FO formulas, or QBFs. This notwithstanding, under some circumstances we may achieve propositional-like decidability for FO formulas by testing for (un)satisfiability. In the framework of SAT, we can test for either satisfiability or unsatisfiability. Indirect proofs are based on the latter. Consider the following language:

Definition 13.13. The language

$$\overline{SAT} = \{(X, \phi) \,|\, (X \cup \{\phi\}) \notin SAT\}$$

for a–possibly empty–set of formulas X and a formula ϕ is the *complement of SAT*.

For simplicity, let us assume that X is empty, so that we have the case $\phi \in \overline{SAT}$. Then, by Proposition 4.39.2, $\neg\phi \in VAL$. Testing for the validity of $\neg\phi$ is, however, not an advantage over testing for the validity of ϕ: Testing for validity remains cumbersome and inefficient (cf. Part IV).

But by Proposition 4.39.1, we also have

$$\phi \in VAL \text{ iff } \neg\phi \in \overline{SAT}.$$

13.1. SAT and refutation proofs

Recall $DT_\models$ as formulated in Theorem 4.41: Testing for unsatisfiability to prove validity is based on this version of DT. In effect, given this formulation of $DT_\models$, we may reformulate VAL in terms of the complement of SAT in the following way:

Definition 13.14. The language

$$VAL_{ref} = \{(X, \phi) \mid (X \cup \{\neg\phi\}) \notin SAT^\neg\}$$

for a–possibly empty–set X of formulas and a formula ϕ, and the specified language

$$SAT^\neg = \{(X, \phi) \mid X \models_\mathcal{M} \neg\phi\}$$

is called *refutation validity*.

We remark that $SAT^\neg$ is *not* a new language; it is merely the version of SAT in which we consider the negation of ϕ. We can say that $SAT^\neg$ just is the language SAT for instances of decision problems formulated by means of negation, namely with refutation in view. Neither is VAL_{ref} a *new* language, but only the formal specification of validity by refutation. In fact, we have the following equivalence:

Proposition 13.15. *The languages* $\overline{SAT^\neg}$ *and* VAL_{ref} *are equivalent.*

Proof: Left as an exercise.

This equivalence between VAL_{ref} and $\overline{SAT^\neg}$ lies at the heart of the notion of an indirect proof. To see how, let us revisit the group theory $\Theta_\mathcal{G}$ for a group a pair $\mathcal{G} = (G, \star)$. Let the statements (1)-(3) of Exercise 2.10.4, known as the axioms of associativity, of the identity element, and of the inverse element, respectively, be the axioms of this theory. We shall denote these three axioms by θ_i, $i = 1, 2, 3$. Given $x, y, z \in G$, we wish to know if the following formula belongs to $\Theta_\mathcal{G}$:[3]

$$\phi = \forall x \forall y \forall z \left[(x \star z = y \star z) \to x = y\right].$$

In other words, we wish to know whether ϕ is a valid formula of, or "holds in," $\Theta_\mathcal{G}$, i.e.

$$\Theta_\mathcal{G} \overset{?}{\models} \phi \quad \equiv \quad (\Theta_\mathcal{G} \cup \{\neg\phi\}) \overset{?}{\in} \overline{SAT^\neg}.$$

Equivalently, we wish to know whether $(\theta_1 \wedge \theta_2 \wedge \theta_3) \to \phi$, for $\theta_1, \theta_2, \theta_3 \in \Theta_\mathcal{G}$, is a valid formula, which, if so, allows us to decide that $\phi \in \Theta_\mathcal{G}^*$.

---
[3]The fact that this formula belongs to $CL^=$ is negligible for our purposes.

13. The satisfiability problem, or SAT

Because, as said above, validity testing is less efficient than testing for (un)satisfiability, relying on Theorem 4.41 and Proposition 13.14 we ask instead whether the formula $\theta_1 \wedge \theta_2 \wedge \theta_3 \wedge \neg\phi$ is unsatisfiable, i.e. $(\Theta_{\mathcal{G}} \cup \{\neg\phi\}) \stackrel{?}{\in} \overline{SAT^{\neg}}$. Theorem 4.53 gives us a decision procedure in more than one way. Firstly, it tells us that if $\Theta_{\mathcal{G}}^* = \{\theta_1, \theta_2, \theta_3, \neg\phi\}$ is an unsatisfiable set of formulas, then the formula $(\theta_1 \wedge \theta_2 \wedge \theta_3) \to \phi$ is indeed valid, and we have $\Theta_{\mathcal{G}} \models \phi$; otherwise, $\Theta_{\mathcal{G}}^*$ is satisfiable and we have $\Theta_{\mathcal{G}} \not\models \phi$. Secondly, because (un)satisfiability is the counterpart of (in)consistency, if we have an adequate deductive system at hand, then we can simply apply some proof calculus to decide algorithmically whether $\Theta_{\mathcal{G}} \vdash \phi$. This means that a decision problem that is expressed in a model-theoretical way can find a solution by proof-theoretical means, namely a refutation proof.

Refutation proofs are often indifferently called also *indirect proofs* and/or *reductio (ad absurdum) proofs*. Because there are in fact nuances to be taken into consideration in this terminology (cf. Exercise 13.3), we shall consider refutation proofs as a class of proofs including both indirect proofs and *reductio (ad absurdum)* proofs, where these might be identical (always, or in some contexts). We shall actually favor the expression "indirect proofs" but shall disambiguate if necessary. Recall the general notion of a logical proof in Definition 4.16; the following elaboration specifies the notion of indirect proofs.

Definition 13.16. An *indirect proof* $\blacksquare \in \mathcal{P}$ for some proof system $\mathcal{P}$ and some decision problem $DP(\Theta, \phi)$ where $\Theta = \{\theta_1, ..., \theta_k\}$ is a set of FO formulas and ϕ is a FO formula consists in a sequence

1.	θ_1
2.	θ_2
$\vdots$	$\vdots$
$k.$	θ_k
$(k+1).$	$\theta_{k+1} = \neg\phi$
$(k+2).$	$\theta_{k+2} = \mathbf{r}_l(\theta_i, ..., \theta_j), i, j \leq k+1$
$\vdots$	$\vdots$
$(n-1).$	$\theta_{n-1} = \mathbf{r}_l(\theta_i, ..., \theta_j), i, j \leq n-2$
$n.$	$\vdash \perp$

where $\theta_i \in F_{L1}$ for $1 \leq i \leq n-1$, $\mathbf{r}_l \in |RI| = o$ for some finite $1 \leq l \leq o$, $\mathbf{r}_l(\theta_i, ..., \theta_j)$ for possibly $i = j$ denotes the application of rule $\mathbf{r}_l$ on the formula(s) $\theta_i, ..., \theta_j$, and $\perp$ denotes a contradiction.

Remark 13.17. Just as in the case of direct proofs (cf. Remark 10.8), one may consider the solution for $DP(\Theta, \phi)$ in the form $\models \Theta \to \theta$, in which case the indirect proof ■ takes the form

$$\nvDash \neg(\Theta \to \phi)$$

or, equivalently (cf. Theorem 4.24),

$$\nvDash \theta_1 \wedge \ldots \wedge \theta_k \wedge \neg\phi.$$

Above, we remarked that an indirect proof may not always be synonymous with a proof by *reductio ad absurdum*. In effect, our definition (Def. 13.16) does not suit the notion of a proof by *reductio ad absurdum* if this, as is typical, includes assumptions or hypotheses other than the negation of ϕ (cf. Def. 10.6; see Exercise 13.3). Indeed, our definition of indirect proof leaves no room for these, as for each θ_j, $1 \leq j \leq n-1$ and $j \neq k+1$, we have it that θ_j is either a premise or the result of applying some rule of inference **r** on one or more θ_i for $i < j$.

Example 13.18. The proof in Figure 12.1.4 is a proof by *reductio ad absurdum*. The proofs in Chapters 14 and 15 are indirect proofs.

Remark 13.19. Our definition of indirect proof (Def. 13.16) guarantees that this is an algorithmic procedure. Indeed, the sole assumption or hypothesis permissible–i.e. $\neg\phi$–is fixed and predetermined.

As in the case of direct proofs, variations in the form given in Definition 13.16 are acceptable. In particular:

Remark 13.20. Steps $k+2$ to $n-1$ in Definition 13.16 may be constituted by formulas θ_{jl} denoting the j-th formula in the l-th branch. When the latter is the case, we speak of a proof as a *(labeled) tree*.

Example 13.21. Proofs in the analytic tableaux calculus have the form of a downwards-growing labeled binary tree (see Chapter 15).

▶ *Do Exercises 13.3-4.*

13.2. The complexity of *SAT*

The importance of *SAT* for classical logic, both in theoretical and practical terms, lies in its being a measure of the complexity of logical problems formulated as computational problems. Recall the discussion in Section 1.4 on decidability and tractability. The following result gives a first boundary for the *SAT* from the viewpoint of tractability.

13. The satisfiability problem, or SAT

Lemma 13.22. *SAT is a **NP** problem.*

Proof: (Sketch) Recall Definition 1.18. Let us consider a Boolean formula in the form

$$\phi = \bigwedge_i \left(\bigvee_j L_{ij} \right).$$

This is an abbreviation of a CNF formula (see Section 2.4.5), i.e. conjunctions of disjunctions of (negated) atoms. An example of a formula in this form is

$$\phi = (x_1 \vee x_3 \vee \neg x_4) \wedge \neg x_2 \wedge (\neg x_1 \vee x_2).$$

Because these are Boolean variables, it is easy to devise a deterministic algorithm: we take all the possible truth-value assignments of the variables $x_1, ..., x_n$ and evaluate ϕ for each of the assignments. As there are 2^n different assignments, this exhaustive algorithm is of exponential-time complexity, i.e. in the **EXPTIME** class. However, we can improve on this by means of a non-deterministic polynomial-time algorithm that guesses a truth-value assignment to ϕ and returns "accept" if the assignment satisfies ϕ. Hence, *SAT* is a **NP** problem. **QED**

This result having been established, it might in fact be very difficult, if not wholly impossible, to solve in practice a computational problem formulated as an instance of *SAT*, as this is an intractable problem in the **NPC** class, i.e. the class of the **NP**-complete problems. This is a fundamental result (cf. Theorem 13.23 below), as it gives us a measure of the *difficulty of solving* a large collection of computational problems: if *SAT* (or *3-SAT*, for that matter) is polynomial-time reducible to some decision problem, then we know that this is also a **NP**-complete problem. So far, thousands of computational problems have been shown to be in the **NPC** class via this reduction. But first we have to prove that *SAT* is **NP**-complete. This was originally done in Cook (1971).[4]

Theorem 13.23. *(Cook-Levin Theorem) SAT is **NP**-complete.*

Proof: Recall Definition 1.18: This dictates that the first step in proving that a computational problem is in the **NP**-complete class is to prove that it is a **NP** problem. This we did for Lemma 13.22. Let now L be any language in the **NP** class. We aim to show that

$$L \preceq_P SAT.$$

---
[4] Independently, and a little later on, also by L. Levin, which explains the common coinage of the theorem as Cook-Levin Theorem. We focus on Cook's proof.

13.2. The complexity of SAT

We consider a non-deterministic Turing machine M_T that decides L in polynomial time (i.e. time n^k for some constant k); so we have $L = L(M_T)$. We abbreviate M_T as M. The alphabet of L is $\Sigma = \{\neg, \wedge, \vee, x, 0, 1\}$ possibly with parentheses. For convenience, we shall write $\bar{x}$ instead of $\neg x$. Recall from Definition 1.16 that $L \preceq SAT$ means that $w \in L$ iff $f(w) \in SAT$. We aim to show that we can construct a function

$$f_L : \Sigma_L^* \longrightarrow CNF_L$$

such that (i) for every $w \in \Sigma^*$, w is accepted by M iff $f_L(w) = \phi \in CNF_L$, ϕ is a (Boolean) formula in CNF, is satisfiable, and (ii) the corresponding function $g_M : \Sigma^* \to \{\neg, \wedge, x, 1\}^*$,

$$g_M(w) = \phi \in L$$

is computable in polynomial time.

Part (i) is the fastidious one; (ii) is easily verifiable. The idea of the proof is based on the fact that there will be satisfying assignments of $f_L(w)$ iff there are accepting configurations of M on w, so that in fact the reduction of L to SAT is carried out via what we can call "computation histories." In order to achieve this objective we have to both describe the computations of M by Boolean variables and express accepting states of M on w by Boolean formulas.

But first of all, we need a *tableau* for M, i.e. a $n^k \times n^k$ table each row of which is one of the n^k configurations $C_i = Q \cup \Gamma \cup \{\#\}$ of a branch of the computation of M on w for $|w| = n$ (cf. Fig. 13.2.1).[5] Let C_1 be the start configuration. Let each cell (i, j) of the $(n^k)^2$ cells of the tableau have its contents represented by $|C|$ Boolean variables of ϕ

$$\{x_{i,j,\sigma} | \sigma \in C\}$$

indicating that cell (i, j) contains σ if $x_{i,j,\sigma} = 1$. Given $1 \leq i, j \leq n^k$, there will be $|C| \cdot (n^k)^2$ Boolean variables.

A tableau is said to be an *accepting tableau* if there is a row that is an accepting configuration $C_{p(|w|) = n^k}$, and $f_L(w)$ is satisfiable iff there is a "computation history" (a computation branch) of consecutive configurations of M

$$C_1, C_2, ..., C_{n^k}$$

where two configurations are said to be consecutive if we have $C_{i-1} \vdash_M C_i$ (or $C = C'$). In order to produce an accepting tableau for ϕ we have

---
[5]Not to be confused with a *tableau* as in the *tableaux calculus* of Chapter 15.

13. The satisfiability problem, or SAT

C_1	#	q_0	w_1	w_2	...	w_n	ϵ	...	ϵ	#
C_2	#	q_1								#
	#									#
C_{n^k}	#									#
	1	2								n^k

Figure 13.2.1.: A tableau for the Turing machine M.

to produce the corresponding CNF of ϕ

$$CNF(\phi) = \phi_{cell} \wedge \phi_{start} \wedge \phi_{move} \wedge \phi_{accept}.$$

The first thing to consider in this correspondence between the tableau and $CNF(\phi)$ is to make sure that for each assignment there must be exactly one symbol $\sigma \in C$ such that $x_{i,j,\sigma} = 1$. This is expressible in Boolean terms as follows:

$$\phi_{cell} = \bigwedge_{\substack{1 \leq i \\ j \leq n^k}} \left[\left(\bigvee_{\sigma \in C} x_{i,j,\sigma} \right) \wedge \left(\bigwedge_{\substack{\sigma, \tau \in C \\ \sigma \neq \tau}} (\overline{x}_{i,j,\sigma} \vee \overline{x}_{i,j,\tau}) \right) \right]$$

where $\left(\bigvee_{\sigma \in C} x_{i,j,\sigma} \right)$ means that at least for one of the variables it is the case that $x_{i,j,\sigma} = 1$ and $\bigwedge_{\sigma,\tau \in C; \sigma \neq \tau} (\overline{x}_{i,j,\sigma} \vee \overline{x}_{i,j,\tau})$ expresses the fact that for no more than one variable it is the case that $x_{i,j,\sigma} = 1$.

The formula ϕ_{start} guarantees that the first row of the tableau is the starting configuration. This it expresses as

$$\phi_{start} = x_{1,1,\#} \wedge x_{1,2,q_0} \wedge x_{1,3,w_1} \wedge x_{1,4,w_2} \wedge \ldots$$

13.2. The complexity of SAT

$$\wedge x_{1,n+2,w_n} \wedge x_{1,n+3,\epsilon} \wedge \ldots$$

$$\wedge x_{1,n^k-1,\epsilon} \wedge x_{1,n^k,\#}$$

We now require a formula that guarantees that there occurs an accepting configuration in the tableau. This is the formula

$$\phi_{accept} = \bigvee_{\substack{1 \leq i \\ j \leq n^k}} x_{i,j,q_a}$$

where q_a denotes the accepting state. Obviously, if for some cell (i,j) we have a variable x_{i,j,q_a} such that $x_{i,j,q_a} = 1$, then we have an accepting configuration.

We spoke above of a "computation history." This entails that in the sequence of consecutive configurations of M

$$C_1, C_2, \ldots, C_{n^k}$$

we have it that C_{i+1} follows *legally* from C_i. In terms of the tableau for M, this means that each 2×3 *window* of cells (see center of the tableau in Fig. 13.2.1) strictly follows–or does not violate–the transition function of M. In order to verify that each row of the tableau corresponds to a configuration that preserves this legality we have the formula

$$\phi_{move} = \bigwedge_{\substack{1 \leq i < n^k \\ 1 < j < n^k}} \mathcal{W}_{i,j}$$

where $\mathcal{W}_{i,j}$ denotes the valid 2×3 window with top-middle cell at (i,j) and

$$\mathcal{W}_{i,j} =$$

$$\bigvee_{(a_1,\ldots,a_6) \in Val} \left(x_{i,j',a_1} \wedge x_{i,j,a_2} \wedge x_{i,j'',a_3} \wedge x_{i+1,j',a_4} \wedge x_{i+1,j,a_5} \wedge x_{i+1,j'',a_6} \right)$$

where Val is the set of 6-tuples of valid assignments to the cells of $\mathcal{W}$, and $j' = j-1, j'' = j+1$.

This done, we now have to prove that each of the four formulas above can be expressed by a formula of size $\mathcal{O}(n^{2k})$ and can be constructed in polynomial time from w. That is, we have to show that the "computation history" above can be abbreviated as

$$C_1 \vdash^*_M C_{n^k}$$

13. The satisfiability problem, or SAT

for $n^k = p(|w|)$ for a given formula w. This is left as an exercise.

▶ *Do Exercises 13.5-13.*

13.3. Herbrand's Theorem and the SAT

As anticipated in Section 9.2 above, Herbrand semantics is of particular interest when testing for (un)satisfiability. Recall that the SAT is a decision problem defined in the framework of classical logic, and more specifically so in the context of computational problems. As our focus is on refutation proofs, we shall be more concerned with unsatisfiability. This sanctions our application of Herbrand semantics, as it is unsatisfiability-preserving for formulas in SNF.

In effect, by Proposition 4.40, Theorem 9.9 assures us of finite satisfiability (cf. Def. 4.69) for formulas of $L1_{ff}$. Although this holds only for this function-free fragment of L1, on which for instance deductive databases are founded, it is an important result in terms of satisfiability, as infinite satisfiability is of little practical use, if not altogether undesirable. Say we consider models as states of some database; clearly, infinite states simply cannot be stored, let alone manipulated. The importance of unsatisfiability for automated theorem proving resides precisely in this: if a given ground formula ϕ (a set of ground formulas X) is *not* unsatisfiable, then it can be infinitely satisfiable; but if it is unsatisfiable, then we can construct a finite proof–for instance, a closed semantic tree–of its unsatisfiability. This, provided we choose to work with Herbrand semantics.

But the employment of Herbrand semantics has further advantages, as we discuss below.

We begin with a very general result on unsatisfiability-preservation, introducing then Herbrand's Theorem and briefly discussing its impact in the context of the SAT. We then elaborate on versions of this theorem that gives us a semantical foundation for proof-theoretical based unsatisfiability testing.

Theorem 13.24. *Let C_ϕ be a set of clauses that represents a SNF of a FOL formula ϕ. Then ϕ is unsatisfiable iff C_ϕ is unsatisfiable.*

Proof: Left as an exercise.

This theorem requires that we specify when C is unsatisfiable. This was done for satisfiability by J. Herbrand (1930) in a theorem that can be rephrased for unsatisfiability in two equivalent versions for practical purposes. Herbrand's original theorem, in an already simplified formulation, runs as follows:

> Let Θ be a theory axiomatized by exclusively universal formulas. Suppose that $\Theta \models \forall x \exists y_1, ..., y_k \, (P(x,y))$ where $P(x,y)$ is a quantifier-free formula. Then, there is a finite sequence $t_{ij} = t_{ij}(x)$ of terms, $1 \leq i \leq r$ and $1 \leq j \leq k$, such that
>
> $$\Theta \vdash \forall x \left(\bigvee_{i=1}^{r} P(x, t_{i1}, ..., t_{ik}) \right).$$

In other words, a closed FOL formula ϕ is satisfiable iff all its sets of ground clauses are truth-functionally satisfiable. Reformulated in terms of unsatisfiability, this theorem states that a closed formula ϕ in SNF is unsatisfiable iff there is a finite number of clause instances of ϕ whose conjunction is unsatisfiable in terms of truth-functionality, i.e. propositionally.

In Section 9.2, we treated this "propositionalization" rather informally, but now, in light of Herbrand's Theorem, we can give it a more formal treatment.

Definition 13.25. Let C be a set of skolemized FO formulas. We define the *Herbrand expansion* of C as the set

$$HE_C = \{\phi\,[x_1/t_1, ..., x_n/t_n] \,|\, \forall x_1, ..., \forall x_n\,(\phi) \in C, t_i \in H_C\}.$$

Clearly, HE_C contains solely ground atoms, and thus C is amenable to a propositional interpretation via a transformation $\tau(C^g)$ of C (cf. Def. 4.73). Indeed, $HE(C)$ *just is* the ground extension $GE(C)$ of Section 4.2.2.1. Formally, this means that we have now a H-interpretation $HI_C = \tau(C^g)$. This guarantees, by Lemma 4.76, that C is finitely satisfiable iff $\tau(C^g)$ is finitely satisfiable.

But, as we know, HE_C only guarantees finite satisfiability in case of a function-free fragment of L1, as HE_C is finite iff $Fun(C) = \emptyset$. What about unsatisfiability? It so happens that if C is unsatisfiable, we can always find a *finite* subset of HE_C that is unsatisfiable, i.e. Herbrand semantics is propositional-compact with respect to unsatisfiability.

With this in mind, we can actually improve on the compactness results of Section 4.2.2.1 as follows:

Theorem 13.26. *(Compactness of propositional logic) A set X of propositional formulas is unsatisfiable iff at least one finite subset X' of X is unsatisfiable.*

13. The satisfiability problem, or SAT

Proof: Left as an exercise.

To the obvious question of how to find such a counter-model we have an easy answer: by the application of the rules of inference of a refutation calculus such as resolution or analytic tableaux (see Chapters 14 and 15).

Summing up, two fundamental results are obtained from Herbrand's Theorem: firstly, and as seen, it gives us a means to treat FOL formulas as propositional formulas, the advantage being that there are decision procedures for the latter; secondly, but equally importantly, it provides us with a proof-theoretical means to test for unsatisfiability, the advantage being that proofs are by definition finite objects.

We now elaborate on the above by considering the two aforementioned equivalent versions of Herbrand's Theorem. Version 1 is easily understandable, while version 2 requires the new notion of a semantic tree. However, we shall use this notion in the proof of version 1, reason why we give it only after giving the proof for version 2.

Theorem 13.27. *(Herbrand, 1930 - version 1) A set C of clauses is unsatisfiable iff there is a finite unsatisfiable set C' of ground instances of C.*

Proof: Given below.

Although not problem-free, Herbrand's results are fundamental in more than one way. Firstly, this theorem tells us that in order to verify whether C is unsatisfiable we need only focus on the H-interpretations, as a set of clauses C is unsatisfiable iff C is false under all H-interpretations, i.e. iff C has no Herbrand model $H\mathcal{M}_C$. This, in turn, means that we need only consider the Herbrand universe H_C (vs. all possible domains).

We can now provide the following central result:

Theorem 13.28. *A set C of clauses is unsatisfiable iff C is false under all H-interpretations.*

Proof: Obvious from the above. **QED**

Given n elements in $H(C)$, there will generally be 2^n H-interpretations. This, however, is where Herbrand's results might be problematic, as there are infinitely many H-interpretations for $n = \infty$. It so happens that the first requirement of an algorithm is that the number of steps that constitute it be finite. In order to organize all H-interpretations in a systematic way we can apply the notion of a semantic tree.

13.3. Herbrand's Theorem and the SAT

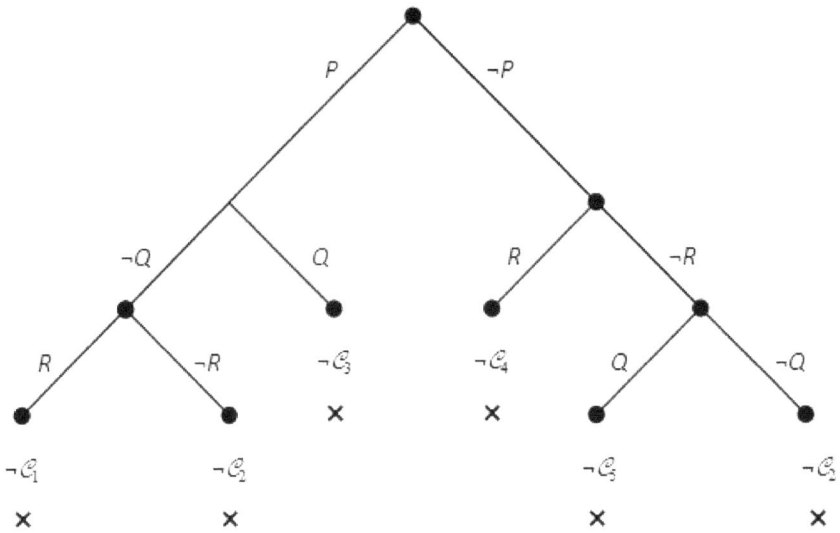

Figure 13.3.1.: Closed semantic tree of $C = \{C_1, C_2, C_3, C_4, C_5\}$ in Example 13.31.

Definition 13.29. A *semantic tree* $\mathcal{T}$ for a set C of clauses, denoted by $\mathcal{T}_C$, is a (downwards-growing) *labeled binary tree* in which each branch is attached with a finite set of (negations of) atoms from $H(C)$ in such a way that:

1. For each node N there are only finitely many immediate edges $L_1, ..., L_n$ from N. For $\mathcal{B}_i$ the conjunction of all the literals in the set attached to $L_i, i = 1, ..., n$, $\mathcal{B}_1 \vee \mathcal{B}_2 \vee ... \vee \mathcal{B}_n$ is a valid formula.

2. For each node N, let $I(N)$ be the union of all the sets attached to the edges of the branch of $\mathcal{T}_C$ down to and including N. Then $I(N)$ does not contain any complementary pair.

Definition 13.30. Let $H(C) = \{A_1, A_2, ..., A_k, ...\}$. A semantic tree $\mathcal{T}_C$ is *complete* iff, for every tip node (i.e. leaf) N, $I(N)$ contains either A_i or $\neg A_i$ for $i = 1, 2, ...$ N is a *failure node* if $I(N)$ falsifies some ground instance of some $\mathcal{C}$ in C, but $I(N')$ does not falsify any ground instance of some $\mathcal{C}$ in C for every ancestor node N' of N. A branch of a semantic tree $\mathcal{T}_C$ is *closed* iff it terminates at a failure node; otherwise, it is said to be *open*. A semantic tree $\mathcal{T}_C$ is closed iff each of its branches is closed.

13. The satisfiability problem, or SAT

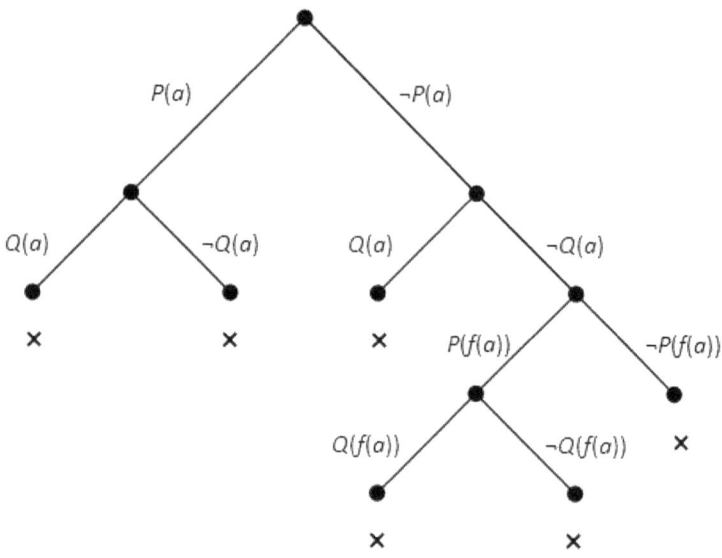

Figure 13.3.2.: A closed semantic tree.

If $I(N)$, which is in fact a partial interpretation for C (i.e., $I(N)$ can be seen as an assignment of truth values to ground atoms of $H(C)$), falsifies C, then we can stop expanding nodes from N. This means that if C is unsatisfiable, its semantic tree $\mathcal{T}_C$ cannot fail to be finite.

Example 13.31. Let $C_1 = Q \vee \neg R$, $C_2 = Q \vee R$, $C_3 = \neg P \vee \neg Q$, $C_4 = P \vee \neg R$, and $C_5 = P \vee \neg Q \vee R$. The atom set of $C = \{C_1, C_2, C_3, C_4, C_5\}$ is $A(C) = \{P, Q, R\}$. The above conditions 13.29.1-2 are satisfied for C. Figure 13.3.1 shows the closed semantic tree of C. A closed branch is marked with ×.

We now give an example in which a semantic tree is more clearly associated to the Herbrand basis $H(C)$ of a set of clauses C.

Example 13.32. Let $C = \{\neg P(x) \vee Q(x), P(f(a)), \neg Q(z)\}$. Then,

$H_C = \{a, f(a), f(f((a))), ...\}$
$H(C) = \{P(a), Q(a), P(f(a)), Q(f(a)), ...\}$.

C is unsatisfiable. Figure 13.3.2 shows the semantic tree for C. A failure node is marked with ×.

Theorem 13.33. *(Herbrand, 1930 - version 2)* A set C of clauses is unsatisfiable iff corresponding to every complete semantic tree of C, there is a finite closed semantic tree.

Proof: The proof follows immediately from the above:

($\Rightarrow$) Assume C is unsatisfiable and $\mathcal{T}_C$ a complete semantic tree for C. For every branch there is the set of labels $I(N)$, which is an interpretation, because the tree is complete. Hence, $I(N)$ falsifies some ground instance $\mathcal{C}'$ of a clause $\mathcal{C} \in C$. Since there are only finitely many literals in $\mathcal{C}'$, there must be a failure node N in a finite distance from the root. Since every branch terminates at a failure node, there is a corresponding closed semantic tree $\mathcal{T}'_C$ that is finite.

($\Leftarrow$) Assume that for every complete semantic tree $\mathcal{T}_C$ there is a corresponding finite closed tree $\mathcal{T}'_C$. Then, every branch terminates at a failure node and hence every interpretation $I(N)$ falsifies C. Hence, C is unsatisfiable. **QED**

Note how version 2 of Herbrand's Theorem can be applied in the context of refutation proofs: Obtaining $H(C)$ from the negation of some formula ϕ (or from a set of premises X and the negation of the conclusion ϕ), we prove the validity of C iff we can construct a closed semantic tree for C. As stated above, a proof of version 1 can be greatly simplified if we already can apply the important definition of semantic tree:

Proof: (Herbrand's Theorem, version 1). ($\Rightarrow$) Assume C is unsatisfiable and $\mathcal{T}_C$ is a complete semantic tree for C. Then, by Herbrand's Theorem, version 2, there is a finite closed semantic tree $\mathcal{T}'_C$ of C. Let C' be the set of ground instances of clauses that are falsified at all failure nodes of $\mathcal{T}'_C$: C' is finite and is falsified by every interpretation. It follows that C' is unsatisfiable.

($\Leftarrow$) The proof is by contraposition. **QED**

Remark 13.34. Version 1 of Herbrand's Theorem can be turned into a proof procedure by successively generating the sets $C'_0, C'_1, \ldots$ in which C'_i is the set of all the ground instances of clauses of C, and by testing them for unsatisfiability by means available to the propositional calculus: the theorem tells us that for some finite N there is a set C'_N that is unsatisfiable if C is unsatisfiable. Gilmore (1960) was the first to implement this idea with the *multiplication method*: as each C'_i is generated, it is multiplied out into a DNF; any conjunction in the DNF containing a complementary pair is then removed, and if some C'_i is found to be empty, a proof for its unsatisfiability has been found. However, this method is highly inefficient, as for, say, a set of ten two-literal ground clauses, there are 2^{10} conjunctions.

▶ *Do Exercises 13.14-16.*

13. The satisfiability problem, or SAT

Exercises

Exercise 13.1. Determine whether the following Boolean formulas are satisfiable by applying Boolean algebra as in Example 13.6.

1. $(x_1 \lor x_3 \lor \neg x_4) \land \neg x_2 \land (\neg x_1 \lor x_2)$
2. $(p_2 \lor \neg p_2 \lor p_3) \land p_1 \land (\neg p_1 \lor \neg p_3)$
3. $p \lor (q \land \neg r) \lor (\neg p \land \neg q \land r)$
4. $(x_2 \land \neg x_1) \lor (x_1 \lor (\neg x_2 \land x_3))$
5. $x_2 \land x_1 \land (x_3 \lor \neg x_4) \land \neg x_2$
6. $(p \lor \neg q) \land (\neg p \lor \neg q) \land (p \lor q) \land (\neg p \lor q)$

Exercise 13.2. Give some thought to the following statements with respect to the concept of satisfiability:

1. Interest in Satisfiability is expanding for a variety of reasons, not in the least because nowadays more problems are being solved faster by SAT solvers than other means. This is probably because Satisfiability stands at the crossroads of logic, graph theory, computer science, computer engineering, and operations research. Thus, many problems originating in one of these fields typically have multiple translations to Satisfiability and there exist many mathematical tools available to the SAT solver to assist in solving them with improved performance. (Franco & Martin, 2009)

2. For many years, SAT solvers were better at solving satisfiable instances than unsatisfiable ones. This is not true anymore. The success of SAT solvers can be largely attributed to their ability to learn from wrong assignments, to prune large search spaces quickly, and to focus first on the "important" variables, those variables that, once given the right value, simplify the problem immensely. All of these factors contribute to the fast solving of both satisfiable and unsatisfiable instances. (Kroening & Strichman, 2008)

Exercise 13.3. Comment on the following statements on indirect proofs and proofs by *reductio (ad absurdum)*:

1. Now consider, say, a four-premise argument, and assume we know three of its premises are true. If we derive a contradiction from that set of four premises, we have proved the fourth premise in

the set is false (since at least one member of the set is false, and we assume the other three are true). So we also have proved the negation of the fourth premise is true (since the negation of a false sentence is true). We have here the main ideas behind the rule of indirect proof. (Indirect proofs also are known as *reductio ad absurdum* proofs, because in an indirect proof an assumption is "reduced to absurdity" by showing that it implies a contradiction.) (Hausman, Kahane, & Tidman, 2010)

2. Classical logic contains the principle of indirect proof: If $\neg A$ leads to a contradiction, A can be inferred. Axiomatically expressed, this principle is contained in the law of double negation, $\neg\neg A \to A$. The law of excluded middle, $A \vee \neg A$, is a somewhat stronger way of expressing the same principle. ... More generally, the inference pattern, if something leads to a contradiction the contrary follows, is known as the principle of *reductio ad absurdum*. Dictionary definitions of this principle rarely make the distinction into a genuine indirect proof and a proof of a negative proposition: If A leads to a contradiction, then $\neg A$ can be inferred. Mathematical and even logical literature are full of examples in which the latter inference, a special case of a constructive proof of an implication, is confused with a genuine *reductio*. A typical example is the proof of irrationality of a real number x: Assume that x is rational, derive a contradiction, and conclude that x is irrational. The fallacy in claiming that this is an indirect proof stems from not realizing that to be an irrational number is a negative property: There do not exist integers n, m such that $x = n/m$. (Negri & von Plato, 2001; adapted notation)

3. The application of *reductio* argument is contrasted with purely mechanical brute algorithmic inferences as an art requiring skill and intelligent intervention in the choice of hypotheses and attribution of contradictions deduced to a particular assumption in a contradiction's derivation base within a *reductio* proof structure. (Jacquette, 2008)

4. It is a curiosity about *reductio* reasoning, and one rich with philosophical significance, especially for mathematical applications, that the contradiction obtaining in an indirect proof framework can be attributed to any of the assumptions on which the contradiction rests. Logicians sometimes speak of "blaming" the contradiction on a particular chosen assumption in the argument, thereby reducing it to an absurdity and supporting the deduction of its negation.

13. The satisfiability problem, or SAT

> Ordinarily, one expects that the contradiction in a *reductio* argument is to be blamed on the *reductio* hypothesis, but there is no imperative necessitating this choice. (Jacquette, 2008)

Exercise 13.4. Show that whatever can be proven by an indirect proof can be so by a slightly longer conditional proof.

Exercise 13.5. Complete the proof of Lemma 13.22 and do so by making a Turing machine feature in it.

Exercise 13.6. With respect to the proof above of Theorem 13.23:

1. Provide the details of the construction and verification of the set Val.

2. Prove, or research into the proof of, (ii).

Exercise 13.7. Research into, and give the details of, the proof that the *general SAT*, i.e. the SAT for Boolean formulas not necessarily in CNF, is **NP**-complete.

Exercise 13.8. It can be shown that $CIRCUIT\text{-}SAT \preceq_P SAT$ (cf. Fig. 2.3.4). Research into the CIRCUIT-SAT and

1. give the essentials of the straightforward proof that this is an **NP**-complete problem.

2. give the proof that $CIRCUIT\text{-}SAT \preceq_P SAT$.

Exercise 13.9. Research into, and give the details of, the proof that *3-SAT* is **NP**-complete by the reduction

$$SAT \preceq_P 3\text{-}SAT.$$

Exercise 13.10. Research into, and give the details of, the proofs of the following statements:

1. $2\text{-}SAT \in \mathbf{P}$.

2. $HORN\text{-}SAT \in \mathbf{P}$.

3. $DNF\text{-}SAT \in \mathbf{P}$.

4. $3\text{-}SAT$ is **NP**-complete.

5. $k\text{-}SAT$, $k \geq 3$, is **NP**-complete.

6. *QBF-SAT* is **PSPACE**-complete.

Exercise 13.11 If $DNF\text{-}SAT \in \mathbf{P}$, but *(CNF-)SAT* is **NP**-complete, why is the Boolean satisfiability problem not typically formulated in terms of DNFs?

Exercise 13.12. Research into the *maximum satisfiability problem*, abbreviated MAX-SAT:

1. Give a formal definition of this problem and its complexity class.

2. Is *MAX-SAT* a decision problem?

Exercise 13.13. The *DUAL-HORN-SAT* is the satisfiability problem for dual-Horn formulas. Show that $DUAL\text{-}HORN\text{-}SAT \in \mathbf{P}$.

Exercise 13.14. Prove Theorem 13.24.

Exercise 13.15. Determine the (un)satisfiability of the following sets of clauses by constructing the corresponding semantic trees:

1. $\{\|P, Q, \neg R\|, \|P, Q\|, \|P, \neg Q\|, \|\neg P\|, \|S, P, Q\|\}$

2. $\left\{ \begin{array}{l} \|R(a), S(a)\|, \|\neg R(a), S(a)\|, \\ \|R(a), \neg S(a)\|, \|\neg R(a), \neg S(a)\| \end{array} \right\}$

3. $\{\|P, \neg Q, R\|, \|Q, R\|, \|\neg P, R\|, \|Q, \neg R\|, \|\neg Q\|\}$

4. $\left\{ \begin{array}{l} \|P(a), \neg Q(f(a))\|, \|P(x), R(b)\|, \\ \|\neg Q(f(a)), R(b)\|, \|\neg P(g(a)), Q(y)\|, \\ \|Q(y), \neg R(z)\|, \|\neg P(g(a)), \neg R(z)\| \end{array} \right\}$

Exercise 13.16. Show that the set of unsatisfiable FO formulas is recursively enumerable under Herbrand semantics.

14. The resolution calculus

Resolution is arguably the most successful refutation calculus in the field of ATP with classical logic, being now implementable in a plethora of fully automated provers. Conceived in the 1960s, the resolution calculus is based on the inference rule of *binary resolution*, which tests for (un)satisfiability of a given set of clauses by deleting pairs of complementary literals occurring in two distinct clauses of the set. This is Algorithm 14.1. This algorithm is tree-based, in order to emphasize the relevance of Herbrand's Theorem in resolution proofs (cf. Section 13.3 above), but it is not necessary actually to construct a tree, as we shall see.

However, this is but the core of a calculus that in the course of time came to include many refinements. These refinements account for some specific changes in mostly Steps 2 and 3 in the general algorithm for resolution (Algorithm 14.1). Below, we discuss both the notion of refinement and some of the most important refinements.

Although making proofs "by hand" is a fundamental exercise to acquire a satisfactory knowledge of resolution and its refinements, reliable proofs of complex theories can only be obtained by means of an automatic prover. We give the basics of the workings of Prover9/Mace4 and give both examples and exercises for resolution with this fully automated prover. This is a two-component prover, with Prover9 producing a proof by refutation and Mace4 generating a counter-model on the same input; one can use only one of these, or both, especially when Prover9 appears not to stop on a given input. Prover9/Mace4 is a free software available at

https://www.cs.unm.edu/~mccune/prover9/download/

We give here the main aspects of both the resolution principle and some important refinements, and we also elaborate on *paramodulation*, the extension of the resolution calculus for CL$^=$. Chang & Lee (1973), which contains a large selection of exercises, is a classic reference for the resolution calculus; Fitting (1996) is also a reference book with many exercises. Readers seeking a comprehensive theoretical approach to this calculus can benefit from Leitsch (1997) and Bachmair & Ganzinger (2001).

14. The resolution calculus

Algorithm 14.1 Binary resolution.

- **Input:** A set of clauses C

- **Output:** An inverted binary tree $\mathcal{T}_C$ for C that constitutes a proof of the (un)satisfiability of C

Steps:

1. Input the (factorized) clauses $C_1, ..., C_n \in C$, for finite $n \geq 2$, as the LEAVES of an inverted binary tree.

2. Resolve a pair of clauses C_i, C_j for $1 \leq i \leq n, 1 \leq j \leq n$, and $i \neq j$, for some l, m such that $L_l \in C_i$ and $L_m \in C_j$ are complementary literals.

3. The derived clause C_{n+1}–called resolvent–is a NODE in the tree and is added to the search space of Step 2, so that now we have $C_1, ..., C_{n+1}$.

4. Repeat Steps 2 and 3 until (i) the empty clause is derived or (ii) there are no more pairs of clauses to be resolved, in which cases the ROOT of the tree is respectively (i) □ or (ii) some non-empty resolvent C_{n+r} for some r.

    a) If (i), then C is unsatisfiable.
    b) Otherwise, C is satisfiable.

14.1. The resolution principle

14.1.1. The resolution principle for propositional logic

The resolution principle is an extension of the one-literal rule first devised by Davis & Putnam (1960). Hence, a brief discussion of this inference rule is called for.

Definition 14.1. Given a set of clauses C, if there is a unit ground clause $\mathcal{C} = \|L\|$ (or $\mathcal{C} = \|\neg L\|$) in C we can obtain C' from C by (i) deleting all the ground clauses in C that contain L (or $\neg L$) and/or (ii) deleting $\neg L$ (respectively, L) from all the clauses in C. This inference rule is called *one-literal rule*.

Proposition 14.2. *Let C' be obtained by an application of the one-literal rule to a unit clause L in a set of clauses C. Then,*

1. *If C' is empty, then C is satisfiable.*

2. *If C' is not empty, we obtain a set C'' from C' by deleting $\neg L$ from C'. C'' is unsatisfiable iff C is.*

Proof: Left as an exercise. (Hint: See Example 14.3.)

Example 14.3. Let $C = \{\neg P \vee Q \vee \neg R, \neg P \vee \neg Q, P, R, U\}$. We have the set $C' = \{\neg P \vee Q \vee \neg R, \neg P \vee \neg Q, R, U\}$ by applying the one-literal rule for P, and then $C'' = \{Q \vee \neg R, \neg Q, R, U\}$. We have $C \equiv_{sat} C''$. We repeat the rule for R in C'' and obtain the set $C''' = \{Q, \neg Q, U\}$. We have $C \equiv_{sat} C'''$. As C''' contains the empty clause ($Q \wedge \neg Q = \square$; cf. Def. 7.3), C is unsatisfiable because C''' is unsatisfiable.

Note in this Example 14.3 that a set of clauses is unsatisfiable iff it contains two complementary unit clauses $\|L\|$ and $\|\neg L\|$; thus, given any set of clauses
$$C = \{\|L\|, \|\neg L\|, \mathcal{C}_1, ..., \mathcal{C}_n\}$$
we need not go through steps 1 and 2 of Proposition 14.2, as C is obviously unsatisfiable.

By extending the one-literal rule to *any* pair of clauses, we obtain the *resolution principle*.

Definition 14.4. For any two clauses $\mathcal{C}_1$ and $\mathcal{C}_2$ and two complementary literals $L_1 \in \mathcal{C}_1$ and $L_2 \in \mathcal{C}_2$, we delete L_1 and L_2 from $\mathcal{C}_1$ and $\mathcal{C}_2$,

14. The resolution calculus

respectively, i.e.

$$(\text{del}) \quad \underbrace{(C_1 - L_1)}_{C'_1} \cup \underbrace{(C_2 - L_2)}_{C'_2}$$

and construct the disjunction of the remaining clauses C'_1, C'_2. The constructed clause, $C'_1 \vee C'_2$, is called a *resolvent* of C_1 and C_2, and the literals L_1, L_2 are the *literals resolved upon*.

Theorem 14.5. *(Binary resolution for CPL) A resolvent $C = C'_1 \vee C'_2$ of two clauses $C_1 = C'_1 \vee L$ and $C_2 = C'_2 \vee \neg L$ is a logical consequence of $C_1 \wedge C_2$, i.e.*[1]

$$(\text{res}) \quad \frac{\begin{array}{c} C'_1 \vee L \\ C'_2 \vee \neg L \end{array}}{C'_1 \vee C'_2}.$$

Proof: Let $C_1 = L \vee C'_1$, $C_2 = \neg L \vee C'_2$, and $C = C'_1 \vee C'_2$. Supposing that C_1 and C_2 are both true in an interpretation $\mathcal{I}$, their resolvent C must also be true in $\mathcal{I}$. Obviously, either L or $\neg L$ is false in $\mathcal{I}$. Assume L is false in $\mathcal{I}$; then C_1 must not be a unit clause, otherwise it would be false in $\mathcal{I}$. Hence, C'_1 must be true in $\mathcal{I}$, and the resolvent $C'_1 \vee C'_2$ is true in $\mathcal{I}$. Assume $\neg L$ is false in $\mathcal{I}$ and proceed in the same way. Hence, $C'_1 \vee C'_2$ is true in $\mathcal{I}$. **QED**

Recall the definition of contradiction (Def. 4.36.2). Recall also that the empty clause is always false (Def. 7.3), thus equating with a contradiction. Given this, we have from Theorem 4.46 that a set of clauses is unsatisfiable iff we can derive from it the empty clause.

Definition 14.6. A *resolution deduction* of C from a set of clauses C, denoted by $C \vdash_{res} C$, is a finite sequence $C_1, C_2, ..., C_k$ of clauses such that each C_i is either a clause in C or a resolvent of clauses preceding C_i, and $C_k = C$. We call the deduction of the empty set $\square$ from C a *refutation*, or *proof* of C.[2]

We can represent a resolution deduction by means of a *deduction*, or *refutation, tree*. This, an upwards-growing tree, is essentially a reverted semantic tree.

---

[1] An alternative form for this rule, more reminding of **del**, is

$$\frac{C'_1 \vee L \quad C'_2 \vee \neg L}{C'_1 \vee C'_2}.$$

[2] Compare with Def. 13.16.

14.1. The resolution principle

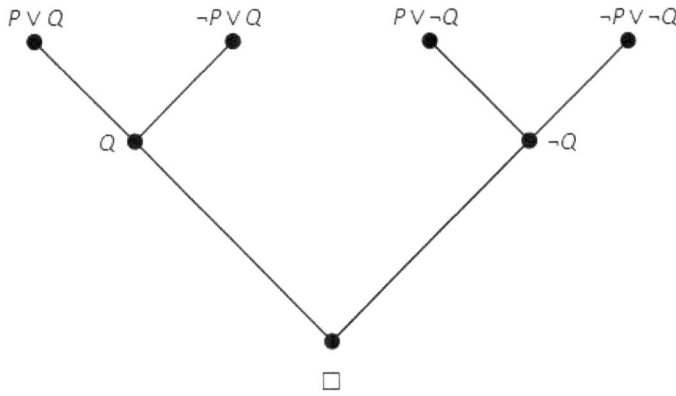

Figure 14.1.1.: A refutation tree.

Example 14.7. Let $C = \{P \vee Q, \neg P \vee Q, P \vee \neg Q, \neg P \vee \neg Q\}$. Figure 14.1.1 shows the refutation tree for C.

Example 14.8. Let there be given the formalized propositional argument of Example 12.13. We implement resolution in Prover9/Mace4. The input is entered as follows: on the Prover9 "Formulas" window, write the premises in the "Assumptions" section, and the conclusion in the "Goals" section.[3] Alternatively, because Prover9 is a refutation-based prover, simply enter the negated conclusion as another premise in the "Assumptions" section. Note in Figure 14.1.2 the notation used in Prover9/Mace4, and the fact that every formula must end with a full stop. In Figure 14.1.3., we give the output by Prover9.[4] An analysis of the output of Prover9, which is easily readable, shows the automatic implementation of the resolution principle: the prover deduces the empty clause (denoted by $F) from the set of propositions (premises and conclusion) that constitute the theory and which it clausifies.

Example 14.9. We now wish to test for the validity of a single formula. We can either enter the formula in the "Goals" section, or negate it in the "Assumptions" section of Prover9/Mace4. Figure 14.1.4 shows the proof

---

[3] Note that for Prover9/Mace4 *assumptions* are what we call *premises* (cf. Def. 10.6 and respective footnote).

[4] The first proof given is always the most abbreviated proof. In order to obtain the complete proof (which is also abbreviated) we must click on the button "Reformat..." (proof window, left upper corner); this will open a small window in which we must now click the Option "Expand Proof."

14. The resolution calculus

> **Assumptions:**
> (P|R)->-(Q&T).
> -S->Q.
> P&T.
>
> **Goals:**
> S.

Figure 14.1.2.: A propositional argument as input in Prover9/Mace4.

```
===================== PROOF =========================
% -------- Comments from original proof --------
% Proof 1 at 0.06 (+ 0.05) seconds.
% Length of proof is 11.
% Level of proof is 3.
% Maximum clause weight is 3.
% Given clauses 0.

1 P | R -> -(Q & T) # label(non_clause).  [assumption].
2 -S -> Q # label(non_clause).  [assumption].
3 P & T # label(non_clause).  [assumption].
4 S # label(non_clause) # label(goal).  [goal].
5 -P | -Q | -T. [clausify(1)].
7 S | Q. [clausify(2)].
8 P. [clausify(3)].
9 T. [clausify(3)].
10 -S. [deny(4)].
11A -Q | -T. [resolve(8,a,5,a)].
11 -Q. [resolve(9,a,11A,b)].
12A Q. [resolve(10,a,7,a)].
12 $F. [resolve(11,a,12A,a)].
===================== end of proof =================
```

Figure 14.1.3.: Output by Prover9: A valid propositional argument.

14.1. The resolution principle

by Prover 9 of the formula obtained from the argument of Example 14.7 by one application of the deduction theorem, i.e.

$$[((P \vee R) \to \neg(Q \wedge T)) \wedge (\neg S \to Q) \wedge (P \wedge T)] \to S.$$

Example 14.10. If we input the argument in Examples 2.22 and 2.24 in Prover9/Mace4, Prover9 does not stop. However, we obtain the counter-model from Mace4 in Figure 14.1.5. Note the usefulness of this component of Prover9/Mace4, as the construction of the truth table for this argument would require $2^9 = 512$ rows.[5]

▶ *Do Exercises 14.1-5.*

14.1.2. The resolution principle for FOL

We now consider the resolution principle for FOL. In order to do so we require the definitions of substitution and unification for FOL (cf. Section 2.4.6).

Definition 14.11. A *factor* of a clause C is a clause $C\sigma$, where σ is a MGU of some $C' \subseteq C$. We say that C' *is factorized to* $C\sigma$ and call this *(positive) factorization*. If $C\sigma$ is a unit clause, then it is called a *unit factor* of C.

Remark 14.12. Every clause is a factor of itself:

$$\frac{C}{C\sigma}.$$

Example 14.13. Given the clause $C = P(x) \vee \neg P(f(x)) \vee P(g(y))$ and the substitution $\sigma = \{x \mapsto g(y)\}$ such that $\sigma = mgu(C')$ for $C' = P(x) \vee P(g(y))$, C is factorized to $C\sigma = P(g(y)) \vee \neg P(f(g(y)))$.

Definition 14.14. Let C_1 and C_2 be two clauses (called *parent clauses*) with no variables in common. Let L_1 and L_2 be two complementary literals in C_1 and C_2, respectively. Then the clause

$$\left(C'_1 \vee C'_2\right)\sigma$$

---

[5] Recall that Prove9/Mace4 tests primarily for $\overline{SAT^{\neg}}$, and hence negates the conclusion in the case of an argument. Thus, the counter-model output by Mace4 is actually an interpretation such that for some formula ϕ we have $\neg\phi \notin \overline{SAT^{\neg}} \equiv \phi \in SAT$.

```
====================== PROOF =========================
% -------- Comments from original proof --------
% Proof 1 at 0.01 (+ 0.05) seconds.
% Length of proof is 8.
% Level of proof is 3.
% Maximum clause weight is 3.
% Given clauses 0.

1 (P | R -> -(Q & T)) & (-S -> Q) & P & T -> S
    # label(non_clause) # label(goal).  [goal].
2 -P | -Q | -T. [deny(1)].
4 S | Q. [deny(1)].
5 P. [deny(1)].
6 T. [deny(1)].
7 -S. [deny(1)].
8A -Q | -T. [resolve(5,a,2,a)].
8 -Q. [resolve(6,a,8A,b)].
9A Q. [resolve(7,a,4,a)].
9 $F. [resolve(8,a,9A,a)].
======================= end of proof ==================
```

Figure 14.1.4.: Output by Prover 9: A valid formula.

```
% number = 1 % seconds = 0
% Interpretation of size 2

A : 0
B : 1
C : 1
D : 0
E : 0
F : 0
K : 0
P : 0
R : 1
```

Figure 14.1.5.: Output by Mace4: A counter-model.

obtained from

$$(\text{del}_\sigma) \quad \underbrace{(C_1 - L_1)\sigma}_{C'_1\sigma} \cup \underbrace{(C_2 - L_2)\sigma}_{C'_2\sigma}$$

where σ is a MGU of L_1 and L_2, is called a *binary resolvent* of C_1 and C_2, and the literals L_1, L_2 are the literals resolved upon.

The above definitions allow us to define the various possible resolvents in a FOL resolution.

Definition 14.15. A *resolvent* of (parent) clauses C_1 and C_2 is one of the following binary resolvents:

1. a binary resolvent of C_1 and C_2;

2. a binary resolvent of C_1 and a factor of C_2;

3. a binary resolvent of a factor of C_1 and C_2;

4. a binary resolvent of a factor of C_1 and a factor of C_2.

Theorem 14.16. *(Binary resolution for CFOL). A resolvent* $C = C'_1 \vee C'_2$ *of two (parent) clauses* $C_1 = C'_1 \vee L_1$ *and* $C_2 = C'_2 \vee L_2$ *of FOL is a logical consequence of* C_1 *and* C_2, *if there is a substitution* σ *such that* σ *unifies the pair of complementary literals* L_1 *and* L_2, *i.e.*,

$$(\text{res}_\sigma) \quad \frac{C'_1 \vee L \qquad C'_2 \vee \neg L}{(C'_1 \vee C'_2)\sigma}.$$

Proof: The proof is left as an exercise.

Example 14.17. Let $C_1 = P(x) \vee Q(x)$ and $C_2 = \neg P(z) \vee R(x) \vee \neg P(a)$. In order to apply binary resolution to this pair of clauses we first must rename the variable x in C_2; renaming x as y will do. We next factor $\neg P(z)$ and $\neg P(a)$ in C_2, namely by applying the substitution set $\sigma = \{z \mapsto a\}$. We now have $C_1 = P(x) \vee Q(x)$ and $C_2 = \neg P(a) \vee R(y)$. Given the MGU $\theta = \{x \mapsto a\}$, we then have the binary resolvent $Q(a) \vee R(y)$. Obviously, the set $C = (C_1, C_2)$ is satisfiable. Figure 14.1.6 shows the refutation failure in a deduction tree.

319

14. The resolution calculus

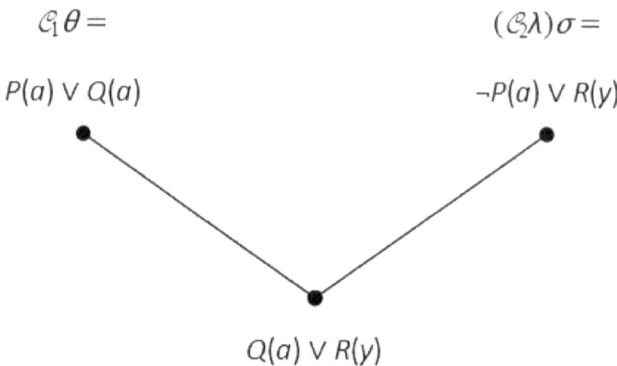

Figure 14.1.6.: A resolution refutation-failure tree.

We now show the usefulness of the automatic prover Prover9/Mace4 for resolution proofs in CFOL.

Example 14.18. We show how Prover9/Mace4 implements resolution in theorem proving in FOL by beginning with a basic geometric property (cf. Example 2.25). We introduce the input in Prover9/Mace4 as in Figure 14.1.7. The proof is given in Figure 14.1.8.

Assumptions:

% Definition of a trapezoid
all x all y all u all v (T(x,y,u,v)->P(x,y,u,v)).
% Alternate interior angles of parallel lines are equal
all x all y all u all v (P(x,y,u,v)->E(x,y,v,u,v,y)).
% Trapezoid in consideration
T(a,b,c,d).

Goals:

% Alternate interior angles formed by a diagonal of a trapezoid are equal
E(a,b,d,c,d,b).

Figure 14.1.7.: Input in Prover9/Mace4: A FO theory.

Example 14.19. Let the following argument formalized over L1 be given:

14.1. The resolution principle

```
===================== PROOF =========================
% -------- Comments from original proof --------
% Proof 1 at 0.00 (+ 0.06) seconds.
% Length of proof is 10.
% Level of proof is 4.
% Maximum clause weight is 0.
% Given clauses 0.


1 (all x all y all z all u (T(x,y,z,u) -> P(x,y,z,u)))
    # label(non_clause).  [assumption].
2 (all x all y all z all u (P(x,y,z,u) -> E(x,y,u,z,u,y)))
    # label(non_clause).  [assumption].
3 E(a,b,d,c,d,b) # label(non_clause) # label(goal).
    [goal].
4 T(a,b,c,d).   [assumption].
5 -T(x,y,z,u) | P(x,y,z,u).  [clausify(1)].
6 P(a,b,c,d).   [resolve(4,a,5,a)].
7 -P(x,y,z,u) | E(x,y,u,z,u,y).  [clausify(2)].
8 E(a,b,d,c,d,b).  [resolve(6,a,7,a)].
9 -E(a,b,d,c,d,b).  [deny(3)].
10 $F.  [resolve(8,a,9,a)].
====================== end of proof ==================
```

Figure 14.1.8.: Output by Prover9.

$$\begin{array}{rl} 1. & \forall x \left(F\left(x\right) \to \exists y \left(G\left(y\right) \wedge H\left(x,y\right)\right) \wedge \exists y \left(G\left(y\right) \wedge \neg H\left(x,y\right)\right)\right) \\ 2. & \underline{\exists x \left(J\left(x\right) \wedge \forall y \left(G\left(y\right) \to H\left(x,y\right)\right)\right)} \\ 3. & \exists x \left(J\left(x\right) \wedge \neg F\left(x\right)\right) \end{array}$$

We want to know whether the argument is valid. The repetition of atoms in premise 1 makes this a somehow complex theory to prove "by hand," reason why Prover9/Mace4 comes in handy. In merely 14 steps, and by applying binary resolution alone, we have a proof of the validity of the given argument (Fig. 14.1.9).

The two rules of inference above, binary resolution and (positive) factoring, describe in an essential way the resolution calculus. In order to prove the completeness of the resolution principle we require also a theorem known as the lifting lemma:

Theorem 14.20. *(Lifting lemma) Let $\mathcal{C}'_1, \mathcal{C}'_2$ be instances of $\mathcal{C}_1$ and $\mathcal{C}_2$,*

321

```
===================== PROOF =========================
% -------- Comments from original proof --------
% Proof 1 at 0.00 (+ 0.06) seconds.
% Length of proof is 14.
% Level of proof is 5.
% Maximum clause weight is 0.
% Given clauses 0.

1 (all x (F(x) -> (exists y (G(y) & H(x,y))) & (exists y
(G(y) & -H(x,y)))))
    # label(non_clause).  [assumption].
2 (exists x (J(x) & (all y (G(y) -> H(x,y))))) # la-
bel(non_clause).  [assumption].
3 (exists x (J(x) & -F(x))) # label(non_clause) # la-
bel(goal).
[goal].
4 -J(x) | F(x).  [deny(3)].
7 -F(x) | G(f2(x)).  [clausify(1)].
8 -F(x) | -H(x,f2(x)).  [clausify(1)].
10 J(c1).  [clausify(2)].
12 -J(x) | G(f2(x)).  [resolve(4,b,7,a)].
13 -J(x) | -H(x,f2(x)).  [resolve(4,b,8,a)].
15 -G(x) | H(c1,x).  [clausify(2)].
16 G(f2(c1)).  [resolve(12,a,10,a)].
17 -H(c1,f2(c1)).  [resolve(13,a,10,a)].
19 H(c1,f2(c1)).  [resolve(16,a,15,a)].
20 $F.  [resolve(19,a,17,a)].
===================== end of proof =================
```

Figure 14.1.9.: Output of Prover9: A valid FO argument.

respectively. If C' is a resolvent of C'_1 and C'_2, then there exists a resolvent C of C_1 and C_2 such that C' is an instance of C.

Proof: (Sketch) Let

$$C' = \left(C'_1\gamma - L'_1\gamma\right) \cup \left(C'_2\gamma - L'_2\gamma\right)$$

where γ is a MGU of L'_1 and L'_2. We need to show that C' is an instance of C,

$$C = ((C_1\lambda)\sigma - L_1\sigma) \cup ((C_2\lambda)\sigma - L_2\sigma)$$

where σ is a MGU of L_1 and L_2 and λ_i is a MGU for $\{L_i^1, ..., L_i^{r_i}\}$ with $L_i = L_i^1 \lambda_i, i = 1, 2$, if $r_i > 1$; if $r_i = 1$, then let $\lambda_i = \epsilon$ and $L_i = L_i^1 \lambda_i$. In order to do so it suffices to show that there is a substitution θ such that $C'_1 = C_1\theta$, $C'_2 = C_2\theta$ and $(\lambda \circ \sigma) \leq_s (\theta \circ \gamma)$. **QED**

Theorem 14.21. *(Completeness of the resolution principle) A set C of clauses is unsatisfiable iff there is a resolution deduction of the empty clause $\square$ from C, written*

$$C \vdash_{res} \square.$$

Proof: (Sketch) ($\Rightarrow$) Given that the empty clause can be deduced from any unsatisfiable set of clauses, the search for one proceeds by saturating the clause set, i.e. by systematically and exhaustively applying the inference rules until the empty clause is derived. In terms of a semantic tree, this means that the tree for a set C consisting only of the root node is generated, after, by Theorem 14.20, a process of obtaining subsequently smaller trees $\mathcal{T}'_C, \mathcal{T}''_C, ...$ for $C \cup \{\mathcal{C}\}$, where $\mathcal{C}$ is a resolvent of clauses C_1 and C_2. The root node is generated only when $\square$ is derived. Therefore, there is a deduction of $\square$ from C.
($\Leftarrow$) Suppose there is a deduction of $\square$ from C. Let $\mathcal{D}_1, ..., \mathcal{D}_k$ be the resultants in the deduction. Assume C is satisfiable. Then there is a model $\mathcal{M}$ of C. By Theorems 14.5 and 14.16, $\mathcal{M}$ satisfies $\mathcal{D}_1, ..., \mathcal{D}_k$. But this is impossible, because one of these resolvents is $\square$. Therefore, C must be unsatisfiable. **QED**

Definition 14.22. Let now Ψ be a complete resolution prover. Applying Ψ to a set of clauses C will produce *one* of the following results:

1. Derivation of $\square$, i.e. $\square \in \Psi(C)$. Obviously, C is unsatisfiable.

2. $\Psi(C)$ is finite (i.e. given C, Ψ is verified to terminate) and does not contain a refutation of C. Given that Ψ is complete, we know that C is satisfiable.

3. $\Psi(C)$ is infinite and does not contain a refutation of C. Given C, it is verified that Ψ does not terminate. Given that Ψ is complete, C must be satisfiable.

▶ *Do Exercises 14.6-13.*

14.2. Resolution refinements

Although binary resolution is indeed a very efficient proof procedure, it often produces redundant resolvents. In order to minimize this problem we apply resolution refinements. We next give some of the most important aspects of some refinements. We begin with a more extensive discussion of a specific refinement, to wit, semantic resolution, and then proceed with briefer treatments of further refinements, leaving many of their aspects as exercises.

The following will be our core definition of a resolution refinement.

Definition 14.23. Let Res_{rf} be a mapping from the set $\mathscr{C}$ of all finite sets of clauses to the set Res of all resolution deductions (i.e. Res-deductions). Let further res denote a ground resolution. We say that Res_{rf} is a *resolution refinement* iff for every set of clauses $C \in \mathscr{C}$,

1. we have $Res_{rf}(C) \subseteq Res(C)$;
2. the set $\{res | res \in Res_{rf}(C)\}$ is decidable;
3. there is an algorithm Ψ that constructs $Res_{rf}(C)$;
4. if $C_1 \subseteq C_2$, then we have $Res_{rf}(C_1) \subseteq Res_{rf}(C_2)$.

Definition 14.24. A resolution refinement Res_{rf} is said to be *complete* if, given an arbitrary unsatisfiable set of clauses C, $Res_{rf}(C)$ contains a refutation of C.

Recall Definition 14.22.

Definition 14.25. Let Res_{rf} be a complete resolution refinement. Given a set of clauses C as input, there are three possible outputs:

1. $Res_{rf}(C)$ terminates with $\square$, i.e. $\square \in Res_{rf}(C)$. Obviously, C is unsatisfiable.

2. $Res_{rf}(C)$ terminates, but does not produce $\square$. Given that Res_{rf} is complete, we know that C is satisfiable.

3. $Res_{rf}(C)$ does not terminate. $Res_{rf}(C)$ is infinite and $\square \notin Res_{rf}(C)$. Then C must be satisfiable, but $Res_{rf}(C)$ does not allow us to detect this property.

In practical terms, the production of irrelevant resolvents can be blocked by dividing a set of clauses C into two sets C_1 and C_2 such that resolution can only be carried out between a clause of C_1 and a clause of C_2. Alternatively, or together with this, the same objective can be obtained by ordering the predicate symbols of a set of clauses C.

14.2.1. Semantic resolution

Semantic resolution is a particularly interesting resolution refinement, because it joins the two methods mentioned in the paragraph immediately above. It owes its coinage to the fact that the very first step is the setting of an interpretation $\mathcal{I}$. We begin by discussing this aspect and then move on to explain how predicate-symbol ordering optimizes this refinement.

Example 14.26. Let there be given the clauses $C_1 = P \vee Q$, $C_2 = R \vee P$, $C_3 = P \vee \neg R \vee \neg Q$, and $C_4 = \neg P$ such that $C = \{C_1, C_2, C_3, C_4\}$. Let now $\mathcal{I} = \{\neg P, \neg Q, \neg R\}$. Then, C_3 and C_4 are satisfied by $\mathcal{I}$, while C_1 and C_2 are falsified by $\mathcal{I}$. We accordingly divide C into $\mathcal{C}_1 = \{C_1, C_2\}$ and $\mathcal{C}_2 = \{C_3, C_4\}$, and stipulate that resolution can only be carried out between these two sets of clauses. One possible resolution proof runs as shown:

Clause		Set	$\text{Res}(\mathcal{C}_1, \mathcal{C}_2)$
C_1	$P \vee Q$	$\mathcal{C}_1$	—
C_2	$R \vee P$	$\mathcal{C}_1$	—
C_3	$P \vee \neg R \vee \neg Q$	$\mathcal{C}_2$	—
C_4	$\neg P$	$\mathcal{C}_2$	—
C_5	Q	$\mathcal{C}_1$	(C_1, C_4)
C_6	R	$\mathcal{C}_1$	(C_2, C_4)
C_7	$P \vee \neg Q$	$\mathcal{C}_2$	(C_2, C_3)
C_8	P	$\mathcal{C}_1$	(C_5, C_7)
C_9	$\square$	—	(C_8, C_4)

Note that every new resolvent is integrated in either of the sets $\mathcal{C}_1, \mathcal{C}_2$. The column with these sets helps us to keep track of the resolution restriction with respect to these two sets.

14. The resolution calculus

We formalize this procedure for FOL, where, of course, atoms are predicates. For convenience, we work with 0-ary predicates, but recall that Herbrand semantics allows us to treat FO formulas as propositional formulas, namely in what we can call ground resolution.

Definition 14.27. Given a set of clauses C and the set $Pred(C)$ of predicates of C, let $\mathcal{I} = (\mathcal{D}, \Theta, \delta)$ be an interpretation such that for every m-place predicate $P(x_1, ..., x_m) \in Pred(C)$ we have

$$\Theta(P)(val_\mathcal{I}(a_1), ..., val_\mathcal{I}(a_m)) = \mathtt{f}$$

for every $a_1, ..., a_m \in \mathcal{D}$. Then, all the positive clauses are false in $\mathcal{I}$ and the remaining clauses are true in $\mathcal{I}$.

We can now define semantic resolution as follows:

Definition 14.28. Let C be a set of clauses and $\mathcal{I}$ an interpretation for C. Let $\mathcal{C}_1$ and $\mathcal{C}_2$ be clauses in the signature of C such that at least one of $\{\mathcal{C}_1, \mathcal{C}_2\}$ is false in $\mathcal{I}$. Then, we say that all the resolvents of $\mathcal{C}_1$ and $\mathcal{C}_2$ are *I-resolvents*.

More specifically, we shall consider the set $C = (E \cup N)$ where, given an interpretation $\mathcal{I}$ for a set of clauses, N is the set of true clauses in $\mathcal{I}$ and E the set of false clauses in $\mathcal{I}$. This allows us to select some clause $\mathcal{N} \in N$, called the *nucleus*, and a set of $\mathcal{E}_i \in E$ *electrons* (or *satellites*) between which alone resolution can be carried out.

Definition 14.29. Given a set of clauses C, let $\mathcal{I}$ be an interpretation for C. A finite set of clauses of C constitutes an *I-clash* $\Xi_\mathcal{I} = (\mathcal{N}; \mathcal{E}_1, ..., \mathcal{E}_n)$, written $\Xi_\mathcal{I}(C)$, iff

1. $\mathcal{E}_1, ..., \mathcal{E}_n$ are false in $\mathcal{I}$;

2. Let $\mathcal{C}_1 = \mathcal{N}$. For every $i = 1, ..., n$ there is a resolvent $\mathcal{C}_{i+1}$ of $\mathcal{C}_i$ and $\mathcal{E}_i$;

3. $\mathcal{C}_{n+1}$ is false in $\mathcal{I}$.[6]

$\mathcal{C}_{n+1}$ is called the *I-resolvent* of $\Xi_\mathcal{I}(C) = (\mathcal{N}; \mathcal{E}_1, ..., \mathcal{E}_n)$.

While this proof procedure is complete and contributes to a reduction of redundant resolvents, it alone does not suffice for an optimal reduction. Additionally, we can stipulate an ordering of the predicate symbols of a set of clauses.

---
[6] Recall that the empty clause is always false in any interpretation.

14.2. Resolution refinements

Example 14.30. Let there be given the set of clauses of Example 14.26. We stipulate the ordering of the predicates of C

$$R > Q > P.$$

We further stipulate that in the resolution of a clause $\mathcal{E}$ with a nucleus, the resolved-upon literal from $\mathcal{E}$ contain the largest predicate symbol in that clause. This stipulation blocks the resolution of $\mathcal{C}_4 = \mathcal{N}$ with either possible electron $\mathcal{C}_1, \mathcal{C}_2$, as in neither is P the largest predicate symbol.

As said above, it is the composition of both an interpretation and an ordering of predicates that makes semantic resolution an optimal refinement of resolution. We redefine the above Definition 14.29, now taking predicate ordering into consideration.

Definition 14.31. Given a set of clauses C, let $\mathcal{I}$ be an interpretation and $\mathscr{P}$ an ordering of $P, Q, ... \in Pred(C)$. A finite set of clauses of C constitutes a *PI-clash* $\Xi_{\mathscr{P}\mathcal{I}} = (\mathcal{N}; \mathcal{E}_1, ..., \mathcal{E}_n)$ iff

1. $\mathcal{E}_1, ..., \mathcal{E}_n$ are false in $\mathcal{I}$;

2. Let $\mathcal{C}_1 = \mathcal{N}$. For every $i = 1, ..., n$ there is a resolvent $\mathcal{C}_{i+1}$ of $\mathcal{C}_i$ and $\mathcal{E}_i$;

3. The resolved-upon literal of $\mathcal{E}_i$ contains the highest predicate symbol in $\mathcal{E}_i$, $i = 1, ..., n$;

4. $\mathcal{C}_{n+1}$ is false in $\mathcal{I}$.

$\mathcal{C}_{n+1}$ is called the *PI-resolvent* of $\Xi_{\mathscr{P}\mathcal{I}}(C) = (\mathcal{N}; \mathcal{E}_1, ..., \mathcal{E}_n)$.

Example 14.32. Given the data of Examples 14.26 and 14.30, we can proceed as shown below in order to obtain a PI-refutation of C. Figure 14.2.1 shows the corresponding deduction tree.

Clause		Type	Ordering	Res($\mathcal{N}, \mathcal{E}$)
C_1	$P \vee Q$	$\mathcal{E}$	$Q > P$	–
C_2	$R \vee P$	$\mathcal{E}$	$R > P$	–
C_3	$P \vee \neg R \vee \neg Q$	$\mathcal{N}$	$R > Q > P$	–
C_4	$\neg P$	$\mathcal{N}$	P	–
C_5	$P \vee \neg R$	$\mathcal{N}$	$R > P$	(C_3, C_1)
C_6	P	$\mathcal{E}$	P	(C_5, C_2)
C_7	$\square$	–	–	(C_4, C_6)

14. The resolution calculus

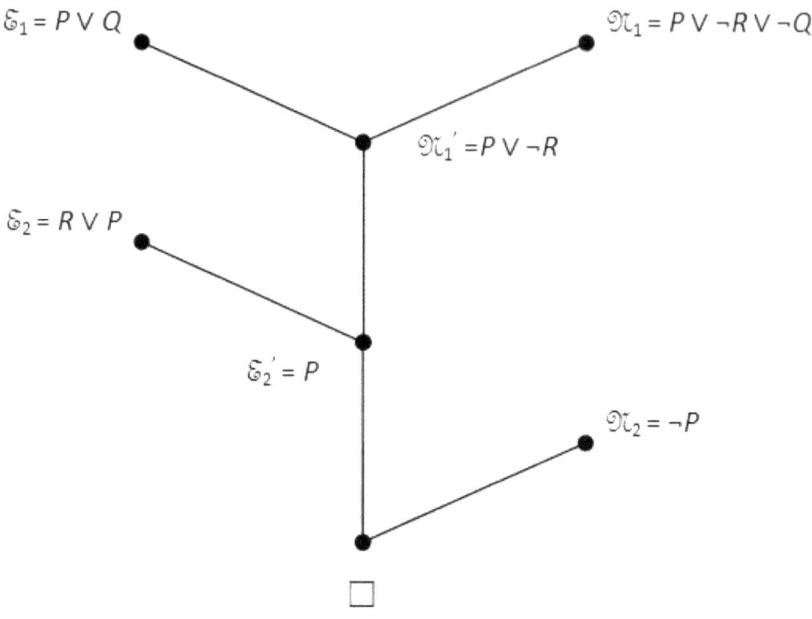

Figure 14.2.1.: A PI-resolution tree.

$\Xi_{\mathscr{PI}}(C) = (C_3; C_1, C_2)$ is indeed a PI-clash of C, as it satisfies all the conditions in Definition 14.31. Note that P, the resolvent of this PI-clash, is false in $\mathcal{I}$. Note also that neither of $(C_3; C_1)$ or $(C_3; C_2)$ alone is a PI-clash of C, as the respective resolvents are not PI-resolvents, i.e. they are not false in $\mathcal{I}$. Now, it is important to keep track of whether the (resolvent) clauses are nuclei or electrons, and an additional column with the stipulated ordering is useful to fulfill the ordering restrictions. Compare the abbreviated proof with that of Example 14.26.

Definition 14.33. Given a set of clauses C, let $\mathcal{I}$ be an interpretation for it and let $\mathscr{P}$ be an ordering of its predicate symbols. We say that a deduction from C is a *PI-deduction* iff each clause in the deduction is either a clause in C or a PI-resolvent.

Semantic resolution is a complete proof procedure. Because by applying any interpretation $\mathcal{I}$ and any ordering $\mathscr{P}$ we can always obtain a PI-deduction of the empty clause from an unsatisfiable set of ground clauses, we prove the completeness of semantic resolution via a proof of PI-resolution.

Theorem 14.34. *(Completeness of PI-resolution) If $\mathscr{P}$ is an ordering of predicate symbols in an unsatisfiable finite set C of clauses and if $\mathcal{I}$*

14.2. Resolution refinements

is an interpretation on C, then there is a PI-deduction of $\square$ from C, written
$$C \vdash_{PIres} \square.$$

Proof: (Chang & Lee, 1973)[7] The proof is by induction on the number of atoms of C. Let C be an unsatisfiable set of ground clauses. Let $At(C) = \{P\}$. Then, $C = \{P, \neg P\} = C'$. Clearly, the resolvent from C' is $\square$, regardless of the interpretation $\mathcal{I}$ (i.e. $\mathcal{I} = \{P\}$ or $\mathcal{I} = \{\neg P\}$). Therefore, either P or $\neg P$ is false in $\mathcal{I}$ and $\square$ is equally false in $\mathcal{I}$, and we have it that $\square$ is a PI-resolvent.

We thus showed that the theorem holds for $n = 1$. Assume now that the theorem holds for $|At(C)| = i$, $1 \leq i \leq n$. In order to complete the induction, we shall consider that $|At(C)| = n+1$. We start by searching for a unit clause $\mathcal{C} = L$ that is false in $\mathcal{I}$.

(i) C contains a unit clause $\|L\|$ that is false in $\mathcal{I}$. Then, by deleting in C the clauses containing L and by deleting $\neg L$ from the remaining clauses of C, we obtain C'. By the one-literal rule (cf. Def. 14.1), C' is unsatisfiable. C' contains n or fewer than n atoms, so by the induction hypothesis there is a PI-deduction of $\square$ from C'. Let us denote this PI-deduction by D'. We can obtain from C a PI-deduction of $\square$ from D': it suffices to replace every clash sequence $\left(\mathcal{N}'; \mathcal{E}'_1, ..., \mathcal{E}'_q\right)$, in which $\mathcal{N}', \mathcal{E}'_1, ..., \mathcal{E}'_q$ are clauses connected to the initial nodes of D' and $\mathcal{N}'$ was obtained from $\mathcal{N}$ by deleting $\neg L$, by the PI-clash sequence $\left(\mathcal{N}, L; \mathcal{E}'_1, ... \mathcal{E}'_q\right)$. If $\mathcal{E}'_i$ was obtained from $\mathcal{E}_i$ by deleting $\neg L$ in it, we add the PI-clash $(L; \mathcal{E}_i)$ above the node of $\mathcal{E}'_i$. In this way, we obtain a PI-deduction of $\square$ from C.

(ii) C does not contain a unit clause $\|L\|$ that is false in $\mathcal{I}$. Then, we can obtain a PI-deduction D' of $\square$ from C', in which C' is obtained by applying the one-literal rule to a literal L that is the symbol of the lowest predicate in some set $\{B, \neg B\} \subseteq At(C)$ and is false in $\mathcal{I}$. Clearly, C' is unsatisfiable. C' contains n or fewer than n atoms, so by the induction hypothesis there is a PI-deduction D' of $\square$ from C'. Replace now again literal L in the clauses from which it was firstly removed and denote by D_1 the deduction obtained from D' by means of this operation: D_1 remains a PI-deduction, given that L contains the lowest predicate symbol and is false in $\mathcal{I}$. It is evident that either $D_1 = \square$ or $D_1 = L$. In the first case, the proof is finished. In the second case, by (i) we obtain

---
[7]Actually, this is a proof of a lemma in Chang & Lee (1973), in which the statement of Theorem 14.34 is both a lemma and a theorem. We think that this proof suffices to prove the completeness of PI-resolution for CFOL, namely with both Herbrand's Theorem and the lifting lemma as a theoretical support.

14. The resolution calculus

a PI-deduction $D_2 = \square$ from $C \cup \{L\}$ in which L is a unit clause and false in $\mathcal{I}$. By the combination of D_1 and D_2 we obtain a PI-deduction of $\square$ from C. **QED**

Herbrand's Theorem (version 1; Theorem 13.27) and the lifting lemma (Theorem 14.20) assure us that this completeness result of PI-resolution holds for any FO unsatisfiable set of clauses, i.e. a set of non-ground clauses.

If we stipulate an interpretation in which every literal is (not) the negation of an atom, then every electron and every PI-resolvent contains only positive (negative, respectively) atoms. With this strategy we obtain *hyper-resolution*, a specific case of semantic resolution. Besides the restriction concerning the signs of the clash sequence (see Def. 14.31), hyper-resolution imposes other restrictions on the space of the selection of the clauses to resolve, blocking in this way the production of many redundant clauses and their addition to the search space.

Definition 14.35. Let $\mathcal{C}$ be any non-positive clause and let $\mathcal{D}_1, ..., \mathcal{D}_n$ be positive clauses, $\mathcal{C}, \mathcal{D}_1, ..., \mathcal{D}_n \in C$. Then, $\Xi = (\mathcal{C}; \mathcal{D}_1, ..., \mathcal{D}_n)$ is a *clash sequence* in which $\mathcal{C}$ is the nucleus and $\mathcal{D}_1, ..., \mathcal{D}_n$ are the electrons (or satellites). Let $\mathcal{C}_0 = \mathcal{C}$ and $\mathcal{C}_{i+1} \in Res(\{\mathcal{C}_i, \mathcal{D}_{i+1}\})$ for $i = 1, ..., n-1$. If $\mathcal{C}_n$ is defined and positive, then we say that $\mathcal{C}_n$ is a *hyper-resolvent of* Ξ.

Example 14.36. Let $C = \{\mathcal{C}_1, \mathcal{C}_2, \mathcal{C}_3, \mathcal{C}_4\}$, $\mathcal{C}_1 = P(a, b)$, $\mathcal{C}_2 = P(b, a)$, $\mathcal{C}_3 = \neg P(x, y) \vee \neg P(y, z) \vee P(x, z)$, and $\mathcal{C}_4 = \neg P(a, a)$. In the following resolution refutation (Fig. 14.2.2), one of the resolving clauses is always positive. In effect, it is a resolution refutation of $\Xi = (\mathcal{C}_3; \mathcal{C}_1, \mathcal{C}_2)$, in which $\mathcal{C}_1$ and $\mathcal{C}_2$ are the electrons. We say that $\mathcal{C}_6 = P(a, a)$ is a hyper-resolvent of $\Xi = (\mathcal{C}_3; \mathcal{C}_1, \mathcal{C}_2)$ insofar as we can say that $\mathcal{C}_5 = \neg P(x, b) \vee P(x, a)$ is an intermediate result with respect to $\mathcal{C}_6$. Note that in $\mathcal{C}_5$ the negative literal belongs to the nucleus and the positive literal is in fact the electron $(\mathcal{C}_1)\lambda$ (or $(\mathcal{C}_2)\lambda$) for $\lambda = \{b \mapsto a\}$. Clearly, we have $\mathcal{I} = \{\neg P(a, a), \neg P(a, b), \neg P(b, a)\}$.

This is an example of *positive hyper-resolution*, because all the electrons and all the hyper-resolvents are positive. Thus, the interpretation for C contains only negated atoms. In the case of *negative hyper-resolution*, the resolvents and the electrons are negative, and thus the interpretation does not contain negated atoms. In either case, it is all about imposing an interpretation when resolving a set of clauses, reason why hyper-resolution is in fact a kind of semantic resolution.

Hyper-resolution is also a kind of *macro-resolution*, i.e. the contraction of a sequence of resolution steps into a single inference. In Example

14.2. Resolution refinements

C_1 $P(a,b)$
C_2 $P(b,a)$
C_3 $\neg P(x,y) \vee \neg P(y,z) \vee P(x,z)$
C_4 $\neg P(a,a)$
C_5 $\neg P(x,b) \vee P(x,a)$ Resolvent of $(C_2, C_3)\sigma$, $\sigma = \{y \mapsto b, z \mapsto a\}$
C_6 $P(a,a)$ Resolvent of $(C_1, C_5)\theta$, $\theta = \{x \mapsto a\}$
C_7 $\square$ Resolvent of C_4 and C_6

Figure 14.2.2.: Hyper-resolution of $\Xi = (C_3; C_1, C_2)$.

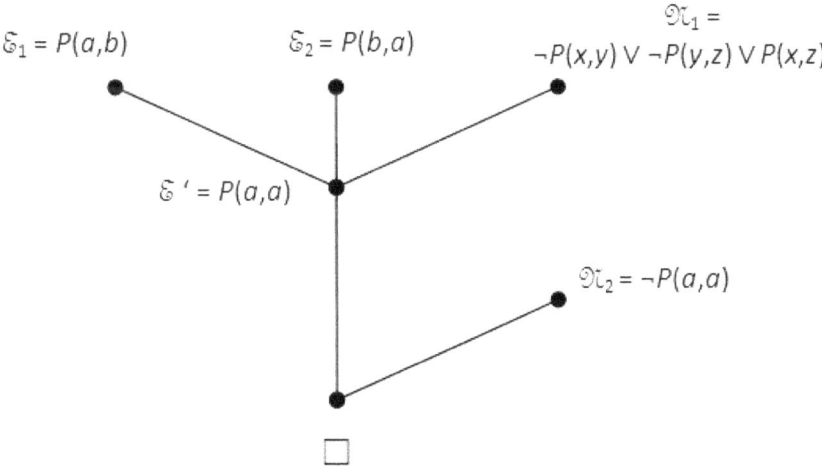

Figure 14.2.3.: A hyper-resolution deduction tree.

14. The resolution calculus

Assumptions:
```
% axioms for a partial order
x <= x.
x <= y & y <= x -> x=y.
x <= y & y <= z -> x <= z.
% lattice axioms
x <= 1.
z <= x ^ y <-> z <= x & z <= y.
0 <= x.
x v y <= z <-> x <= z & y <= z.
% distributivity
(x v y) ^z = (x^z) v (y^z).
```

Goals:
```
% commutativity of meet
x ^ y = y ^ x.
```

Figure 14.2.4.: Theory of distributive lattices and commutativity of meet: Input in Prover9/Mace4.

14.36, the *macro-resolvents* of C are C_6 and C_7. See Figure 14.2.3 for the hyper-resolution deduction tree of Example 14.36; note how the resolvent C_5 does not appear in the tree.

Example 14.37. Let there be given the theory of distributive lattices. We wish to know whether meet is a commutative operation in such a lattice. The input in Prover9/Mace4 is shown in Figure 14.2.4. As can be easily seen in the output by Prover 9 (Fig. 14.2.5), hyper-resolution accounts for most of the proof.

Hyper-resolution can also include a predicate-symbol ordering. We leave this as exercises.

▶ *Do Exercises 14.14-21.*

14.2.2. Linear resolution: Logic programming (II)

An important set of resolution refinements cluster around *linear resolution*. This is a particularly relevant refinement, as it is at the very core of logic programming in Prolog, of which we already gave an introductory

14.2. Resolution refinements

```
======================= PROOF =========================
% -------- Comments from original proof --------
% Proof 1 at 0.05 (+ 0.09) seconds.
% Length of proof is 14.
% Level of proof is 5.
% Maximum clause weight is 11.
% Given clauses 46.


1 x <= y & y <= x -> x = y # label(non_clause).  [assumption].
3 x <= y ^ z <-> x <= y & x <= z # label(non_clause).  [assumption].
5 x ^ y = y ^ x # label(non_clause) # label(goal).  [goal].
6 x <= x.  [assumption].    7 -(x <= y) | -(y <= x) | y = x.  [clausify(1)].
10 -(x <= y ^ z) | x <= y.  [clausify(3)].
11 -(x <= y ^ z) | x <= z.  [clausify(3)].
12 x <= y ^ z | -(x <= y) | -(x <= z).  [clausify(3)].
19 c2 ^ c1 != c1 ^ c2.  [deny(5)].
20 x ^ y <= x.  [hyper(10,a,6,a)].
21 x ^ y <= y.  [hyper(11,a,6,a)].
57 x ^ y <= y ^ x.  [hyper(12,b,21,a,c,20,a)].
484 x ^ y = y ^ x.  [hyper(7,a,57,a,b,57,a)].
485 $F.  [resolve(484,a,19,a)].
======================= end of proof ==================
```

Figure 14.2.5.: Proof by Prover9 of the commutativity of meet in a distributive lattice.

14. The resolution calculus

discussion in Section 5.3.2.[8] As a matter of fact, deduction in Prolog applies a refinement of linear resolution itself, known as SLD resolution. We first give the relevant definitions and then move on to the application of these specific resolution refinements to Prolog.

Definition 14.38. Given a set of clauses C, we say that a clause $\mathcal{C}$ is a *linear resolution deduction* from C, and write $C \vdash_{lres} \mathcal{C}$, if there is a sequence of pairs $(\mathcal{C}_0, \mathcal{D}_0), ..., (\mathcal{C}_n, \mathcal{D}_n)$ such that $\mathcal{C} = \mathcal{C}_{n+1}$ and (i) $\mathcal{C}_0$, called the *starting clause*, and the $\mathcal{D}_i$ are elements of C or of some $\mathcal{C}_j, j < i$, (ii) each $\mathcal{C}_{i+1}, i \leq n$, is a resolvent of $\mathcal{C}_i$ and $\mathcal{D}_i$. We correspondingly have the deduction tree

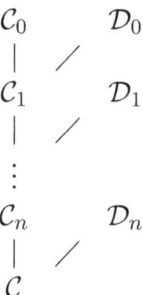

The elements of C are called the *input clauses*, the $\mathcal{C}_i$ are the *center clauses* and the $\mathcal{D}_i$ the *side clauses*. If $\mathcal{C} = \square$, we say that there is a linear-resolution refutation of C, and write $C \vdash_{lres} \square$.

Example 14.39. Let $C = \{\|p, q\|, \|p, \neg q\|, \|\neg p, q\|, \|\neg p, \neg q\|\}$ be given. Figure 14.2.6 shows the linear-resolution refutation of this obviously unsatisfiable set of clauses.

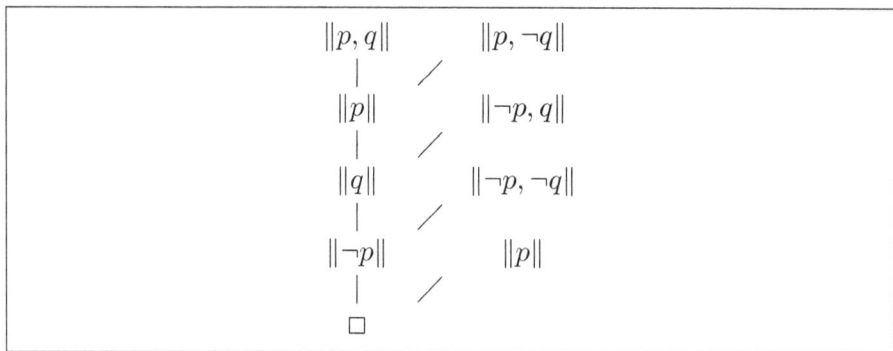

Figure 14.2.6.: A linear-resolution refutation tree.

---

[8] For the sake of simplicity, we consider here Prolog as a synonym for logic programming, but in fact Prolog is just a language (family)–albeit the most important one–in this programming paradigm.

14.2. Resolution refinements

Definition 14.40. Linear resolution is called *linear input resolution* (abbr.: *LI resolution*) if, given a sequence of pairs $(C_0, D_0), ..., (C_n, D_n)$, then all the D_i are variants of clauses in C.

Definition 14.41. Let $C = \|\neg L_1, ..., \neg L_n\|$, $D = \|M, \neg M_1, ..., \neg M_m\|$ be ordered clauses. The following rule of inference, where $\sigma = mgu(M, L_i)$ and the resolvent is an ordered (or definite) clause, is called *linear definite resolution* (abbr.: *LD resolution*):

$$\frac{\neg L_1 \vee ... \vee \neg L_n \qquad M \vee \neg M_1 \vee ... \vee \neg M_m}{(\neg L_1 \vee ... \vee \neg L_{i-1} \vee \neg M_1 \vee ... \vee \neg M_m \vee \neg L_{i+1} \vee ... \vee \neg L_n)\sigma}$$

Definition 14.42. The following inference rule, where $\sigma = mgu(M, L_i)$ and r is a selection rule or function, is called *selective linear definite resolution* (abbr.: *SLD resolution*):

$$\frac{r(\neg L_1 \vee ... \vee \neg L_n) \qquad M \vee \neg M_1 \vee ... \vee \neg M_m}{(\neg L_1 \vee ... \vee \neg L_{i-1} \vee \neg M_1 \vee ... \vee \neg M_m \vee \neg L_{i+1} \vee ... \vee \neg L_n)\sigma}$$

We retake now the discussion from Section 5.3.2. In this, we focused on the expressiveness of **L1**, more specifically so of the subset thereof that constitutes **Prolog**. We now concentrate on deduction in Prolog, from both the semantical and syntactical perspectives; we show the adequateness of Prolog as a deductive system by showing how linear resolution, and in particular SLD resolution, concretizes the meaning of some Prolog program π. We shall follow a circular path: We begin with some basic syntax (Def.s 14.43-4) and then move on to semantic aspects (Def.s 14.46-8) after connecting these two perspectives (Def. 14.45); this done, we build up from the concept of *reduction* to full-blown SLD resolution, showing that this is complete with respect to Prolog, thus concretizing the meaning of some program π.

Recall that given a Prolog program π and some query G, the solution to G is a fact that is a (common) instance of G. This solution is obtained via the application of a rule known as universal *modus ponens*:

Definition 14.43. From the rule $\mathbf{r} = A \leftarrow B_1, ..., B_n$ and the facts $B'_1., ..., B'_n$. we can deduce A' if $A' \leftarrow B'_1, ..., B'_n$ is an instance of $\mathbf{r}$. We call this rule *universal modus ponens (UMP)*.

The next definition establishes the relation between UMP and logical consequence in Prolog:

14. The resolution calculus

Definition 14.44. A goal G is a logical consequence of a program π, denoted by $\pi \vdash G$, if there is a clause in π with a ground instance $A \leftarrow B_1, ..., B_n$, $n \geq 0$, such that $\pi \vdash B_i$ for all $0 < i \leq n$ and A is an instance of G. In other words, G is a logical consequence of π iff G can be deduced from π by a finite number of applications of UMP.[9]

We shall consider the empty clause $\square$ as a goal. We now give the "general" algorithm of Prolog:

Definition 14.45. An *abstract interpreter* Ψ for a Prolog program is an algorithm that takes as input a program π and a goal G, answering *true* (or *yes*) if G is a logical consequence of π and *false* (or *no*) otherwise. In the first case we say that Ψ performs a *true-computation*, and in the second case we say that Ψ performs a *false-computation*.[10]

Given a Prolog program π, a true-computation is the case iff, for finite $n \geq 1$, there are ground goals $(G = \{B_1, ..., B_n\}) \subseteq \pi$ that constitute the meaning of π:

Definition 14.46. The *meaning of a program* π, denoted by $M(\pi)$, is the set of unit ground goals $G = \{B_1, ..., B_n\}$ such that for all $0 < i \leq n$ we have
$$\pi \vdash B_i.$$

1. A program π is said to be *correct* with respect to some intended meaning M iff $M(\pi) \subseteq M$, and it is said to be *complete* with respect to some intended meaning M iff $M \subseteq M(\pi)$.

2. A program π is *adequate*, i.e. both correct and complete, with respect to some intended meaning M iff $M = M(\pi)$.

---

[9] More specifically, G is an existentially quantified goal. See above.

[10] Recall the contents of Section 1.4 above. Given some computing device M (for instance, a Turing machine or a digital computer), the relationship between logical deduction and computation is established as
$$C_0 \vdash^*_M C_f \quad \text{iff} \quad \pi \Vdash^*_M G$$
for the initial and final configurations of M, denoted respectively by C_0 and C_f. We can read the above as follows: the goal G is a logical consequence of program π obtained in zero or more (denoted by *) applications of UMP by M iff from an initial configuration M stops in an accepting configuration C_f after zero or more steps. In other words, given π and the goal G as input to the computing device M, G is a logical consequence of program π iff there is a finite sequence of configurations of M that ends in a "Yes" or "True" configuration. See Augusto (2020a) for a comprehensive elaboration on *deductive computation*.

14.2. Resolution refinements

Informally, a program π is correct iff it does not "say" unintended "things," and it is complete if every "thing" that is intended can be "said." The meaning of a basic program built up solely of ground facts is the program itself. Put differently, the program "means" just what it "says." The meaning of a regular logic program (i.e. a logic program comprising rules) contains explicitly whatever the program states implicitly.

Example 14.47. The meaning of the program *Fatherhood* (Example 5.24), $M(Fatherhood)$, just is the program itself. If we add to this program the rule `parent(X,Y) ← father(X,Y).`, then $M(Fatherhood)$ additionally contains all goals of the form `parent(X,Y)` for every pair (X,Y) such that `father(X,Y).` is in the program.

This said, it should be obvious that the *intended meaning* of a program π–a set $MI(\pi)$ of unit ground goals–is intuitively given by the choice of names in the program. This allows a semantics of quasi-truth values in the following way:

Definition 14.48. Given a program π, we say that a ground goal G is *true* with respect to $MI(\pi)$ if $G \in MI(\pi)$; otherwise, we say that G is *false*.

We now expand on the abstract interpreter Ψ of Definition 14.45, namely as a search algorithm when computing a goal. The interaction with Prolog by the user is by means of *queries*: If one wants to know whether G is a goal of some program π, one asks the Prolog interpreter consulting π

$$\pi \stackrel{?}{\vdash} G$$

which in practical terms is just entering the expression "G?" in some Prolog compiler (typically as "? – G.").[11]

Definition 14.49. We call *resolvent* the current (usually conjunctive) stage of a computation of the abstract interpreter Ψ. The *empty resolvent* (or *empty clause*), denoted by $\square$, is the clause with empty head and empty body, i.e. $\leftarrow$. The sequence of resolvents produced during a computation is called the *trace* of Ψ.

Example 14.50. To the program *Fatherhood* (Example 5.24) add the rule

`daughter(X,Y) : −father(Y,X), female(X).`

Input the goal `? − daughter(X, harry).`, we obtain two replies, to wit, `X = mary.` and `X = jane.`. Figure 14.2.7 shows the trace by SWI-Prolog

---

[11] We shall be using SWI-Prolog, freely available at https://www.swi-prolog.org/.

14. The resolution calculus

```
[trace] ?- daughter(X,harry).
   Call:  (8) daughter(_2768, harry) ?  creep
   Call:  (9) father(harry, _2768) ?  creep
   Exit:  (9) father(harry, louis) ?  creep
   Call:  (9) female(louis) ?  creep
   Fail:  (9) female(louis) ?  creep
   Redo:  (9) father(harry, _2768) ?  creep
   Exit:  (9) father(harry, mary) ?  creep
   Call:  (9) female(mary) ?  creep
   Exit:  (9) female(mary) ?  creep
   Exit:  (8) daughter(mary, harry) ?  creep
X = mary ;
   Redo:  (9) father(harry, _2768) ?  creep
   Exit:  (9) father(harry, jane) ?  creep
   Call:  (9) female(jane) ?  creep
   Exit:  (9) female(jane) ?  creep
   Exit:  (8) daughter(jane, harry) ?  creep
X = jane.
```

Figure 14.2.7.: Trace by SWI-Prolog.

for this goal. However, the trace output by SWI-Prolog shows that the first attempt to produce a reply failed, because father(harry, louis). does not match any fact female(louis). in the program (see line 6 of the trace in Figure 14.2.7).[12] Formally, we have

$$Fatherhood \not\vdash \texttt{female(louis)}$$

though we have

$$Fatherhood \vdash \texttt{father(harry, louis)}.$$

Only for Mary and Jane do we have instances of the rule above when we apply the substitutions $\sigma = \{Y = \texttt{harry}\}$ and $\theta = \{X = \texttt{mary}\}$ or $\theta = \{X = \texttt{jane}\}$.

Definition 14.51. Given a Prolog program π and a goal G, the replacement of G by the body of an instance of a clause $\mathcal{C} \in \pi$ whose head is identical to G is called a *reduction*. A reduction is *ground* if both the

---
[12] This shows an important feature of Prolog search (*depth-first with backtracking*) to be discussed below, which should be carefully taken into consideration when writing a Prolog program.

goal G and the instance of the clause $\mathcal{C}$ are ground. The goal replaced in a reduction is said to be *reduced* and we say that the new goals are *derived*.

If the goal G is not deducible from program π, then Ψ may fail to terminate. We give this reduction procedure by Ψ in Algorithm 14.2. Note that each iteration of the "while loop" is a single application of UMP, i.e. a reduction. It is easy to see that reduction is the basic computational step in Prolog. The selection of the goal to be reduced and the order of the reductions thereof is arbitrary, as all the goals in a given resolvent must be reduced. The selection of a clause and a suitable instance thereof is non-deterministic but critical.

The output "True" by the compiler actually is a proof that the query follows from the program. Such a proof is implicitly represented in the trace of a query, but we can represent it explicitly in the form of a tree.

Definition 14.52. A *(reduction) proof tree* is a down-growing tree whose nodes represent the goals that are reduced during a computation, there being a directed edge from a node to each node that corresponds to a derived goal of a reduced goal. In a proof tree, the number of nodes corresponds to the number of reduction steps in a computation. The root of a proof tree for a simple query is the query itself. The proof tree for a conjunctive query is the collection of all the proof trees for its individual goals.

Strictly considered, a proof tree for some rule $A \leftarrow B_1, ..., B_n$ is an elementary tree of the form

for $n = 0$, and of the form

for $n > 0$. When, as in Figure 14.2.8, the search process is included in the tree, this is more properly called a *search tree*. But, in fact, a search tree in Prolog coincides with a proof tree, reason why we shall relax this distinction.

Example 14.53. Figure 14.2.8 shows the (failed) proof tree for the first attempt in Example 14.50.

14. The resolution calculus

Algorithm 14.2 Reduction.
- **Input:** a Prolog program π and a goal G
- **Output:** *True* or *False*

Initialize the resolvent to G
while resolvent $\neq \square$ **do**
    choose a goal A from the resolvent
    choose a ground instance of a clause $A' \leftarrow B_1, ..., B_n \in \pi$ s.t. $A = A'$
       if no such goal and clause exist, leave the while loop
    replace A by $B_1, ..., B_n$ in the resolvent
if resolvent $= \square$, **then** output *true*, **else** output *false*

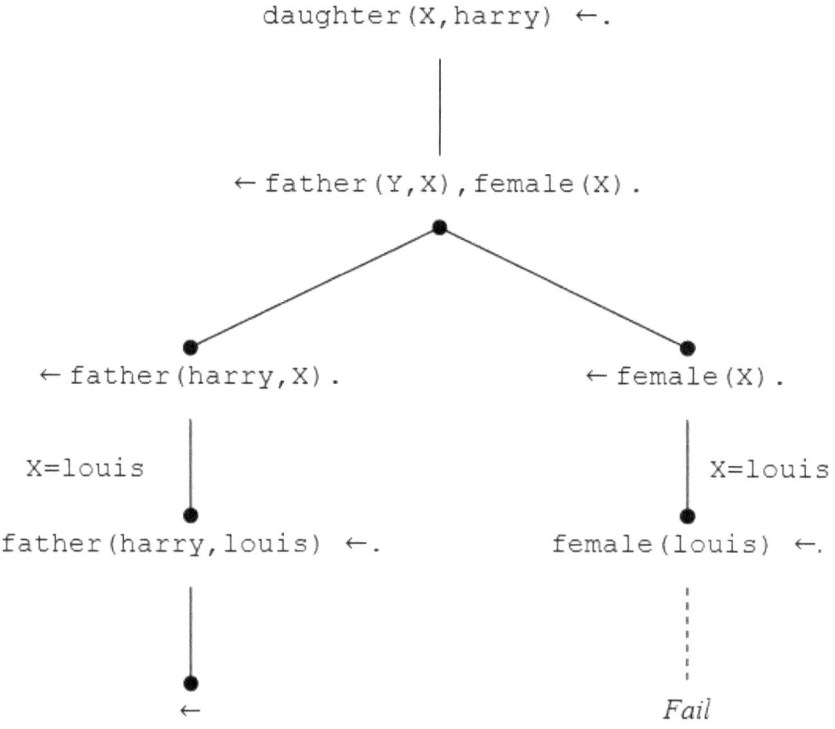

Figure 14.2.8.: A failed proof tree.

However, knowledge of the resolution calculus and a little thought will reveal that reduction, a generalization of UMP, can be re-expressed in the terms of this calculus. Indeed, it is easy to see that we can apply resolution to find a contradiction from the combination of a *goal clause*, with solely negative literals,[13] and a fact (in a rule), a positive literal. It is important to remark that a goal clause $\mathcal{G}$ is not a *program clause*, which can be only either a rule or a fact. This should be born in mind when considering $\pi \cup \{\mathcal{G}\}$, i.e. when we add a goal clause $\mathcal{G}$ to a Prolog program π. The goal clause $\mathcal{G} = \|\neg q_1, ..., \neg q_n\|$ is added to the program, in order to test if the conjunctive query $q_1 \wedge ... \wedge q_n$ follows from it, and this is the case iff $\pi \cup \{\mathcal{G}\}$ is unsatisfiable. This is so iff we can deduce the *empty goal clause* $\square$ from $\pi \cup \{\mathcal{G}\}$ by an application of resolution. More specifically, we refer here to *linear input resolution* (cf. Def. 14.40), as this has been proven complete for Horn clauses. As seen above, LI resolution, in turn, is a refinement of linear resolution, and we begin by giving important results for Prolog with relation to it.

Example 14.54. Consider the program *Fatherhood* and the goal of Example 14.50. Figure 14.2.9 shows a possible way to prove that X = mary (for simplicity, we omit renaming of variables). Figure 14.2.10 shows the corresponding LI-resolution proof (for convenience, only the numbers of the clauses in the resolution are shown).

Lemma 14.55. *Let π be a Prolog program and $\mathcal{G} = \|\neg q_1, ..., \neg q_n\|$ a goal clause. Then, all the q_i are consequences of π iff $\pi \cup \{\mathcal{G}\}$ is unsatisfiable.*

Proof: Left as an exercise.

Theorem 14.56. *(Refutation completeness of linear resolution for Horn clauses) If C is an unsatisfiable set of Horn clauses, then there is a linear resolution proof that is a refutation of C, i.e. $C \vdash_{lres} \square$.*

Proof: (Idea) Assume that C is finite and proceed by induction on the elements of C. **QED**

Although LI resolution is the general resolution rule for Prolog, when given as input the query $q_1, ..., q_n$? a Prolog interpreter actually searches for a *SLD-resolution proof of* $\square$ (cf. Def.14.42), in turn a case of LD resolution (cf. Def.14.41). We now define these two resolution refinements with respect to Prolog.

---

[13] A goal clause always corresponds to the body of a rule, i.e. it corresponds to the literals on the right of $\leftarrow$. Cf. Def.s 5.15 and 5.17.

1. ← daughter(X, harry).   Goal
2. daughter(X, Y) ← father(Y, X), female(X).   Rule
3. father(harry, mary) ← .   Fact
4. female(mary) ← .   Fact
5. ← father(harry, X), female(X).   Resolution (1, 2)
                                                $\sigma = \{Y = \text{harry}\}$
6. ← father(harry, mary).   Resolution (4, 5)
                                                $\theta = \{X = \text{mary}\}$
7. ← .   Resolution (3, 6)

Figure 14.2.9.: A successful reduction interpreted as a resolution proof.

14.2. Resolution refinements

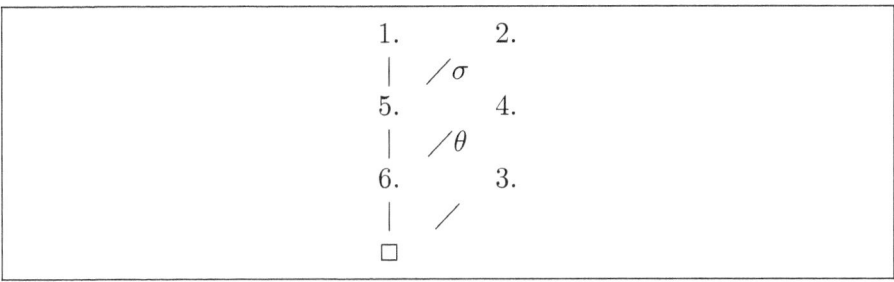

Figure 14.2.10.: A LI-resolution proof tree.

Definition 14.57. Let $\pi \cup \{\mathcal{G}\}$ be given as a set of ordered clauses. Then, a LD-resolution refutation of $\pi \cup \{\mathcal{G}\}$, denoted by $\pi \cup \{\mathcal{G}\} \vdash_{ldres} \Box$, is a sequence
$$(\mathcal{G}_0, \mathcal{C}_0), ..., (\mathcal{G}_n, \mathcal{C}_n)$$
where the $\mathcal{G}_i, \mathcal{C}_i$, $0 \leq i \leq n$, are ordered clauses, such that $\mathcal{G}_0 = \mathcal{G} = \|\neg p_1, ..., \neg p_n\|$ and $\mathcal{G}_{n+1} = \Box$. More specifically, we have
$$\mathcal{G}_i = \|\neg p_{i,1}, ..., \neg p_{i,n(i)}\|, |\mathcal{G}_i| = n(i)$$
are the goal clauses, and for the $\mathcal{C}_i \in \pi$ we have
$$\mathcal{C}_i = \|q, \neg q_{i,1}, ..., \neg q_{i,m(i)}\|, |\mathcal{C}_i| = m(i) + 1 \text{ or } 1 \text{ if } \mathcal{C}_i = \|q\|.$$
Then, for each $i < n$ there is a resolution rule
$$\frac{\mathcal{G}_i \quad \mathcal{C}_i}{\mathcal{G}_{i+1}}$$
where $\mathcal{G}_{i+1} = \|\neg p_{i,1}, ..., \neg p_{i,k-1}, \neg q_{i,1}, ..., \neg q_{i,m(i)}, \neg p_{i,k+1}, ..., \neg p_{i,n(i)}\|$, an ordered clause with $|\mathcal{G}_{i+1}| = (n(i) - 1) + m(i)$, is the resolvent.

Lemma 14.58. *For a Prolog program π and $\mathcal{G} = \|\neg q_1, ..., \neg q_n\|$ a goal clause, if $\pi \cup \{\mathcal{G}\}$ is an unsatisfiable set of ordered clauses, then there is a LD-resolution refutation of $\pi \cup \{\mathcal{G}\}$ beginning with $\mathcal{G}$.*

Proof: Left as an exercise.

We obtain SLD resolution by introducing a *selection rule* r to choose the literal $p_i \in \mathcal{G}_i$ to be resolved upon with LD resolution, so that $r(\mathcal{G}_i)$ is the literal resolved upon in the $(i+1)$-th step of the proof.

14. The resolution calculus

Definition 14.59. The LD-resolution rule

$$\frac{r(\mathcal{G}_i) \quad \mathcal{C}_i}{\mathcal{G}_{i+1}}$$

where r is a selection rule, is called a *SLD-resolution rule*.

In Prolog, the selection rule simply chooses the leftmost literal in the goal clause to be resolved upon.

Theorem 14.60. *(Completeness of SLD resolution for Prolog) Given a Prolog program π and a goal clause $\mathcal{G} = \|\neg q_1, ..., \neg q_n\|$, if $\pi \cup \{\mathcal{G}\}$ is an unsatisfiable set of ordered clauses, then there is a selection rule r such that there is a SLD-resolution refutation of $\pi \cup \{\mathcal{G}\}$ via r beginning with $r(\mathcal{G})$.*

Proof: (Sketch). Lemma 14.57 guarantees us that there is a LD-resolution refutation of $\pi \cup \{\mathcal{G}\} \notin SAT$ beginning with $\mathcal{G}$. We only need to prove that there is a SLD-resolution refutation of $\pi \cup \{\mathcal{G}\} \notin SAT$ via $r(\mathcal{G})$. The proof is by induction on the length of $\mathcal{G}$. If $|\mathcal{G}| = 1$, then $\mathcal{G}_0$ is a unit clause and $r(\mathcal{G}_0)$ is irrelevant. We now let $(\mathcal{G}_0, \mathcal{C}_0), ..., (\mathcal{G}_n, \mathcal{C}_n)$ be a LD-resolution refutation of $\pi \cup \{\mathcal{G}_0\} \notin SAT$ and we suppose that the selection rule r chooses the literal $\neg p_{0,k} \in \mathcal{G}_0$. Because $\pi \cup \{\mathcal{G}_0\} \notin SAT$, we must have $\mathcal{G}_{n+1} = \square$, and hence there must be some $j < n$ at which we resolve on $\neg p_{0,k}$. If $j = 0$, we are done. If $j \geq 1$, then there must be some $\mathcal{C}$ that is a resolvent of $\mathcal{G}_0$ and $\mathcal{C}_j$. Then, there must be a LD-resolution refutation of length $n-1$ of $\pi \cup \{\mathcal{C}\}$ beginning with $\mathcal{C}$. By induction, this refutation can be replaced by a SLD-resolution refutation via r. We add this refutation onto the single step resolution of $\mathcal{G}_0$ and $\mathcal{C}_j$ obtaining the SLD-resolution of $\pi \cup \{\mathcal{G}\} \notin SAT$ via $r(\mathcal{G})$ beginning with $\mathcal{G} = \mathcal{G}_0$. **QED**

We now elaborate on how SLD resolution corresponds to the search process in Prolog when a query is entered as input. A Prolog proof tree corresponds to the search process known as *depth-first search with backtracking*: by "depth-first search" it is meant that, given a finitely branching tree, all the descendants of a node are checked before their siblings on the right of the tree and no edge is traversed more than once; if a *fail* leaf is encountered, then the search "backtracks" to the immediate ancestor of this leaf and the depth-search process is resumed. If a success leaf (denoted by $\square$) is found, the search stops until we prompt the search to proceed further by means of an *expand* "command" that makes the search retake.[14] The search is considered successful if at least

---
[14] In the SWI-Prolog interpreter, we simply enter ";".

14.2. Resolution refinements

one □-resolvent is found on the tree; otherwise, the search fails and "false" is the output to the query.

Put briefly, given a Prolog program π and a goal clause $\mathcal{G}$, every branch of a complete Prolog proof tree is either a successful SLD-resolution proof or a failed SLD-resolution proof. For this reason, we–perhaps less properly–refer to this tree also as a *SLD-resolution tree*.

Example 14.61. Let there be given the following Prolog program π_1:

1. $p(X, Y) \leftarrow q(X, Z), r(Z, Y)$
2. $p(X, X) \leftarrow s(X)$
3. $q(X, b)$
4. $q(b, a)$
5. $q(X, a)$
6. $r(b, a)$
7. $s(X) \leftarrow t(X, a)$
8. $s(X) \leftarrow t(X, b)$
9. $s(X) \leftarrow t(X, X)$
10. $t(a, b)$
11. $t(b, a)$

Our query is $?-p(X, X)$, i.e. $\|\neg p(X, X)\|$, where we use the symbol $\neg$ for convenience.[15] The depth-first algorithm starts by checking premise 1 and then moves to premise 2. This is so because conjunction is not commutative in Prolog, reason why the order of the premises is crucial. Beginning with premise 1, we have it that there is a successful SLD-resolution proof when we apply SLD resolution to the premises 1, 3, and 6, in this exact order, with, after renaming of variables, substitutions $\sigma = \{X_1 \to X, Y \to X\}$, $\theta = \{X_2 \to X, Z \to b\}$, and $\lambda = \{X \to a\}$ (cf. Figure 14.2.11). Applying SLD resolution to the sequences 1 and 4 or 1 and 5 will not produce successful proofs. The search starting by checking premise 2 produces two successful proofs and a failure. Figure 14.2.12 shows the complete proof tree for $\pi_1 \cup \{\|\neg p(X, X)\|\}$ with the further substitutions $\omega = \{X \to b, Z \to a\}$, $\varsigma = \{Z \to a\}$, and $\mu = \{X \to b\}$; renaming of variables was omitted and ε denotes the empty substitution.

Example 14.62. Figure 14.2.13 shows how SWI-Prolog answers the query $?-p(X, X)$ when given program π_1 and how it answers the request to produce a trace of some instantiations. In the first case, given the input (query) $?-p(X, X)$, SWI-Prolog gives the first answer, to wit, X = a. Asked to provide more answers by means of the prompt ";",

---

[15] Explicit negation is a tricky thing in Prolog. See Augusto (2020a).

14. The resolution calculus

$$
\begin{array}{ll}
\|\neg p(X,X)\| & 1.\ \|p(X_1,Y), \neg q(X_1,Z), \neg r(Z,Y)\| \\
\quad | \ /\sigma & \\
\|\neg q(X,Z), \neg r(Z,X)\| & 3.\ \|q(X_2,b)\| \\
\quad | \ /\theta & \\
\|\neg r(b,X)\| & 6.\ \|r(b,a)\| \\
\quad | \ /\lambda & \\
\quad \square &
\end{array}
$$

Figure 14.2.11.: A SLD-resolution proof.

SWI-Prolog gives the replies X = b and X = a, finally replying that there are no more instantiations (denoted by false). Asked to output traces of ? − p(X, X), ? − p(a, X), and ? − p(b, X), SWI-Prolog does so, in each case adding that X = a. Compare these with the proof tree in Figure 14.2.12. Figure 14.2.14 shows both the case of a successful instantiation ? − p(b, b) and a failed instantiation ? − p(c, d). In these last traces, "redo" indicates backtracking.

If you check now the trace of Example 14.50 for the reply X = mary., you should notice that it corresponds in fact to a SLD-resolution. Moreover, it is now an easy matter, given the complete trace for the given goal, to construct the complete SLD-resolution tree for this goal.

▶ *Do Exercises 14.22-38.*

14.3. Paramodulation

As seen in Section 7.2, equality is a fundamental relation in many theories, namely in mathematical theories, and we often need to extend CL to $CL^=$. Thus, one can say that a proof calculus for CFOL is not entirely adequate, and an automated prover associated to it is not full-blown, if it is not capable of tackling equality. *Paramodulation* is the way conceived to introduce equality in the resolution calculus. We give here the essentials of this proof technique; a deeper treatment of paramodulation can be found in, for example, Nieuwenhuis & Rubio (2001).

Suppose you are given the clauses $\mathcal{C}_1 = P(a)$ and $\mathcal{C}_2 = (a = b)$. Then, we should be able to derive $\mathcal{C} = P(b)$, i.e. we should be able to have $\{\mathcal{C}_1, \mathcal{C}_2\} \vdash_{res} \mathcal{C}$, by applying a simple substitution of b for a. Obvious as this is, the resolution calculus above cannot handle this without being

14.3. Paramodulation

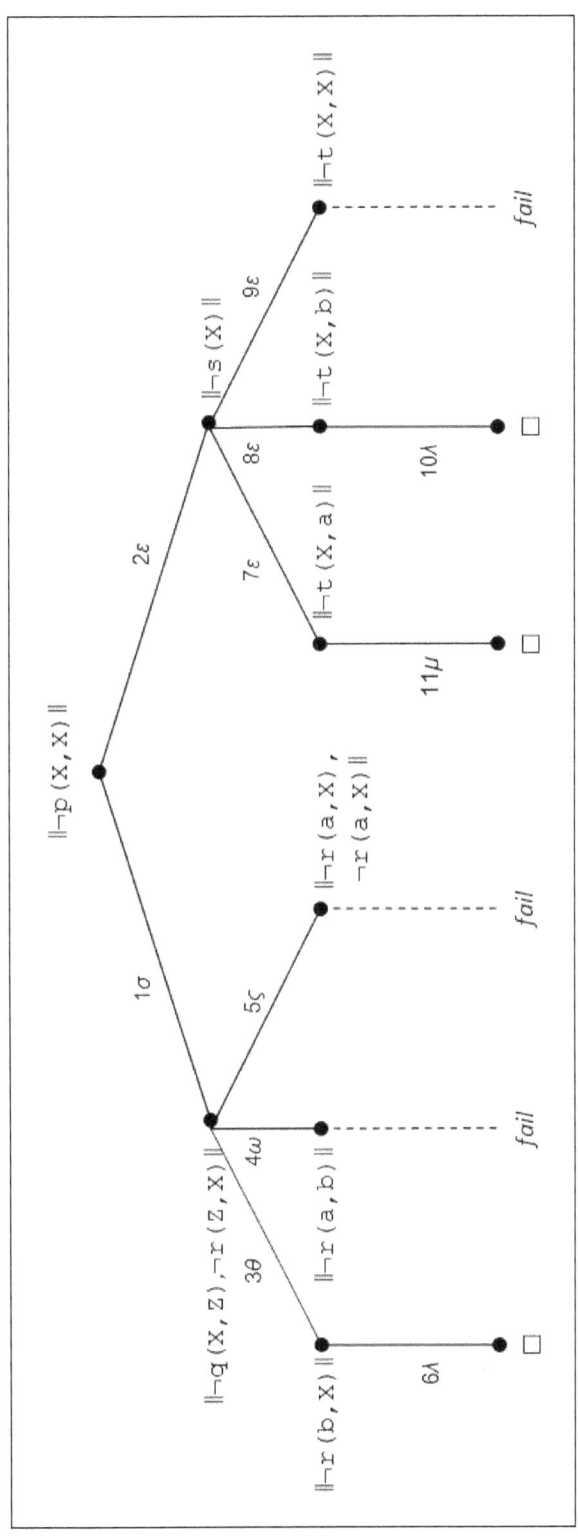

Figure 14.2.12.: A complete proof tree.

14. The resolution calculus

```
?- p(X,X).
X = a ;
X = b ;
X = a ;
false.

?- trace.
true.

[trace]  ?- p(X,X).
   Call:  (8) p(_2768, _2768) ?  creep
   Call:  (9) q(_2768, _2986) ?  creep
   Exit:  (9) q(_2768, b) ?  creep
   Call:  (9) r(b, _2768) ?  creep
   Exit:  (9) r(b, a) ?  creep
   Exit:  (8) p(a, a) ?  creep
X = a .

[trace]  ?- p(a,X).
   Call:  (8) p(a, _2770) ?  creep
   Call:  (9) q(a, _2986) ?  creep
   Exit:  (9) q(a, b) ?  creep
   Call:  (9) r(b, _2770) ?  creep
   Exit:  (9) r(b, a) ?  creep
   Exit:  (8) p(a, a) ?  creep
X = a .

[trace]  ?- p(b,X).
   Call:  (8) p(b, _2770) ?  creep
   Call:  (9) q(b, _2986) ?  creep
   Exit:  (9) q(b, b) ?  creep
   Call:  (9) r(b, _2770) ?  creep
   Exit:  (9) r(b, a) ?  creep
   Exit:  (8) p(b, a) ?  creep
X = a .
```

Figure 14.2.13.: SWI-Prolog answering a query and outputting traces for some "true" instantiations.

```
[trace] ?- p(b,b).
   Call:  (8) p(b, b) ? creep
   Call:  (9) q(b, _2950) ? creep
   Exit:  (9) q(b, b) ? creep
   Call:  (9) r(b, b) ? creep
   Fail:  (9) r(b, b) ? creep
   Redo:  (9) q(b, _2950) ? creep
   Exit:  (9) q(b, a) ? creep
   Call:  (9) r(a, b) ? creep
   Fail:  (9) r(a, b) ? creep
   Redo:  (9) q(b, _2950) ? creep
   Exit:  (9) q(b, a) ? creep
   Call:  (9) r(a, b) ? creep
   Fail:  (9) r(a, b) ? creep
   Redo:  (8) p(b, b) ? creep
   Call:  (9) s(b) ? creep
   Call:  (10) t(b, a) ? creep
   Exit:  (10) t(b, a) ? creep
   Exit:  (9) s(b) ? creep
   Exit:  (8) p(b, b) ? creep
true.

[trace] ?- p(c,d).
   Call:  (8) p(c, d) ? creep
   Call:  (9) q(c, _2950) ? creep
   Exit:  (9) q(c, b) ? creep
   Call:  (9) r(b, d) ? creep
   Fail:  (9) r(b, d) ? creep
   Redo:  (9) q(c, _2950) ? creep
   Exit:  (9) q(c, a) ? creep
   Call:  (9) r(a, d) ? creep
   Fail:  (9) r(a, d) ? creep
   Redo:  (8) p(c, d) ? creep
   Fail:  (8) p(c, d) ? creep
false.
```

Figure 14.2.14.: SWI-Prolog traces of a "true" and a "false" instantiation.

14. The resolution calculus

augmented with a *generalization* of substitution that, in turn, requires a clear definition of equality substitution.

Definition 14.63. Given $\mathcal{C}(t)$, a clause $\mathcal{C}$ in which a term t occurs, and a unit clause $(t = s)$, we can derive a new clause $\mathcal{C}(s)$ by substituting s for a single occurrence of t. We call this derivation rule *equality substitution*:

$$\textbf{(eqSub)} \quad \frac{\mathcal{C}(t) \quad t = s}{\mathcal{C}(s)}.$$

Definition 14.64. Consider the clauses $C_1 = L(t) \vee C_1'$, where $L(t)$ denotes that the term t occurs in the literal L, and $C_2 = (r = s) \vee C_2'$ where r and s are any terms. Then, *paramodulation* is the inference rule

$$\textbf{(par)} \quad \frac{L(t) \vee C_1' \quad (r = s) \vee C_2'}{L(s) \vee C_1' \vee C_2'}$$

and the *paramodulant* of C_1 and C_2 is the clause

$$L(s) \vee C_1' \vee C_2'.$$

Example 14.65. Given the clauses $C_1 = P(a) \vee Q(b)$ and $C_2 = (a = b) \vee R(b)$, we can derive the paramodulant $P(b) \vee Q(b) \vee R(b)$.

The above is a definition of *ground paramodulation*. This is now generalized to formulas with individual variables in the following way:

Definition 14.66. Let there be given the clauses $C_1 = L(t) \vee C_1'$, where $L(t)$ denotes that the term t occurs in the literal L, and $C_2 = (r = s) \vee C_2'$ where r and s are any terms. Let further $\sigma = mgu(t, r)$, C_1 and C_2 have no variables in common. Then, *binary paramodulation* is the inference rule

$$\textbf{(par}_\sigma\textbf{)} \quad \frac{L(t) \vee C_1' \quad (r = s) \vee C_2'}{(L(s\sigma) \vee C_1' \vee C_2')\sigma}$$

and the *binary paramodulant* of C_1 and C_2 is the clause

$$\left(L(s\sigma) \vee C_1' \vee C_2'\right)\sigma.$$

We say that L and $r = s$ are the literals *paramodulated upon* in the paramodulation from C_2 to C_1.

14.3. Paramodulation

Example 14.67. Given the clauses $C_1 = P(x) \vee Q(x)$ and $C_2 = (a = b)$, by applying the substitution $\sigma = \{x \mapsto a\}$ we can obtain the paramodulants $P(b) \vee Q(a)$ or $P(a) \vee Q(b)$.

Example 14.67 shows an important feature of paramodulation, to wit, substitutions are applied in one argument position at a time. In this example, we have

$$(P(a) \vee Q(x))\sigma = \underbrace{P(b)}_{L\sigma(r\sigma)} \vee \underbrace{Q(a)}_{C'_1\sigma} \vee \underbrace{\square}_{C'_2\sigma}$$

or

$$(P(x) \vee Q(a))\sigma = \underbrace{P(a)}_{C'_1\sigma} \vee \underbrace{Q(b)}_{L\sigma(r\sigma)} \vee \underbrace{\square}_{C'_2\sigma}.$$

Example 14.68. A slightly more complex example can be given by the paramodulation from $C_2 = (f(g(b)) = a) \vee R(g(c))$ to $C_1 = P(g(f(x))) \vee Q(x)$. Following Definition 14.66, we make t be $f(x)$, $L(t)$ is $P(g(f(x)))$, r is $f(g(b))$; the MGU of r and t is $\sigma = \{x \mapsto g(b)\}$. We apply the paramodulation inference rule and obtain the paramodulant $P(g(a)) \vee Q(g(b)) \vee R(g(c))$ in the following two steps:

$$\underbrace{P(g(f(g(b))))}_{L(r\sigma)} \vee Q(x) \vee R(g(c))$$

$$\underbrace{P(g(a))}_{L\sigma(r\sigma) = (L(s\sigma))\sigma} \vee \underbrace{Q(g(b))}_{C'_1\sigma} \vee \underbrace{R(g(c))}_{C'_2\sigma}$$

We now further specify the definition of paramodulant:

Definition 14.69. A paramodulant of (parent) clauses C_1 and C_2 is one of the following binary paramodulants:

1. A binary paramodulant of C_1 and C_2;
2. A binary paramodulant of a factor of C_1 and C_2;
3. A binary paramodulant of C_1 and a factor of C_2;
4. A binary paramodulant of a factor of C_1 and a factor of C_2.

14. The resolution calculus

However, it is not often the case that one is confronted with the simplicity above; more often than not, there are more than one equality literals and/or more than one literal to be paramodulated upon, so that strategies are required such as ordering of the equality literals and simultaneous paramodulation.

Definition 14.70. The rule of inference

$$\text{(orPar)} \quad \frac{\begin{array}{c} L(t) \vee C_1' \\ (r = s) \vee C_2' \end{array}}{\left(L(s\sigma) \vee C_1' \vee C_2' \right) \sigma}$$

where $\sigma = mgu(r, t)$, $L\sigma \not< L\sigma(s\sigma)$, $(r\sigma = s\sigma) \not< M$ for all the literals $M \in \left(C_2'\right)\sigma$ and $L\sigma \not< M$ for all the literals $M \in \left(C_1'\right)\sigma$, is called *ordered paramodulation*.

Definition 14.71. The rule of inference is called *simultaneous paramodulation*.

$$\text{(smPar)} \quad \frac{\begin{array}{c} L(t) \vee S(t) \\ (r = s) \vee C \end{array}}{(L(s\sigma) \vee S(s\sigma) \vee C)\sigma}$$

Example 14.72. Let there be given the clauses $C_1 = P(f(x, g(x))) \vee Q(x)$ and $C_2 = (a = b) \vee (g(a) = a) \vee (f(a, g(a)) = b)$. It will be useful to indicate the constituents of these clauses as follows:

$$\underbrace{C_1}_{1} = \underbrace{P(f(x, g(x)))}_{a(1,2)} \vee \underbrace{Q(x)}_{b}$$

and

$$\underbrace{C_2}_{2} = \underbrace{(a = b)}_{a} \vee \underbrace{(g(a) = a)}_{b} \vee \underbrace{(f(a, g(a)) = b)}_{c}$$

We may obtain the following sample of all the paramodulants of the paramodulation from C_2 into C_1:

By an application of paramodulation with the MGU $\sigma_1 = \{x \mapsto a\}$ we obtain the paramodulants

para[2(a),1(a,1)]: $P(f(b, g(a))) \vee Q(a) \vee (g(a) = a) \vee (f(a, g(a)) = b)$
para[2(a),1(a,2)]: $P(f(a, g(b))) \vee Q(a) \vee (g(a) = a) \vee (f(a, g(a)) = b)$
para[2(a),1(a,1-2)]: $P(f(b, g(b))) \vee Q(a) \vee (g(a) = a) \vee (f(a, g(a)) = b)$
para[2(a),1(b)]: $P(f(a, g(a))) \vee Q(b) \vee (g(a) = a) \vee (f(a, g(a)) = b)$
para[2(a),1(a,b)]: $P(f(b, g(b))) \vee Q(b) \vee (g(a) = a) \vee (f(a, g(a)) = b)$

14.3. Paramodulation

We can obtain a factor of $C_2' = (g(a) = b) \vee f(a, g(a)) = b)$ by the transitivity of the equality relation on $g(a) = a$ and $a = b$, which gives us $g(a) = b$. Applying paramodulation with the MGU $\sigma_2 = \{x \mapsto g(a)\}$, we obtain the paramodulants

para[2'(a),1(a,1)]: $P(f(b, g(g(a)))) \vee Q(g(a)) \vee (f(a, g(a)) = b)$
para[2'(a),1(a,2)]: $P(f(g(a), g(b))) \vee Q(g(a)) \vee (f(a, g(a)) = b)$
para[2'(a),1(a,1-2)]: $P(f(b, g(b))) \vee Q(g(a)) \vee (f(a, g(a)) = b)$
para[2'(a),1(b)]: $P(f(g(a), g(g(a)))) \vee Q(b) \vee (f(a, g(a)) = b)$
para[2'(a),1(a,b)]: $P(f(b, g(b))) \vee Q(b) \vee (f(a, g(a)) = b)$

C_2' can further be factorized into $C_2'' = (g(a) = f(a, g(a)))$, which in turn gives us simpler paramodulants.

Example 14.73. Given the group theory $\Theta_{\mathcal{G}}$, we prove in Prover9/Mace4 that if $x \star x = e$ for all x in a group $\mathcal{G} = (G, \star)$, then $\mathcal{G}$ is a commutative group. Figure 14.3.1 shows the input, and Figure 14.3.2 shows the proof ("! =" stands for "$\neq$").

Assumptions:
f(e,x)=x.
f(x,e)=x.
f(x,f(y,z))=f(f(x,y),z).
f(x,x)=e.
f(a,b)=c.

Goals:
c=f(b,a).

Figure 14.3.1.: Theory of commutative groups: Input in Prover9/Mace4.

▶ *Do Exercises 14.39-44.*

Exercises

Exercise 14.1. Prove Proposition 14.2.

14. The resolution calculus

```
===================== PROOF =========================
% -------- Comments from original proof --------
% Proof 1 at 0.01 (+ 0.05) seconds.
% Length of proof is 15.
% Level of proof is 5.
% Maximum clause weight is 11.
% Given clauses 9.


1  c = f(b,a) # label(non_clause) # label(goal).  [goal].
2  f(e,x) = x.  [assumption].
3  f(x,e) = x.  [assumption].
4  f(x,f(y,z)) = f(f(x,y),z).  [assumption].
5  f(f(x,y),z) = f(x,f(y,z)).  [copy(4),flip(a)].
6  f(x,x) = e.  [assumption].
7  f(a,b) = c.  [assumption].
8  c = f(a,b).  [copy(7),flip(a)].
9  f(b,a) != c.  [deny(1)].
10 f(b,a) != f(a,b).  [copy(9),rewrite([8(4)])].
11 f(x,f(x,y)) = y.  [para(6(a,1),5(a,1,1)),rewrite([2(2)]),
     flip(a)].
12 f(x,f(y,f(x,y))) = e.  [para(6(a,1),5(a,1)),flip(a)].
15 f(x,f(y,x)) = y.  [para(12(a,1),11(a,1,2)),rewrite([3(2)]),
     flip(a)].
19 f(x,y) = f(y,x).  [para(15(a,1),11(a,1,2))].
20 $F.  [resolve(19,a,10,a)].
===================== end of proof ==================
```

Figure 14.3.2.: Output by Prover9.

14.3. Paramodulation

Exercise 14.2. Test the validity of the following arguments by applying resolution. First prove "by hand" and then in Prover9/Mace4. Compare both proofs, yours and the expanded proof by Prover9.[16]

1. $\{Q \to R, R \to (P \land Q), P \to (Q \lor R)\} \vdash P \leftrightarrow Q$.

2. $\{(\neg S \land V) \to \neg P, P, V\} \vdash S$.

3. $\{\neg R \leftrightarrow (\neg P \lor \neg Q), P \land Q\} \vdash R$.

4. $(P \lor Q) \land (P \lor R) \vdash \neg P \to (Q \land R)$.

5. $\{(P \lor Q) \land (P \lor R), P \to S, Q \to S, P \to T, R \to T\} \vdash S \land T$.

Exercise 14.3. Prove the soundness of the propositional resolution calculus, i.e.

$$\text{if } \vdash_{res} \phi, \text{ then } \models \phi$$

for a formula $\phi \in$ L0.

Exercise 14.4. Prove the completeness of the propositional resolution calculus, i.e.

$$\text{if } \models \phi, \text{ then } \vdash_{res} \phi$$

for a formula $\phi \in$ L0.

Exercise 14.5. Is the propositional resolution calculus strongly complete?

Exercise 14.6. Determine whether the following sets of clauses are satisfiable or unsatisfiable by means of the resolution calculus:

1.
$$C = \left\{\begin{array}{c} \|\neg P(y), Q(z, h(z))\| \\ \|S(z, f(x)), x), \neg P(x)\| \\ \|P(a)\| \\ \|\neg R(x, h(y)), \neg S(g(z), z, a)\| \\ \|\neg T(g(x)), \neg R(x, h(x))\| \\ \|R(x, y), \neg Q(z, y)\| \end{array}\right\}$$

---

[16] Given the first output by Prover9, click the button "Reformat..." on the top left corner of the proof window; next, in "Options" select "Expand proof."

14. The resolution calculus

2.
$$C = \begin{cases} \|P(x), \neg Q(x, f(y)), \neg R(a)\| \\ \|R(x), \neg Q(x, y)\| \\ \|\neg P(x), \neg Q(y, z)\| \\ \|P(x), \neg R(x)\| \\ \|Q(x, y), \neg P(y)\| \\ \|R(f(b))\| \end{cases}$$

Exercise 14.7. Prove the following arguments by applying resolution. First prove "by hand" and then in Prover9/Mace4. Compare both proofs, yours and the expanded proof by Prover9.

1. $\forall x \, (P(x) \to Q(x)) \vdash \forall x \, (P(x)) \to \forall x \, (Q(x))$

2. $\{\forall x \, (F(x)) \to \forall x \, (G(x)), \neg \forall x \, (G(x))\} \vdash \neg \forall x \, (F(x))$

3. $\{\forall x \, (F(x) \to G(x)), \forall x \, (Fx)\} \vdash G(a)$

4. $\{\exists x \, (P \to F(x)), \exists x \, (F(x) \to P)\} \vdash \exists x \, (P \leftrightarrow F(x))$

5.
$$\frac{\begin{array}{c} Q(x) \lor R(x) \\ R(x) \lor P(x) \lor \neg Q(f(x)) \\ R(x) \lor P(x) \lor T(x, y) \\ \neg R(x) \\ Q(a) \end{array}}{\exists x \forall y \, (P(x) \lor T(x, y))}$$

6.
$$\frac{\begin{array}{c} P(f(x)) \to S(g(y), y) \\ S(y, g(y)) \to (P(y) \lor R(y)) \\ Q(x) \to (P(f(x)) \lor R(x)) \\ \neg(\neg Q(x) \land \neg R(x)) \\ \neg S(x, y) \end{array}}{R(x)}$$

7.
$$\frac{\begin{array}{c} P(x) \to P(g(x)) \\ \neg (P(x) \land Q(a, g(x))) \\ Q(g(x), g(y)) \to (\neg (P(x) \land P(y)) \lor Q(x, y)) \\ Q(g(f(x, g(y))), f(g(x), g(y))) \\ P(a) \\ Q(f(x, a), x) \end{array}}{\neg Q(g(a), g(g(a)))}$$

Exercise 14.8. Prove in the FO resolution calculus the following reasoning instances:

14.3. Paramodulation

1. Exercise 2.10.7.

2. Exercise 2.10.8.

3. Exercise 2.10.10.

4. Exercise 2.13.

Exercise 14.9. Prove Theorem 14.16.

Exercise 14.10. With respect to Theorem 14.20:

1. Summarize the idea of the lifting lemma.

2. Complete the given proof.

Exercise 14.11. Complete the proof of Theorem 14.21.

Exercise 14.12. Is it enough to prove the completeness of the FO resolution calculus, or do we have to prove that it is

1. actually refutation-complete?

2. only refutation-complete?

Exercise 14.13. Prove the soundness of the FO resolution calculus.

Exercise 14.14. Apply plain binary resolution without refinements to the set of clauses of Example 14.26. Find the worst scenario, i.e. the resolution proof with the most redundant resolvents.

Exercise 14.15. Apply semantic resolution to the following sets of clauses given the corresponding interpretations:

1.
$$C = \begin{Bmatrix} \neg R(y) \\ \neg S(a) \\ \neg S(b) \\ \neg Q(a) \\ \neg Q(b) \\ P(a) \vee P(b) \\ \neg P(x) \vee Q(x) \vee S(x) \vee R(x) \end{Bmatrix} ;$$

$\mathcal{I} = \{P(x), \neg Q(x), R(x), S(x)\} ; \mathscr{D} = \{a, b\}$.

14. The resolution calculus

2.
$$C = \left\{ \begin{array}{c} P(x) \vee S(x) \vee Q \\ \neg P(a) \\ S(a) \vee \neg Q \\ R \vee \neg S(a) \\ \neg R \vee \neg S(z) \end{array} \right\} ;$$
$$\mathcal{I} = \{\neg P(a), Q, \neg S(a), R\}.$$

3.
$$C = \left\{ \begin{array}{c} T(y) \vee \neg R(a,b) \vee T(x) \\ S \\ P(x) \vee \neg Q(x,y) \vee \neg S \vee P(y) \\ \neg P(x) \\ \neg T(y) \vee \neg T(x) \\ R(x,y) \vee Q(x,y) \end{array} \right\} ;$$
$$\mathcal{I} = \{\neg P(x), \neg Q(x,y), \neg R(x,y), \neg S, \neg T(x)\} ; \mathcal{D} = \{a,b\}.$$

4.
$$C = \left\{ \begin{array}{c} \neg P(y) \vee \neg P(a) \\ Q(b) \vee R(c) \\ R(c) \vee P(z) \vee P(a) \vee \neg Q(z) \vee \neg Q(b) \end{array} \right\} ;$$
$$\mathcal{I} = \{P(x), \neg Q(x), \neg R(x)\} ; \mathcal{D} = \{a,b,c\}.$$

Exercise 14.16. Let $C = \{P, \neg P \vee Q, R \vee \neg P, \neg P \vee \neg Q \vee \neg R\}$. Prove that the set C is unsatisfiable by means of PI-resolution. The following cases should be considered:

1. $\mathcal{I} = \{\neg P, \neg Q, \neg R\}$; $R < Q < P$.
2. $\mathcal{I} = \{P, Q, R\}$; $R < P < Q$.
3. $\mathcal{I} = \{\neg P, \neg Q, R\}$; $P < Q < R$.

Exercise 14.17. Apply PI-resolution to the sets of clauses of Exercise 14.15 by specifying at least two different predicate-symbol orderings for each.

Exercise 14.18. The PI-refutation of Example 14.32 is not a PI-deduction, as the resolvent C_5 is neither a clause in C nor a PI-resolvent. Nevertheless, it can be shown that a PI-deduction is obtainable from C. Show how by appealing to Theorem 14.34.

14.3. Paramodulation

Exercise 14.19. Given the following set of clauses and respective predicate-symbol ordering

$$C = \left\{ \begin{array}{c} Q(x) \vee P(c) \\ \neg P(x) \vee Q(x) \\ \neg R(c) \vee \neg Q(c) \\ R(x) \end{array} \right\},$$

$$\mathscr{P} = Q < P < R$$

obtain:

1. a refutation by positive hyper-resolution.

2. a refutation by negative hyper-resolution.

Exercise 14.20. Given a set of clauses C and a subset $D \subset C$, we say that D is a *set of support* of C if $C - D$ is satisfiable, and we call a resolution of two clauses that are not both from $C - D$ a *set-of-support resolution* (SSR). Then, a *set-of-support deduction* is a deduction in which every resolution is a SSR.

1. Apply SSR to the sets of clauses of Exercise 14.15.

2. Show that SSR is a kind of semantic resolution.

3. Prove the completeness of SSR.

Exercise 14.21. Prove that the following are refinements of resolution:

1. Semantic resolution

2. Hyper-resolution.

3. SSR.

Exercise 14.22. With respect to linear resolution:

1. Show that linear resolution is a resolution refinement.

2. Prove the completeness of linear resolution.

3. Show that the following set of clauses is unsatisfiable by applying linear resolution:

$$C = \left\{ \begin{array}{c} \|P(x), \neg Q(x, f(y)), \neg R(a)\| \\ \|R(x), \neg Q(x, y)\| \\ \|\neg P(x), \neg Q(y, z)\| \\ \|P(x), \neg R(x)\| \\ \|R(f(b))\| \\ \|Q(x, y), \neg P(y)\| \end{array} \right\}$$

Exercise 14.23. With respect to LI resolution:

1. Show that LI resolution is not generally complete.

2. Show that LI resolution is complete for Horn clauses.

3. Show that LI resolution is a resolution refinement.

4. Show that the set of clauses of Exercise 14.22.3 is unsatisfiable by applying LI resolution.

Exercise 14.24. With respect to LD resolution:

1. Show that LD resolution is complete for Horn clauses.

2. Show that LD resolution is a resolution refinement.

3. Prove the unsatisfiability of the set of clauses of Exercise 14.22.3 by applying LD resolution.

Exercise 14.25. With respect to SLD resolution:

1. Elaborate on the SLD resolution rule.

2. Prove that SLD resolution is complete for Horn clauses.

3. Prove that SLD resolution is a LD-resolution refinement.

4. Prove the unsatisfiability of the set of clauses of Exercise 14.22.3 by applying SLD resolution.

Exercise 14.26. Consider the predicates male/1, sibling/2, and parent/2.

1. From these, define a predicate `uncle/2`.

2. Change the program obtained in 1 so that it is

   a) incomplete.
   b) incorrect.

Exercise 14.27. Show that a set of Horn clauses is unsatisfiable iff it contains at least one fact and one goal clause.

Exercise 14.28. With respect to the argument in Exercise 2.13, write it as a Prolog program in SWI-Prolog.

1. Verify if the conclusion holds and construct the corresponding SLD-resolution refutation tree.

2. Input the questions in Exercise 5.16 as queries and obtain the replies thereto.

Exercise 14.29. Draw the search tree for the query `?-p(b,b).` in Figure 14.2.14.

Exercise 14.30. Consider the program *Fatherhood* with the following additional rule and fact:

$$\text{son}(X, Y) \leftarrow \text{father}(Y, X), \text{male}(X).$$

$$\text{son}(\text{louis}, \text{harry}).$$

Given the input `son(louis, harry)?`, provide the trace of an interpreter Ψ and the final output by Ψ.

Exercise 14.31. What is the result in terms of computation of adding a rule of the form $p \leftarrow p$ to a Prolog program?

Exercise 14.32. Given a Prolog program π and a query clause $\mathcal{G}$, exactly one of the five following SLD trees $\mathcal{T}_{\pi,\mathcal{G}}$ for $\pi \cup \{\mathcal{G}\}$ exists. Explain when each SLD tree is the case.

1. $\mathcal{T}_{\pi,\mathcal{G}}$ terminates.

2. $\mathcal{T}_{\pi,\mathcal{G}}$ diverges.

3. $\mathcal{T}_{\pi,\mathcal{G}}$ potentially diverges.

14. The resolution calculus

4. $\mathcal{T}_{\pi,\mathcal{G}}$ gives infinitely many answers.

5. $\mathcal{T}_{\pi,\mathcal{G}}$ fails.

Exercise 14.33. A *breadth-first search* traverses a SLD-resolution tree level by level. What is the computational advantage, in terms of complexity, of a depth-first over a breadth-first search?

Exercise 14.34. Prove the soundness of SLD resolution for Prolog.

Exercise 14.35. Prove Lemmas 14.55 and 14.56.

Exercise 14.36. Complete the proofs of Theorems 14.56 and 14.60.

Exercise 14.37. Another way to prove the completeness of SLD resolution for Prolog is via the *least Herbrand model* for a Prolog program π, denoted by $\underline{HM}_\pi$.

1. Research into this proof and give its details, not forgetting to define $\underline{HM}_\pi$.

2. Explain why it can be stated

$$\underline{HM}_\pi = M(\pi)$$

where $M(\pi)$ denotes the meaning of a Prolog program π (cf. Def. 14.46).

Exercise 14.38. Prolog programs are often *recursive*. Define this property and explain what it entails from the viewpoint of logic programming.

Exercise 14.39. Find the binary paramodulants of the following pairs of clauses:

1. $P(f(x,a),y) \vee R(y)$ and $(f(c,a) = g(b)) \vee R(g(b))$
2. $(f(x) = b) \vee P(x)$ and $Q(f(a)) \vee R$
3. $(f(g,g(x)) = a) \vee P(x)$ and $R(y, f(g(y), z)) \vee S(z)$
4. $(f(g(b)) = a) \vee R(g(c))$ and $P(g(f(x))) \vee Q(x)$
5. $Q(g(a)) \vee (f(h(z),z) = g(z))$, $\neg P(f(x,g(y))) \vee (k(x,g(y)) = h(y))$

Exercise 14.40. Resolution-based paramodulation makes implicit use of the following equality axioms (cf. Section 7.2):

$\mathcal{E}1$ $\forall x \, (x = x)$
$\mathcal{E}2$ $\forall x_i \forall y_i \, [\neg (x_1 = y_1) \vee ... \vee \neg (x_n = y_n) \vee (f(x_1, ..., x_n) = f(y_1, ..., y_n))]$
$\mathcal{E}3$ $\forall x_i \forall y_i \, [\neg (x_1 = y_1) \vee ... \vee \neg (x_n = y_n) \vee \neg P(x_1, ..., x_n) \vee P(y_1, ..., y_n)]$

Show that $\mathcal{E}2$ and $\mathcal{E}3$ above are equivalent to $\mathcal{E}4$ and $\mathcal{E}5$ of Proposition 7.5, respectively.

Exercise 14.41. Given the theory of distributive lattices (cf. Example 14.37), we wish to know whether the following property belongs to a distributive lattice:
$$x = x \wedge y \leftrightarrow x \vee y = y.$$
Prove that it does in Prover9/Mace4 and analyze the proof by Prover9.

Exercise 14.42. Apply ordered paramodulation to the clauses of Example 14.72.

Exercise 14.43. Prove the refutation-completeness of

1. ordered paramodulation.

2. simultaneous paramodulation.

Exercise 14.44. RUE resolution, denoting "resolution with unification and equality," is based on the rule

$$\frac{L(t_1, ..., t_n) \vee \mathcal{C} \qquad \neg L(s_1, ..., s_n) \vee \mathcal{D}}{\sigma \mathcal{C} \vee \sigma \mathcal{D} \vee (t_1 \neq s_1) \vee ... \vee (t_n \neq s_n)}$$

where σ is a substitution.

1. Elaborate on the RUE resolution principle.

2. Prove the soundness of the RUE resolution rule.

3. Prove the completeness of RUE resolution.

15. The analytic tableaux calculus

Not as successful in the field of ATP as the resolution calculus (Chapter 14), analytic tableaux is also a popular refutation calculus. This proof system was firstly conceived as *semantic tableaux* by Beth (1955) and Hintikka (1955), whose concerns were mostly semantical; it was later greatly simplified by Smullyan (1968) into the variant known as *analytic tableaux*. We concentrate on the latter.

Analytic tableaux (henceforth often just: tableaux) is a remarkably efficient indirect proof calculus for classical logic, and some non-classical logics as well, based on labeled binary trees.[1] Briefly, analytic tableaux are implemented by binary trees whose nodes are formulas that are (sub)goals in the proofs; the tree structure concretizes the logical dependence among the (sub)goals. A tree constitutes a proof iff it is a closed tableau, i.e. a tree whose every branch has a contradictory pair of literals.

Although tableaux for CFOL have some additional features with respect to those for CPL, the algorithm to be applied is essentially the same and we give it in this short introduction to this calculus. This is Algorithm 15.1. We refer the reader to the next Sections in this Chapter for the specific jargon in this algorithm.[2]

Recall the definitions above (Section 4.2) of (semi-)decidability. In fact, given a finite set of formulas $X \subseteq F_L$, the tableaux proof is guaranteed to *terminate* in either a closed or an open tableau, being thus indeed a decision procedure–albeit only for propositional logic. If we allow for X being infinite, then the tableau construction is guaranteed

---

[1] Priest (2008) gives tableaux proofs for a large selection of non-classical logics. See Augusto (2020c) for a comprehensive discussion of analytic tableaux for many-valued logics.

[2] While in the general algorithm for resolution (cf. Algorithm 14.1) we emphasized the binary tree implementing the single rule of binary resolution, in this algorithm for analytic tableaux proofs we emphasize the several rules that the binary tree implements. Note that this Algorithm 15.1 is conceived as a refutation proof algorithm, reason why it is considered to terminate successfully only if the tree is closed (cf. Step 4); this notwithstanding, one may apply the tableaux calculus to test for satisfiability.

15. *The analytic tableaux calculus*

Algorithm 15.1 Analytic tableaux proof.

- **Input:** A set of formulas $X \subseteq F_L$
- **Output:** A closed/open tableau $\mathcal{T}_X$ of X

Steps:

1. INITIALIZATION: In the root of a downwards-growing binary tree, put:

    a) $\neg \chi$ if $X = \{\chi\}$;

    b) $\chi_1 \wedge \ldots \wedge \chi_{n-1} \wedge \neg \chi_n$ if $X = \{\chi_1, \ldots, \chi_{n-1}, \chi_n\}$, $\chi_1, \ldots, \chi_{n-1}$ are the premises and χ_n is the conclusion of some formalized argument.

2. Apply the EXPANSION RULES to the sub-formulas on the tree until they can no longer be applied (i.e. there are only literals).

3. CLOSE contradictory branches (i.e. branches containing both L and $\neg L$, where L is a literal).

4. TERMINATE successfully iff all branches are closed; unsuccessfully, otherwise.

---

to terminate only if X is unsatisfiable, running forever if X is satisfiable. Thus, in the latter case we speak of semi-decidability.

Although some proof *assistants* are available, which require interaction between the human computer and the software, there is as yet no full-blown *prover* for this calculus for CFOL. Nevertheless, there are already a few web-based interactive provers that can be used for low-complexity FO theories. For example, Tableaux Package, available at

http://hackage.haskell.org/package/tableaux

and Tree Proof Generator, available at

https://www.umsu.de/logik/trees/.

15.1. Analytic tableaux as a propositional calculus

We now expand on, and formalize, the above introductory remarks and Algorithm 15.1 in terms of a propositional calculus for classical logic.

Definition 15.1. Let the language L0 be given and a non-empty set of formulas $X \subseteq F_{L0}$. A *tableau* is a finite binary tree $\mathcal{T}$ whose nodes are (sub-)formulas from X. A *tableau proof system* (or *calculus*) over X is a set $RT = \{\mathbf{I}, \mathbf{A}, \mathbf{B}, \mathsf{X}\}$ of *initialization rule(s)* $\mathbf{I}$, *expansion rules* $\mathbf{A}$ and $\mathbf{B}$ each over a set $X' \subseteq X$, and *closure rule(s)* X.

For each set X, the transitive closure of the rules in RT defines a set of tableaux constructed with RT for X. We denote the tree for X resulting from the application of the rules in RT over X by $\mathcal{T}_X$.

Definition 15.2. In a tableau $\mathcal{T}_X$ for a set of formulas $X \subseteq F_{L0}$, a branch $\mathcal{B}$ of $\mathcal{T}_X$ is *closed* iff $\mathcal{B} \cup X'$ for $X' \subseteq X$ contains either a pair of literals $(L, \neg L) \in F_{L0}$ or $\bot$. Otherwise, $\mathcal{B}$ is *open*.

1. A tableau $\mathcal{T}_X$ is *closed* iff all its branches are closed; otherwise, it is *open*.

Definition 15.3. A *tableau proof* of a set $X \subseteq F_{L0}$ is a closed tableau $\mathcal{T}_X$.

We now formalize Algorithm 15.1 for CPL.

15. The analytic tableaux calculus

Definition 15.4. Given a set $X \subseteq F_{L0}$, a *tableau for X* is defined as a tableau constructed according to the following rules:

1. *Initialization rule:* The tree $\mathcal{T}_X$ consisting of a single node t is a tableau for X.

2. *Expansion rule:* Let $\mathcal{T}_X$ be a tableau for X, and $\mathcal{B}$ a branch of $\mathcal{T}_X$. Let further χ be a formula in $\mathcal{B} \cup X'$, $X' \subseteq X$. Obtain the tree $\mathcal{T}'_X = \mathcal{T}_X$ by extending $\mathcal{B}$ with n new subtrees whose nodes are the sub-formulas in the extension of the rule instance for χ. Then, $\mathcal{T}'_X$ is a tableau for X.

3. *Closure rule:* Given a tableau $\mathcal{T}_X$ and a branch $\mathcal{B}$ thereof, if we have $(L, \neg L) \in \mathcal{B} \cup X'$, $X' \subseteq X$, L is a propositional atom, then $\mathcal{B}$ is a closed branch of $\mathcal{T}_X$. If all branches of $\mathcal{T}_X$ are closed, then $\mathcal{T}_X$ is a closed tableau for X.

The closing of a branch is typically indicated by the symbol ✗. The notions of unicity of decomposition and immediate sub-formula (cf. Prop. 2.5 and Def. 2.6) allow us to act in such a way as to have a final set of (negated) propositional atoms by proceeding to a step-by-step decomposition of complex formulas. Underlying the construction of the tree is the rewriting of all the formulas into equivalent (negations of) conjunctions and disjunctions in the usual ways, being a tree thus in fact a disjunction of conjunctions. Now, recall that for literals L and $\neg L$ we have $(L \wedge \neg L) = \bot$, and that in turn $(\bot \vee \bot)$ is also a contradiction; this explains why a disjunction of finitely many $\bot$-containing branches constitutes a closed tree, i.e. a contradiction of the negated formula. Recall now the classicality conditions for the $\heartsuit$-consequences in Section 6.1.

Definition 15.5. The expansion rules and the closure rule, expressed in the language of set theory and accounted for in terms of the classical logical consequence relation, are as follows:

1.
$$(\wedge_{RE}) \quad \frac{X \cup \{\phi \wedge \psi\}}{X \cup \{\phi, \psi\}}$$

given that $\wedge$ is classical in terms of $\Vdash$ iff we have $X, \phi \wedge \psi \Vdash \chi$ iff $X, \phi, \psi \Vdash \chi$.

15.1. Analytic tableaux as a propositional calculus

2.
$$(\vee_{RE}) \quad \frac{X \cup \{\phi \vee \psi\}}{X \cup \{\phi\} \mid X \cup \{\psi\}}$$

given that $\vee$ is classical in terms of $\Vdash$ iff we have $X, \phi \vee \psi \Vdash \chi$ iff $X, \phi \Vdash \chi$ and $X, \psi \Vdash \chi$.

3.
$$(\mathsf{X}) \quad \frac{X \cup \{L, \neg L\}}{\mathsf{X}}$$

because classically we have $X, \phi, \neg\phi \Vdash \chi$, i.e. *ex contradictione quodlibet*, or "explosion."

All the expansions over some complex formula ϕ can be reduced to two rules in what is known as the $\alpha\beta$-*classification* (Fig. 15.1.1).[3]

α	α_1	α_2
$A \wedge B$	A	B
$\neg(A \vee B)$	$\neg A$	$\neg B$
$\neg(A \to B)$	A	$\neg B$
$\neg\neg A$	A	A

β	β_1	β_2
$\neg(A \wedge B)$	$\neg A$	$\neg B$
$A \vee B$	A	B
$A \to B$	$\neg A$	B

Figure 15.1.1.: Analytic tableaux expansion rules: $\alpha\beta$-classification.

Definition 15.6. There are thus in effect only two expansion rules, **A** (for α) and **B** (for β), where α and β are the *formula types* in Figure 15.1.1:

$$(\mathbf{A}) \quad \frac{\alpha}{\begin{array}{c}\alpha_1 \\ \alpha_2\end{array}}$$

$$(\mathbf{B}) \quad \frac{\beta}{\beta_1 \mid \beta_2}$$

Informally, these two rules mean that if α is a conjunction of α_1 and α_2, then both these two sub-formulas are logical consequences of α (rule **A**) and thus are nodes in the one and same branch of the tree (i.e.,

---

[3] By convention, doubly negated formulas are treated as formulas of type α, rather than as an elimination rule. One can further treat $\neg\top$ and $\neg\bot$ as type-α formulas, resulting in the nodes labeled with $\bot$ and $\top$, respectively.

15. The analytic tableaux calculus

analytic tableau). Graphically, we have a single branch $\mathcal{B}$ of the form[4]

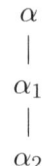

If $\mathcal{B}$ is a disjunction of β_1 and β_2, then these sub-formulas originate two different branches $\mathcal{B}', \mathcal{B}''$ as a logical consequence of β (rule **B**); graphically, we have

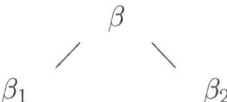

As said above, a tree is a disjunction of conjunctions; this is the result of giving priority to α-type formulas over β-type formulas in terms of formula decomposition.[5]

Theorem 15.7. *For any propositional tableau, after a finite number of steps no more expansion rules will be applicable.*

Proof: (Sketch) The theorem holds assuming that we analyze each formula at most once. Let us begin by assuming that we have a formula with $n = 0$ connectives. Then, this is a propositional atom, and no expansion rules apply. Let us now assume that the theorem holds for any formula with at most n connectives. Then we can prove it for a formula ϕ with $n + 1$ connectives. This can be done in two ways, depending on the type of the formula:

1. ϕ is a formula of type α. We apply rule **A**, and we mark ϕ as analyzed once. Clearly, α_1 and α_2 contain each fewer connectives with relation to ϕ; we apply the inductive hypothesis, and say that a tableau can be built such that each formula is analyzed at most once. After

---

[4] For graphical convenience, this can be abbreviated as

$$\alpha \\ | \\ \alpha_1 \\ \alpha_2$$

or even simpler as

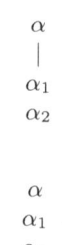

[5] A tableau can be a conjunction of disjunctions, but this typically increases the size of the tableau, and it is thus a less efficient proof procedure.

15.1. Analytic tableaux as a propositional calculus

a finite number of steps, no more expansion rules can be applied. We combine the two proofs, and the theorem is proved for a type-α formula.

2. ϕ is a formula of type β. We apply rule **B**, and we mark ϕ as analyzed once. Clearly, β_1 and β_2 contain each fewer connectives with relation to ϕ; we apply the inductive hypothesis and say that two tableaux can be built for β_1 and β_2 such that each formula is analyzed at most once. After a finite number of steps, no more expansion rules can be applied. We combine the two proofs, and the theorem is proved for a type-β formula. **QED**

Recall the sub-formula property of a sequent-calculus proof (cf. Prop. 12.40). It is evident that this property is at play in the proof above. In effect, we say that an expansion rule is *analytic* whenever its every application yields the sub-formula property. Formalizing this:

Definition 15.8. We say that an expansion rule is *analytic* if every formula that occurs as a conclusion of the rule–i.e. α_i, β_i, for $i = 1, 2$–is a sub-formula of the formula that occurs as the premise of the rule (i.e. α, β).

This shows the interesting relation between the sequent calculus and analytic tableaux: a downward growing tree in the latter is an upward growing tree in the former.

Example 15.9. Figure 15.1.2 shows the tableaux proof of the formula

$$A = ((P \to Q) \land ((P \land Q) \to R)) \to (P \to R).$$

We show the step-by-step application of Algorithm 15.1. We put the negated formula, $\neg A$, at the root of the tree $\mathcal{T}_A$ for A, according to Step 1. By applying first $\to_{def}$ and then DeM$_\lor$ we obtain the formula $((P \to Q) \land ((P \land Q) \to R)) \land \neg (P \to R)$ equivalent to $\neg A$. This is clearly a type-α formula, and we apply the corresponding **A** expansion rule, obtaining two consecutive nodes on the main branch of $\mathcal{T}_A$, to wit, $\alpha_1 = (P \to Q) \land ((P \land Q) \to R)$ and $\alpha_2 = \neg(P \to R)$. These are not atomic formulas, and thus further application of expansion rules is required. With respect to α_1, this is again a type-α formula and we proceed as above. By applying first $\to_{def}$ and then DeM$_\lor$ to $\alpha_2 = \neg(P \to R)$ we obtain $P \land \neg R$, again a type-α formula; we apply again expansion rule **A**, and we obtain two atoms on the main branch of $\mathcal{T}_A$, to wit, P and $\neg R$. We go back to the decomposed formulas $P \to Q$ and $(P \land Q) \to R$ obtained by applying expansion rule **A** to the formula

15. The analytic tableaux calculus

$(P \to Q) \wedge ((P \wedge Q) \to R)$: By applying $\to_{def}$ we obtain the type-β formula $\neg P \vee Q$; the same procedure gives us $\neg(P \wedge Q) \vee R$. Both are type-β formulas and we apply expansion rule **B** until no more expansion rules are applicable. This was Step 2 of Algorithm 15.1. Application of Step 3 closes all the branches of $\mathcal{T}_A$, and the algorithm terminates successfully in this case: Formula A is valid. In Figure 15.1.2, all the steps and corresponding application of the expansion rules are indicated on the left and on the right of the subformulas.[6]

Although analytic tableaux is a proof system, it actually is a hybrid system in the sense that a tableau proof is a *counter-model* with respect to some semantics. In effect, a tableaux proof procedure is so with respect to the negation of the formula one wishes to prove. Informally, a counter-model corresponds to a tree whose branches are partial descriptions of the model, where the $\alpha\beta$-classification is to be understood as follows:

Proposition 15.10. *Under any interpretation $\mathcal{I}$, the following facts clearly hold:*

1. (α_T) α *is true iff* α_1, α_2 *are both true;*
2. (β_T) β *is true iff at least one of* β_1, β_2 *is true.*

Proof: Left as an exercise.

As it is already known, a branch in a tree is said to close if both L and $\neg L$ are in it (i.e. if they are nodes of the same branch), and the tree itself is said to close if all its branches close. We next formalize this, leaving all the proofs as exercises.

Proposition 15.11. *A tableau $\mathcal{T}_X$ for a set of formulas $X \subseteq F_{L0}$ is satisfiable iff there is a model $\mathcal{M}_X$ of X such that for every valuation val of the formulas of X there is a branch $\mathcal{B}$ of $\mathcal{T}_X$ with $\models_{val_\mathcal{M}} \mathcal{B}$. We then say that $\mathcal{M}$ is a model of $\mathcal{T}_X$, and we write $\models_\mathcal{M} \mathcal{T}_X$.*

Proposition 15.12. *It is evident that no model or valuation can satisfy a closed branch, so a closed tree $\mathcal{T}_X$ is a proof of the unsatisfiability of X.*

---

[6]This is merely an auxiliary graphical means, and may be dispensed with.

15.1. Analytic tableaux as a propositional calculus

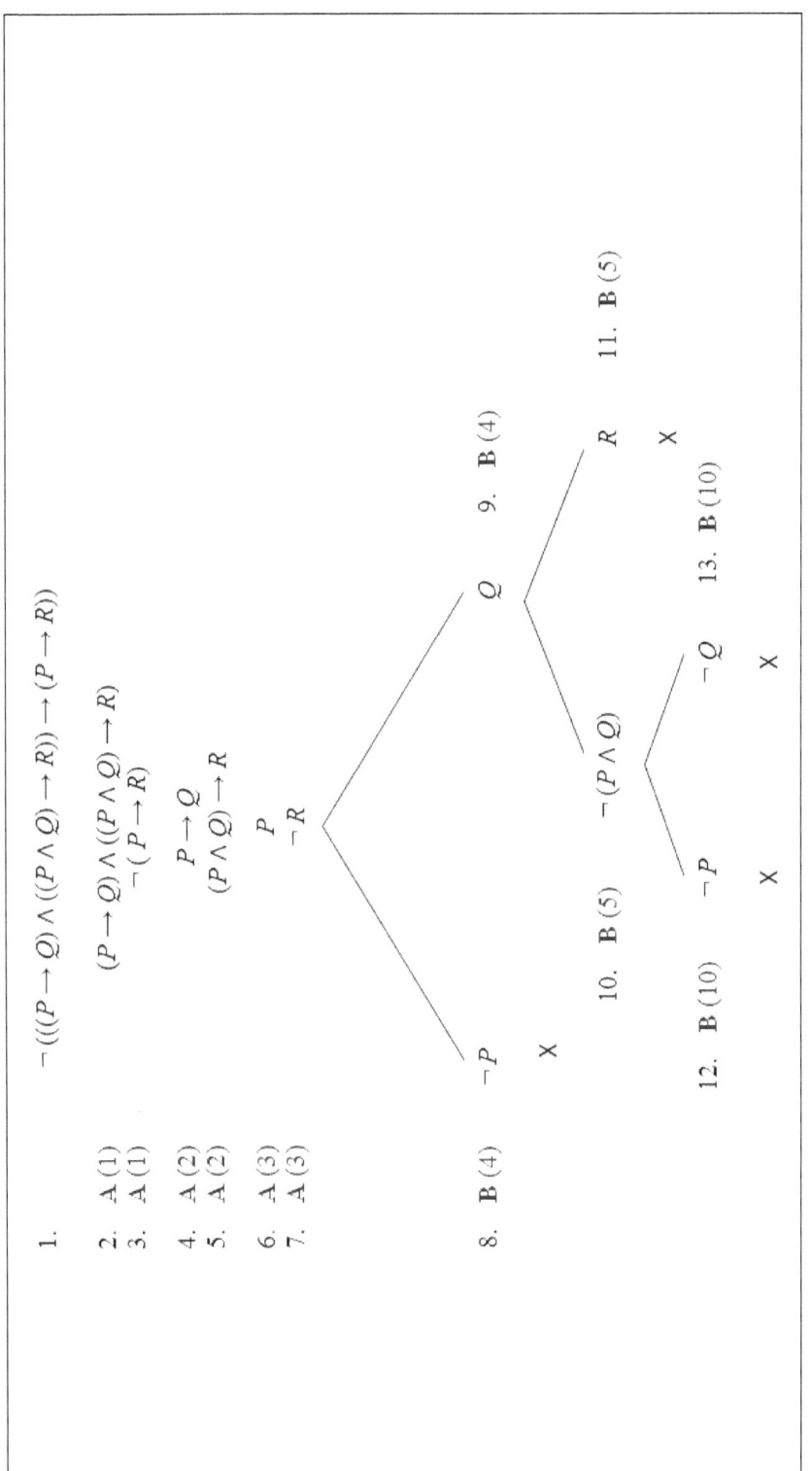

Figure 15.1.2.: A closed propositional tableau.

15. The analytic tableaux calculus

In terms of satisfiability, we have the following facts with respect to the expansion rules for formulas of types α and β:

Proposition 15.13. *Let $X \subseteq F_{L0}$ be a set of formulas. Then the following facts hold:*

1. (α_{sat}) If X is satisfiable and $\alpha \in X$, then $\{X, \alpha_1, \alpha_2\}$ is satisfiable.

2. (β_{sat}) If X is satisfiable and $\beta \in X$, then at least one of $\{X, \beta_1\}$, $\{X, \beta_2\}$ is satisfiable.

Propositions 15.10 and 15.13 are fundamental in that they state that the decomposition of formulas in the propositional tableaux calculus preserves truth and satisfiability.

We now require a few further definitions in order to address the soundness and completeness of the propositional tableaux calculus.

Definition 15.14. *Let $\mathcal{T}$ be a tableau. We say that a branch $\mathcal{B} \in \mathcal{T}$ is* complete *if for every α occurring in $\mathcal{B}$, both α_1 and α_2 occur in $\mathcal{B}$, and for every β occurring in $\mathcal{B}$, at least one of β_1, β_2 occurs in $\mathcal{B}$. We call a tableau $\mathcal{T}$* complete(d) *if every branch of $\mathcal{T}$ is either closed or complete.*

Theorem 15.15. *(Soundness of propositional analytic tableaux) Every formula provable by the analytic tableaux calculus is a tautology.*

Proof: (Sketch) Let X be a set of formulas, possibly a singleton, and let $\mathcal{I}$ be an interpretation. Let $\models_\mathcal{I} \mathcal{T}_X^{(0)}$, $\mathcal{T}_X^{(0)}$ is the completed tableau for X. Then, for any subtree $\mathcal{T}_X^i$, $0 < i \leq n$, we have $\models_\mathcal{I} \mathcal{T}_X^i$ by Propositions 15.10 and 15.13, i.e. any interpretation satisfying all the formulas in X must also satisfy all the formulas contained in at least one complete branch of any completed tableau for X. Thus, for any completed tableau $\mathcal{T}$, and applying induction and the analyticity property of the tableaux calculus (Def. 15.8), if the root is true under an interpretation $\mathcal{I}$, then $\mathcal{T}$ must be true under $\mathcal{I}$. It is evident (by Proposition 15.12) that a closed tableau cannot be true under any interpretation, and therefore its root cannot be true under any interpretation. Hence, every formula provable by the tableaux algorithm must be a tautology. **QED**

Corollary 15.16. *(Consistency of propositional analytic tableaux) The tableau method for classical propositional logic is consistent.*

15.1. Analytic tableaux as a propositional calculus

Proof: From the proof of the soundness theorem, we conclude that no formula and its negation are both provable in the tableaux calculus. **QED**

We turn now to the completeness theorem of analytic tableaux. In order to prove this theorem, a few more fundamental notions are required.

Definition 15.17. Let $X \subseteq F_{L0}$. We say that X is a *downward saturated set* iff

1. if $\alpha \in X$, then $\alpha_1 \in X$ and $\alpha_2 \in X$;
2. if $\beta \in X$, then $\beta_1 \in X$ or $\beta_2 \in X$.

Note that this entails that a tableau $\mathcal{T}_X$ for a downward saturated set X is completed in the sense that every branch $\mathcal{B}$ of $\mathcal{T}_X$ is complete.

Definition 15.18. A downward saturated set that does not contain an (atomic) formula and its negation is called an *(atomic) Hintikka set*.

Lemma 15.19. *(Hintikka's lemma) Every downward saturated set X (whether finite or infinite) is satisfiable.*

Proof: Let X be a Hintikka set. We need to show that some Herbrand interpretation $H\mathcal{I}_X$ of X is a model of X.[7] Let H_X be the set of ground terms in X. Because $H(X)$ by definition (of X as a Hintikka set) does not contain an atomic formula and its negation, it immediately defines a Herbrand interpretation $H\mathcal{I}_X$. Assume that $H\mathcal{I}_\phi = \mathbf{t}$ for all formulas $\phi \in X$ with complexity less than n, where by complexity of a formula χ it is meant the number n of occurrences of sub-formulas in χ. Now consider any complex formula $\psi \in X$ with complexity n. It is evident that the complexity of any tableau sub-formula of ψ is less than n. Clearly, ψ must be a α- or a β-type formula, and we have two cases.

Case 1: If ψ is a α-type formula, then by definition of downward saturation every α_i is in X. By induction, and by $\alpha_\top$ (Prop. 15.10.1), we have $H\mathcal{I}_{\alpha_i} = \mathbf{t}$, $H\mathcal{I}_{\phi,\psi} = \mathbf{t}$, and of course $H\mathcal{I}_X = \mathbf{t}$.

Case 2: If ψ is a β-type formula, then by definition of downward saturation some β_i is in X. By induction, and by $\beta_\top$ (Prop. 15.10.2), we have $H\mathcal{I}_{\beta_i} = \mathbf{t}$, $H\mathcal{I}_{\phi,\psi} = \mathbf{t}$, and of course $H\mathcal{I}_X = \mathbf{t}$. **QED**

The proof of Hintikka's lemma, together with α_{sat} and β_{sat} (Prop.s 15.13.1-2), actually proves the following important theorem:

---
[7] Recall the contents of Section 9.2.

Theorem 15.20. *(Smullyan, 1968) Any complete open branch of any tableau is (simultaneously) satisfiable.*

In turn, this theorem implies the completeness of the propositional tableaux calculus.

Theorem 15.21. *(Completeness of propositional analytic tableaux) (a) If ψ is a tautology, then every complete tableau starting with $\neg\psi$ must close. (b) Every tautology is provable in the tableaux calculus.*

The derivation of (a) from Theorem 15.20 runs as follows: Assume that $\mathcal{T}_\psi$ is a complete tableau starting with $\neg\psi$. By Theorem 15.20, if $\mathcal{T}_\psi$ is open, then $\neg\psi$ is satisfiable. Therefore, ψ cannot be a tautology. Hence, if ψ is a tautology, then $\mathcal{T}_\psi$ must be closed.

The proofs of Theorems 15.20-1 are left as exercises.

▶ *Do Exercises 15.1-6.*

15.2. Analytic tableaux as a FO predicate calculus

The introduction of analytic tableaux rules for the quantifiers gives rise to some problems, in particular because the set of constants is infinite, a problem that actually accounts for the non-terminating character of tableaux proofs in FOL. In effect, Theorem 15.7 does *not* hold for FO tableaux. In any case, these problems can be overcome in several ways, so that the tableaux calculus is sound and complete for FOL. Our elaboration of this FO calculus falls only on the essential points thereof; in particular, we leave the proofs of soundness and completeness as exercises, and we refer the reader interested in applying this calculus to CL$^=$ to Fitting (1996).

Definition 15.22. Let Υ be a signature for the language L1. A *tableau (over Υ)* is a finite tree whose nodes are formulas from F_{L1}. A *tableau proof system* (or *calculus*) over F_{L1} is a set $RT = \{\mathbf{I}, \mathbf{A}, \mathbf{B}, \mathbf{C}, \mathbf{D}, \mathsf{X}\}$ where $\mathbf{I}, \mathbf{A}, \mathbf{B}, \mathsf{X}$ are essentially as in Definition 15.1 and $\mathbf{C}, \mathbf{D}$ are additional *rules of expansion for FO formulas*.

Definitions 15.1-3 are thus generalizable to classical FOL; the same holds for formulas of types α and β (with the proviso that now by "formula" it is understood "closed formula of quantification theory"; cf. Smullyan, 1968). This sets the formal scenario for approaching tableaux for FOL.

15.2. Analytic tableaux as a FO predicate calculus

Just as in the case of propositional formulas, there are only two expansion rules when quantifiers are involved: this is the $\gamma\delta$-classification (Fig. 15.2.1). It is evident that Definition 15.8 generalizes to FOL.

γ	$\gamma(a)$
$\forall x A(x)$	$A(a)$
$\neg \exists x A(x)$	$\neg A(a)$

δ	$\delta(a)$
$\exists x A(x)$	$A(a)$
$\neg \forall x A(x)$	$\neg A(a)$

Figure 15.2.1.: Analytic tableaux expansion rules: $\gamma\delta$-classification.

Definition 15.23. Formulas with the universal quantifier are classified as type γ, and those with the existential quantifier are of type δ. The corresponding tableaux rules for the quantified formulas are

$$(\mathbf{C}) \quad \frac{\gamma}{\gamma(a)}$$

where a is a ground term or a Skolem constant (more strictly: a parameter), and

$$(\mathbf{D}) \quad \frac{\delta}{\delta(a)}$$

where a is new to the branch. By $\gamma(a)$ or $\delta(a)$ it is meant, for a formula ϕ, ϕ_a^x or $\neg\phi_a^x$, i.e. the formula ϕ with x substituted by a.

Just as in the case of the $\alpha\beta$-classification, there is a semantical account for the expansion rules for formulas of types γ and δ that assures us that the decomposition of these latter types of formulas preserves truth and satisfiability. We leave the proofs of the following propositions as exercises.

Proposition 15.24. *Under any interpretation $\mathcal{I}$ for a domain $\mathcal{D}$, the following facts hold:*

1. (γ_T) γ is true iff $\gamma(a)$ is true for every $a \in \mathcal{D}$.

2. (δ_T) δ is true iff $\delta(a)$ is true for at least one $a \in \mathcal{D}$.

In terms of satisfiability, we have the following facts:

Proposition 15.25. *Let $X \in F_{L1}$ be a set of formulas. Then,*

377

15. The analytic tableaux calculus

1. (γ_{sat})   If X is satisfiable and $\gamma \in X$, then for every constant (more strictly: parameter) a the set $\{X, \gamma(a)\}$ is satisfiable.

2. (δ_{sat})   If X is satisfiable and $\delta \in X$, and if a is a constant (more strictly: parameter) that does not occur in any element of X, then $\{X, \delta(a)\}$ is satisfiable.

However, the above classifications are all too general. Importantly, we need to discuss separately tableaux for FOL with and without unification, as the expansion rules for the quantifiers are different for each.

15.2.1. FOL tableaux without unification

To begin with, some useful heuristics: whenever possible, apply propositional rules before quantifier rules; in the latter case, apply the tableau rules on δ-formulas before γ-formulas. "Beyond this, you are on your own" (Fitting, 1996), a remark that works as a reminder that analytic tableaux is *not* a decision procedure for FOL, for the simple reason that no such procedure exists. In particular, misapplication of the tableau rule for γ-formulas may cause the tableau not to close.

Definition 15.26. The analytic tableaux rules for the quantifiers without unification are

$$(\mathbf{C}) \quad \frac{\forall x \gamma(x)}{\gamma(t)}$$

where t is any ground term, and

$$(\mathbf{D}) \quad \frac{\exists x \delta(x)}{\delta(c)}$$

where the constant symbol c is new to the branch. Note that in $\exists x \delta(x)$, the variable x does not occur within the scope of a universal quantifier. Therefore, the Skolemization of $\exists x \delta(x)$ generates solely a constant, i.e. a 0-ary function.

Example 15.27. To prove by means of analytic tableaux the validity of the formula $\forall x (F(x) \land G(x)) \leftrightarrow \forall x (F(x)) \land \forall x (G(x))$ we apply rules **C** and **D**, i.e. the expansion rules for the universal quantifier without unification. See Fig. 15.2.2 for the respective tableau proof.

15.2. Analytic tableaux as a FO predicate calculus

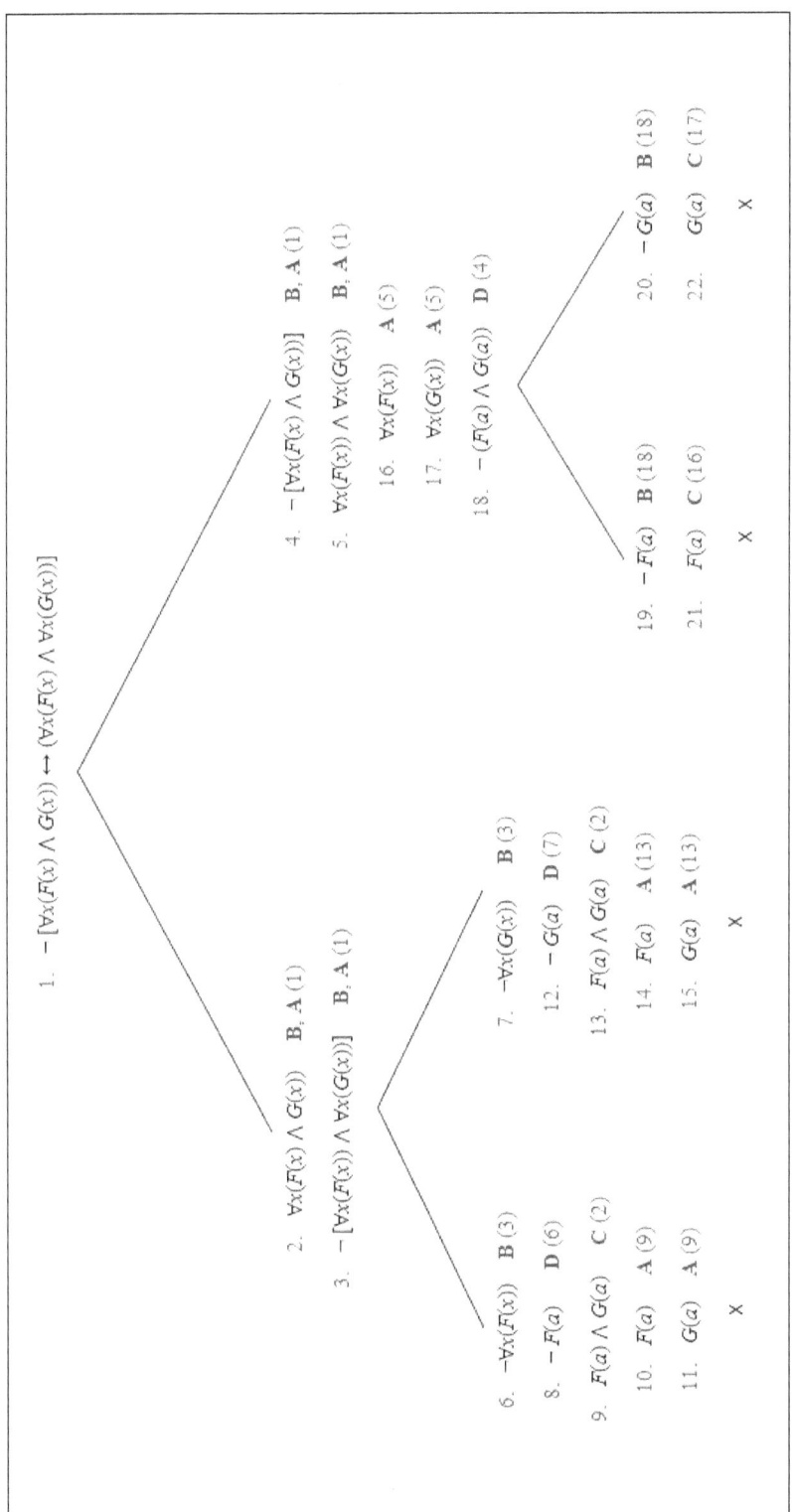

Figure 15.2.2.: A closed FO tableau without unification.

15. The analytic tableaux calculus

Definition 15.28. Given a tree $\mathcal{T}_X$, for $X \subseteq F_{L1}$, and a branch $\mathcal{B}$ thereof, let $(L, \neg L) \in \mathcal{B} \cup X'$, $X' \subseteq X$, L is an atom. Then $\mathcal{B}$ is a closed branch of $\mathcal{T}_X$. If all branches of $\mathcal{T}_X$ are closed, then $\mathcal{T}_X$ is a closed tableau for X.

The proof of soundness of the predicate tableaux calculus without unification is as for the propositional case, but appealing now also to Propositions 15.24-5. In order to prove the completeness of tableaux for FOL we need to extend our definition of a downward saturated set (Def. 15.17) to sets containing γ- and δ-type formulas:

Definition 15.29. Let $X \subseteq F_{L1}$. We say that X is a downward saturated set iff conditions 1 and 2 of Def. 15.17 hold, and additionally

1. if $\gamma \in X$, then $\gamma(t) \in X$ for all $\gamma(t)$ with $t \in H_X$;
2. if $\delta \in X$, then $\delta(c) \in X$ for at least one $c \in Cons \subseteq \Upsilon_X$.

The definition of Hintikka set (Def. 15.18) generalizes immediately to the sets above, and so does Lemma 15.19. Thus, the proof of completeness of the predicate tableaux calculus without unification runs as for the propositional case, with the additional assumptions that $H\mathcal{I}_{\gamma(t)} = \mathsf{t}$ for any $t \in H_X$ and $H\mathcal{I}_{\delta(c)} = \mathsf{t}$ for some constant $c \in Cons \subseteq \Upsilon_X$. In the former case, let $\phi \in X$, ϕ is a formula of type $\gamma = \forall x \phi'$, where ϕ' is an *immediate tableau sub-formula of* ϕ (in this case, $\phi' = \phi[x/t]$); by the property of being a downward saturated set and by the induction assumption, we have $H\mathcal{I}_{\gamma(t)} = \mathsf{t}$ for any term $t \in H_X$. Because H_X is the universe of $H\mathcal{I}_X$, and $H\mathcal{I}_X$ maps any term t to itself (cf. Def. 9.5), for all variable assignments ϖ to H_X we have $H\mathcal{I}_{\phi'}^{\varpi} = H\mathcal{I}_{\phi'[x/\varpi(x)]} = \mathsf{t}$. In the latter case, $\phi \in X$ is a δ-formula, by downward saturation and the induction hypothesis we have it that $H\mathcal{I}_{\delta(c)} = \mathsf{t}$ for some constant $c \in T$ and therefore $H\mathcal{I}_\phi = \mathsf{t}$.

15.2.2. FOL tableaux with unification

Note that when applying rule **C** without unification we have the problem of choosing the ground term t. In effect, any possible ground term can be chosen, or even every one, but this would certainly not be relevant for the tableau proof. The problem is particularly acute in terms of automation, as this is basically impossible without guidance on what term(s) to choose. The strategy then is to delay the choice of the ground

15.2. Analytic tableaux as a FO predicate calculus

term until $\gamma(t)$ allows the closure of *at least* one of the branches of the tree, though evidently a substitution σ that will *simultaneously* close *all* branches is what is required. To do this, we simply use a variable instead of a term, so that the rule now is defined as follows:

Definition 15.30. The expansion rule for a type-γ formula with unification is

$$(\mathbf{C'}) \quad \frac{\forall x \gamma(x)}{\gamma(x')}$$

where x' is a variable not occurring anywhere else in the tableau.

Just as for rule **D**, we apply Skolemization again, but now Skolem terms must not be constants, because the application of unification may create free variables that are implicitly universally quantified. This means that a formula $\exists x \delta(x)$ can now be within the scope of one or more universal quantifiers. According to Step 2 of Algorithm 2.2, we now have:

Definition 15.31. The expansion rule for a type-δ formula with unification is

$$(\mathbf{D'}) \quad \frac{\exists x \delta(x)}{\delta(f(x_1, ..., x_n))}$$

where f is a new function symbol and $x_1, ..., x_n$ are the n free variables of δ.

We can now rephrase the closure rule above (Def. 15.28) for our FO language L* with formulas with free variables as follows:

Definition 15.32. Given a tableau $\mathcal{T}_X$ for $X \subseteq F_{L1}$ and a branch $\mathcal{B}$ thereof, let $L, \neg L \in (\mathcal{B} \cup X)$, L is an atom. Let further L and $\neg L$ be unifiable by means of the MGU σ. Then $\mathcal{B}$ is a closed branch of $\mathcal{T}_{\sigma X}$, the tree for σX constructed by applying σ to all formulas of X. If all branches of $\mathcal{T}_{\sigma X}$ are closed, then $\mathcal{T}_{\sigma X}$ is a closed tableau for σX.

Example 15.33. Figure 15.2.3 shows a tableau with unification for the unsatisfiable set $A = \{\forall x (P(x)), \exists x (\neg P(x) \vee \neg P(f(x)))\}$. Rule D' was not applied, as $\exists x \delta(x)$ does not fall within the scope of a universal quantifier. For further examples, see Fitting (1996).

▶ *Do Exercises 15.7-10*

15. The analytic tableaux calculus

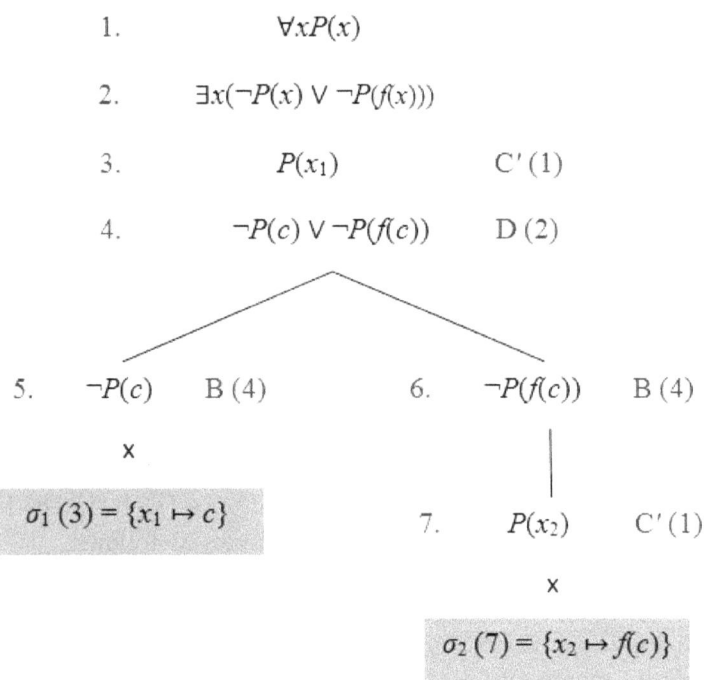

Figure 15.2.3.: A closed FO tableau with unification.

Exercises

Exercise 15.1. Prove the formulas of Exercise 11.2 in the analytic tableaux calculus.

Exercise 15.2. Test the validity of the argument in Example 11.4 in the analytic tableaux calculus.

Exercise 15.3. Test the validity of the arguments of Exercise 14.2 in the analytic tableaux calculus.

Exercise 15.4. Prove Propositions 15.10-13.

Exercise 15.5. Complete the proofs of Theorems 15.7 and 15.15.

Exercise 15.6. Prove Theorems 15.20-1.

Exercise 15.7. Prove the formulas of Exercise 11.9 in the FO analytic tableaux calculus.

Exercise 15.8. Prove in the FO analytic tableaux calculus the arguments of Exercise 14.7.

Exercise 15.9. Prove Propositions 15.24-5.

Exercise 15.10. Prove the soundness and completeness of the FO analytic tableaux calculus.

Bibliography

Bibliographical references

- Aristotle (ca. 350 BC). *Metaphysics*. Trans. by W. D. Ross (1908). Available at http://classics.mit.edu//Aristotle/metaphysics.html.

- Augusto, L. M. (2017). *Logical consequences. Theory and applications: An introduction*. London: College Publications.

- Augusto, L. M. (2018). *Computational logic. Vol. 1: Classical deductive computing with classical logic*. London: College Publications.

- Augusto, L. M. (2020a). *Computational logic. Vol. 1: Classical deductive computing with classical logic*. 2nd ed. London: College Publications.

- Augusto, L. M. (2020b). *Languages, machines, and classical computation*. 2nd ed. London: College Publications.

- Augusto, L. M. (2020c). *Many-valued logics. A mathematical and computational introduction*. 2nd ed. London: College Publications.

- Bachmair, L. & Ganziger, H. (2001). Resolution theorem proving. In A. Robinson & A. Voronkov (eds.), *Handbook of automated reasoning*, vol. 1 (pp. 19-99). Amsterdam: Elsevier; Cambridge, MA: MIT Press.

- Beth, E. W. (1955). Semantic entailment and formal derivability. *Mededlingen der Koninklijke Nederlandse Akademie van Wetenschappen, 18*, 309-342.

- Beth, E. W. (1960). Completeness results for formal systems. In J. A. Todd (ed.), *Proceedings of the International Congress of Mathematicians, 14-21 August 1958* (pp. 281-288). Cambridge: CUP.

- Bloch, E. D. (2011). *Proofs and fundamentals. A first course in abstract mathematics*. New York, etc.: Springer.

Bibliographical references

- Boole, G. (1847). *The mathematical analysis of logic. Being an essay towards a calculus of deductive reasoning.* Cambridge: Macmillan, Barclay, and Macmillan.

- Boole, G. (1854). *An investigation of the laws of thought, on which are founded the mathematical theories of logic and probabilities.* London: Walton and Maberly.

- Boolos, G. S., Burgess, J. P., & Jeffrey, R. C. (2007). *Computability and logic.* 5th ed. Cambridge, etc.: Cambridge University Press.

- Börger, E., Grädel, E., & Gurevich, Y. (2001). *The classical decision problem.* Berlin, etc.: Springer.

- Chang, C.-L. & Lee, R. C.-T. (1973). *Symbolic logic and mechanical theorem proving.* New York & London: Academic Press.

- Chomsky, N. (1957). *Syntactic structures.* The Hague & Paris: Muton.

- Church, A. (1936a). An unsolvable problem of elementary number theory. *American Journal of Mathematics, 2,* 345-363.

- Church, A. (1936b). A note on the Entscheidungsproblem. *Journal of Symbolic Logic, 1,* 40-41.

- Church, A. (1956). *Introduction to mathematical logic.* Princeton, NJ: Princeton University Press.

- Cleave, J. P. (1991). *A study of logics.* Oxford: Clarendon Press.

- Cook, S. A. (1971). The complexity of theorem proving procedures. *Proceedings of the 3rd Annual ACM Symposium of Theory of Computing,* 151-158.

- Curry, H. B. (1963). *Foundations of mathematical logic.* New York, etc.: McGraw-Hill.

- Davis, M. & Putnam, H. (1960). A computing procedure for quantification theory. *Journal of the ACM, 7,* 201-215.

- Enderton, H. B. (2001). *A mathematical introduction to logic.* 2nd ed. San Diego, etc.: Harcourt Academic Press.

- Etchemendy, J. (1999). *The concept of logical consequence.* Stanford: CSLI Publications.

- Fitting, M. (1996). *First order logic and automated theorem proving.* 2nd ed. New York, etc.: Springer.

- Fitting, M. (1999). Introduction. In M. D'Agostino et al. (eds.), *Handbook of tableau methods* (pp. 1-44). Dordrecht: Kluwer.

- Franco, J. & Martin, J. (2009). A history of satisfiability. In A. Biere, M. Heule, H. van Maaren, & T. Walsh (eds.), *Handbook of satisfiability* (pp. 3-74). Amsterdam, etc.: IOS Press.

- Frege, G. (1879). *Begriffsschrift, eine der arithmetischen Formelsprache des reinen Denkens.* Halle a. S.: Louis Nebert. (Engl. trans.: *Begriffsschrift, a formula language, modeled upon that of arithmetic, for pure thought.* Trans. by S. Bauer-Mengelberg. In J. van Heijenoort (ed.), *From Frege to Gödel. A source book in mathematical logic, 1879-1931* (pp. 1-82). Cambridge, MA: Harvard University Press, 1967.)

- Frege, G. (1892). Über Sinn und Bedeutung. *Zeitschrift für Philosophie und philosophische Kritik C*, 25-50.

- Gabbay, D. M. & Woods, J. (2003). *A practical logic of cognitive systems. Vol. 1: Agenda relevance. A study in formal pragmatics.* Amsterdam, etc.: Elsevier.

- Gentzen, G. (1934-5). Untersuchungen über das logische Schliessen. *Mathematische Zeitschrift, 39*, 176-210, 405-431. (Engl. trans.: Investigations into logical deduction. In M. E. Szabo (ed.), *The Collected Papers of Gerhard Gentzen* (pp. 68-131). Amsterdam: North-Holland.)

- Gilmore, P. (1960). A proof method for quantification theory: Its justification and realization. *IBM Journal of Research and Development, 4*, 28-35.

- Gödel, K. (1930). Die Vollständigkeit der Axiome des logischen Funktionkalküls. *Monatshefte für Mathematik, 37*, 349-360. (Engl. trans.: The completeness of the axioms of the functional calculus of logic. In S. Feferman et al. (eds.), *Collected works. Vol. 1: Publications 1929-1936* (pp. 103-123). New York: OUP & Oxford: Clarendon Press, 1986.)

- Gödel, K. (1931). Über formal unentscheidbare Sätze der *Principia Mathematica* und verwandter Systeme, I. *Monatshefte für Mathematik und Physik, 38*, 173-198. (Engl. trans.: On formally undecidable propositions of *Principia Mathematica* and related systems, I. In S. Feferman et al. (eds.), *Collected works. Vol. 1: Publications 1929-1936* (pp. 144-195). New York: OUP & Oxford: Clarendon Press, 1986.)

- Hausman, A., Kahane, H., & Tidman, P. (2010). *Logic and philosophy. A modern introduction.* Boston, MA: Wadsworth, Cengage Learning.

- Henkin, L. (1949). The completeness of the first-order functional calculus. *Journal of Symbolic Logic, 14,* 159-166.

- Herbrand, J. (1930). *Recherches sur la théorie de la démonstration.* Thèses présentées à la Faculté des Sciences de Paris.

- Hilbert, D. & Ackermann, W. (1928). *Grundzüge der theoretischen Logik.* Berlin: Springer.

- Hilbert, D. & Bernays, P. (1934). *Grundlagen der Mathematik. Vol. I.* 1st ed. Berlin & New York: Springer.

- Hintikka, J. (1955). Form and content in quantification theory. *Acta Philosophica Fennica, 8,* 7-55.

- Hurley, P. J. (2012). *A concise introduction to logic.* 11th ed. Boston, MA: Wadsworth.

- Jacquette, D. (2008). Mathematical proof and discovery *reductio ad absurdum. Informal Logic, 28,* 242-261.

- Jaśkowski, S. (1934). On the rules of suppositions in formal logic. *Studia Logica, 1,* 5-32.

- Kalmár, L. (1935). Über die Axiomatisierbarkeit des Aussagenkalküls. *Acta Scientiarum Mathematicarum, 7,* 222-243.

- Kant, I. (1787). *Kritik der reinen Vernunft.* 2. Auflage. Riga: Johan Friedrich Hartknoch. (Engl. trans.: *Critique of pure reason.* Trans. by N. Kemp Smith. London: MacMillan & Co., 1929.)

- Kant, I. (1800). *Logik. Ein Handbuch zu Vorlesungen.* Königsberg: Nicolovius. (Engl. trans.: *Introduction to logic.* Trans. by T. K. Abbott. London: Longmans, Green, & Co., 1885.)

- Kroening, D. & Strichman, O. (2008). *Decision procedures. An algorithmic point of view.* Berlin, Heidelberg: Springer.

- Leitsch, A. (1997). *The resolution calculus.* Berlin, etc.: Springer.

- Łukasiewicz, J. (1929). *Elementy logiki matematycznej.* Warsaw. (Engl. trans.: *Elements of mathematical logic.* Trans. by O. A. Wojtasiewicz. Oxford: Pergamon Press. 1963.)

- Łukasiewicz, J. (1934). Z historii logiki zdan. *Przeglad Filozoficzny, 37*, 417-437. (Engl. trans.: On the history of the logic of propositions. In S. McCall (ed.), *Polish Logic 1920-1939* (pp. 66-87). Trans. by S. McCall and P. Woodruff. Oxford: Oxford University Press. 1967.)

- Łukasiewicz, J. & Tarski, A. (1930). Untersuchungen über den Aussagenkalkül. *Comptes Rendus des Séances de la Société des Sciences et des Lettres de Varsovie, 23*, cl. iii, 39-50. (Engl. trans.: Investigations into the sentential calculus. In A. Tarski, *Logic, semantics, metamathematics: Papers from 1923 to 1938* (pp. 38-59). Trans. by J. H. Woodger. Oxford: Clarendon Press, 1956.)

- Makinson, D. (2008). *Sets, logic, and maths for computing.* London: Springer.

- Martin, N. M. & Pollard, S. (1996). *Closure spaces and logic.* Dordrecht: Kluwer.

- McKeon, M. W. (2010). *The concept of logical consequence. An introduction to philosophical logic.* New York, etc.: Peter Lang.

- Mendelson, E. (2015). *Introduction to mathematical logic.* 6th ed. Boca Raton, FL: Taylor & Francis Group.

- Minsky, M. (1974). A framework for representing knowledge. Report AIM, 306, Artificial Intelligence Laboratory, MIT.

- Negri, S. & von Plato, J. (2001). *Structural proof theory.* New York: Cambridge University Press.

- Nicod, J. G. (1917). A reduction in the number of primitive propositions in logic. *Proceedings of the Cambridge Philosophical Society, 19*, 32-41.

- Nieuwenhuis, R. & Rubio, A. (2001). Paramodulation-based theorem proving. In A. Robinson & A. Voronkov (eds.), *Handbook of automated reasoning*, vol. 1 (pp. 371-443). Amsterdam: Elsevier; Cambridge, MA: MIT Press.

- Post, E. L. (1921). Introduction to a general theory of elementary propositions. *American Journal of Mathematics, 43*, 163-185.

- Prawitz, D. (1965). *Natural deduction. A proof-theoretical study.* Stockholm: Almqvist & Wiksell.

- Priest, G. (2008). *An introduction to non-classical logic. From if to is.* Cambridge: Cambridge University Press.

- Prior, A. (1960). The runabout inference-ticket. *Analysis, 21*, 38-39.

- Quine, W. V. O. (1938). Completeness of the propositional calculus. *Journal of Symbolic Logic, 3*, 37-40.

- Smith, N. J. J. (2012). *Logic. The laws of truth.* Princeton and Oxford: Princeton University Press.

- Smullyan, R. M. (1968). *First-order logic.* Mineola, NY: Dover.

- Stone, M. H. (1936). The theory of representation for Boolean algebras. *Transactions of the American Mathematical Society, 40*, 37-111.

- Tarski, A. (1930). Fundamentale Begriffe der Methodologie der deduktiven Wissenschaften. I. *Monatshefte für Mathematik und Physik, 37*, 361-404. (Engl. trans.: Fundamental concepts of the methodology of the deductive sciences. In A. Tarski, *Logic, semantics, metamathematics: Papers from 1923 to 1938* (pp. 60-109). Oxford: Clarendon Press, 1956.)

- Tarski, A. (1935). Der Wahrheitsbegriff in formalisierten Sprachen. *Studia Philosophica, 1*, 261-405 (Engl. trans.: The concept of truth in formalized languages. In A. Tarski, *Logic, semantics, metamathematics: Papers from 1923 to 1938* (pp. 152-278). Trans. by J. H. Woodger. Oxford: Clarendon Press, 1956.) (Originally published in Polish in 1933.)

- Tarski, A. (1994). *Introduction to logic and to the methodology of deductive sciences.* 4th ed. J. Tarski (ed.). New York & Oxford: Oxford University Press.

- Troelstra, A. S. & Schwichtenberg, H. (2000). *Basic proof theory.* 2nd ed. Cambridge: Cambridge University Press.

- Turing, A. (1936-7). On computable numbers, with an application to the Entscheidungsproblem. *Proceedings of the London Mathematical Society, Series 2, 41*, 230-265.

- van Fraassen, B. (1971). *Formal semantics and logic.* New York: Macmillan.

- Venn, J. (1881). *Symbolic logic.* London: Macmillan and Co.

- Wajsberg, M. (1937). Metalogische Beiträge. *Wiadomości Matematyczne, 43*, 1-38. (Engl. trans.: Contributions to metalogic. In S. McCall (ed.), *Polish Logic 1920-1939* (pp. 285-318). Trans. by S. McCall and P. Woodruff. Oxford: Oxford University Press. 1967.)

- Whitehead, A. N. & Russell, B. (1910). *Principia mathematica.* Vol. 1. Cambridge: Cambridge University Press.

- Whitehead, A. N. & Russell, B. (1912). *Principia mathematica.* Vol. 2. Cambridge: Cambridge University Press.

- Whitehead, A. N. & Russell, B. (1913). *Principia mathematica.* Vol. 3. Cambridge: Cambridge University Press.

Index

This Index is a hybrid of an *Index rerum* and an *Index nominum*. In the first case, we give solely the page for the first definitional occurrence of the term or expression; in the second case, we give all the occurrences of a specific name. In the latter case, only names of (historical) significance for the classical formalization of logic are given.

Index

A
Abstract interpreter, 336
Adequateness of a logical system, 118
Adequateness of a program, 336
Affirming the consequent, 52
Algorithm, Robinson's, 65
Algorithm, Tseitin transformation, 60
Analytic tableaux, 365
Argument, 48
Aristotle, 8, 185, 186, 191
Assumption (in a proof), 229
Automated theorem proving (ATP), 287
Axiom, 106
Axiom schemata, Derived, 242
Axiom system, 237
Axiom system, Church's, 248
Axiom system, Frege-Łukasiewicz's, 238
Axiom system, Frege's, 247
Axiom system, H propositional, 242
Axiom system, Hilbert's, 247
Axiom system, Kalmár's, 247
Axiom system, Nicod's, 248
Axiom system, Tarski-Bernays', 248
Axiom system, Tarski-Bernays-Wajsberg's, 248
Axiom, Logical, 121
Axiom, Non-logical or proper, 121

B
Backtracking, 344
Backus-Naur form, 38
Bernays, P., 226, 248
Beth, E. W., 194, 365
Big-O notation, 20
Boole, G., 8, 9, 27, 79, 150
Boolean expression, 81
Boolean function, 81, 85
Boolean variable, 81

C
Characteristic function, 124
Church, A., 224, 233, 248
Church-Turing Theorem, 225
Clause, 52
Clause, Definite, 52
Clause, Dual-Horn, 52
Clause, Horn, 52
Closure operation, 133
Closure system, 101
Closure, Existential, 39
Closure, Universal, 39
Compactness, 105
Compactness of propositional logic, 301
Complete lattice, 102
Completeness, 116
Completeness of propositional analytic tableaux, 376
Completeness of the resolution principle, 323
Completeness theorem, 195

Consequence operation, 100
Consequence relation, 100
Consistency, 109
Constructive dilemma (CD), 51
Contingency, 113
Contradiction, 113
Contraposition, Law of, 183
Cook-Levin Theorem, 296
Counter-model, 111
Counter-proof, 108

D
Davis, M., 313
De Morgan's laws (DM), 183
Decidable language (or problem), 19
Deduction theorem (DT), 116
Deduction, Resolution, 314
Deduction-Detachment theorem (DDT), 135
Deductive system, 105
Denotation, 204
Denying the antecedent, 52
Derivability, 107
Destructive dilemma (DD), 51
Discharge (of assumptions), 256
Distributive laws, 58
Domain of discourse, 89

E
Entscheidungsproblem, 125, 224
Equality, 186
Equality substitution, 350
Equisatisfiability, 57
Equivalence relation, 191
Euler diagram, 143
Ex contradictione quodlibet (ECQ), 51
Ex falso quodlibet (EFQ), 176
Excluded middle, Principle of (PEM), 178
Existential distribution, 40

Existential fallacy, 171
Explosion, Principle of, 79
Extensionality, Principle of, 84

F
Fact, Prolog, 156
Factor, 317
Factorization, 317
Finite satisfiability, 126
Finite-model property (FMP), 132
Formal language, 3
Formal logic, 3
Frege, G., 10, 83, 204, 237, 238, 247
Fregean axiom, 84
Function (symbol), 36
Functional completeness, 86

G
Generalization rule (GEN), 245
Gentzen, G., 253, 255, 271
Goal clause, 341
Goal clause, Empty, 341
Goal, Prolog, 156
Gödel, K., 117, 193, 195, 224
Ground expression, 34
Ground extension, 127
Ground instance, 63
Ground substitution, 62

H
Henkin, L., 136, 194, 195, 270, 283
Herbrand base, 208
Herbrand instance (H-instance), 208
Herbrand interpretation (H-interpretation), 208
Herbrand model (H-model), 208
Herbrand model, Least, 362
Herbrand satisfiability (H-satisfiability), 209
Herbrand universe, 207

Herbrand, J., 206, 211, 300
Herbrand's Theorem, 301
Hilbert(-style) systems, 237
Hilbert, D., 31, 125, 206, 223, 224, 226, 237, 247
Hintikka set, 375
Hintikka, J., 365
Hintikka's Lemma, 375
Hyper-resolution, 330
Hypothesis (in a proof), 229
Hypothetical syllogism (HS), 51

I
I-clash, 326
Identity of indiscernibles (IdI), 187
Identity, Law of, 183
Induction, Mathematical, 12
Induction, Structural, 12
Inference, 104
Inference operation, 105
Inference relation, 105
Inference rule, 106
Inference system, 105
Interpretation, 89

J
Jaśkowski, S., 253

K
Kalmár, L., 242, 247
Kant, I., 8, 26
Kleene, S., 87

L
Language, First-order (FO), 38
Language, Logical, 33
Language, Object, 7
Language, Propositional, 37
Leibniz, G. W. von, 9
Leibniz's law (LL), 187
Lifting lemma, 321
Lindenbaum's Theorem, 122

Lindenbaum-Tarski algebra, 213
Logic (of a logical system), The, 110, 113
Logic circuit, 162
Logic design, 162
Logic gate, 163
Logic program, 159
Logic, Classical first-order (CFOL), 144
Logic, Classical propositional (CPL), 144
Logical consequence, 100
Logical equivalence, 87, 92
Logical system, 100
Löwenheim-Skolem Theorem, 136
Łukasiewicz, J., 87, 141, 238, 250

M
Macro-resolution, 332
Meaning, 77
Meaning of a program, 336
Meaning of a program, Intended, 337
Meaning, Principle of compositionality of, 84
Metalanguage, 7
Metalogic, 11
Metaproof, 11
Model, 111
Modus ponens (MP), 51
Modus ponens, Universal (UMP), 335
Modus tollendo ponens (TP), 51
Modus tollens (MT), 51
Monotonicity, 104

N
Natural deduction calculus, 253
Negation distribution, 40
Negation law, Double (DN), 177
Nicod, J. G., 248
Non-contradiction, Principle of (PNC), 178

Index

Normal form, Conjunctive (CNF), 57
Normal form, Disjunctive (DNF), 58
Normal form, Negation (NNF), 53
Normal form, Prenex (PNF), 54
Normal form, Skolem (SNF), 54

O

One-literal rule, 313

P

Parameter, 265
Paramodulation, 350
Paramodulation, Binary, 350
Paramodulation, Ordered, 352
Paramodulation, Simultaneous, 352
Peirce arrow, 147
PI-clash, 327
PI-deduction, 328
Post, E., 28, 31
Prawitz, D., 253, 255, 279
Predicate (symbol), 36
Prefix classes, 132
Prior, A., 235
Problem (MAX-SAT), The maximum satisfiability, 309
Problem for 2-CNF formulas (2-SAT), The satisfiability, 290
Problem for 3-CNF formulas (3-SAT), The satisfiability, 291
Problem for DNF formulas (DNF-SAT), The satisfiability, 292
Problem for dual-Horn formulas (DUAL-HORN-SAT), The satisfiability, 309
Problem for Horn formulas (HORN-SAT), The satisfiability, 291
Problem for k-CNF formulas (k-SAT), The satisfiability, 291
Problem for quantified Boolean formulas (QBF-SAT), The satisfiability, 292
Problem, (Logical) Decision, 123
Problem, Computational, 15
Problem, Decision, 19
Problem, Hilbert's tenth, 31
Problem, Post's correspondence, 31
Problem, The acceptance, 30
Problem, The Boolean satisfiability (SAT), 289
Problem, The busy beaver, 31
Problem, The clique, 22
Problem, The decision, 224
Problem, The halting, 30
Problem, The satisfiability (SAT), 123, 288
Problem, The validity (VAL), 123
Problem, The vertex cover, 22
Procedure, Decision, 123
Program clause, 341
Prolog, 155
Prolog program, 159
Proof, 107
Proof by contradiction, 13
Proof calculus, 106
Proof system, 106
Proof, Constructive, 250
Proof, Logical, 11
Proof, Metalogical, 11
Proposition, Categorical, 150
Provability, 107
Prover9/Mace4, 311
Putnam, H., 313

Q

Quantification, 39
Quantification, Trivial, 39

Quantifier (symbol), 39
Quantifier axioms, 245
Quantifier duality, 92
Quantifier reversal, 40
Query, Prolog, 156
Quine, W. V. O., 194

R
Recursion, 362
Recursive language (or set), 19
Recursively enumerable language (or set), 20
Reductio ad absurdum (RA), 183
Reductio ad absurdum proof, 14
Reduction (in LP), 338
Reduction, Ground, 338
Refutation, 108
Refutation completeness, 341
Representation theorem, 214
Resolution principle for FOL, 317
Resolution principle for propositional logic, 313
Resolution refinement, 324
Resolution, Binary, 314
Resolution, Hyper-, 330
Resolution, LD, 335
Resolution, LI, 335
Resolution, Linear, 334
Resolution, Macro-, 330
Resolution, RUE, 363
Resolution, Semantic, 325
Resolution, Set-of-support, 359
Resolution, SLD, 335
Rule, Prolog, 156
Russell, B., 28, 72

S
Satisfiability, 111
Schema, 50
Search, Breadth-first, 362
Search, Depth-first, 344
Semantical correlate, 83

Semantics, 112
Semi-decidable language (or problem), 20
Sequent calculus, 272
Set-of-support deduction, 359
Set-of-support resolution, 359
Sheffer stroke, 147
Skolem constant, 54
Skolem function, 54
Smullyan, R. M., 365, 376
Soundness, 116
Square of opposition, 150
Statement, Prolog, 156
Stone, M. H., 214
Substitution, 62
Substitution principle (SubP), 187
Substitution rule (SUB), 107
Syllogism, Categorical, 148
Syntax, 7
Syntax, Ambivalent, 158

T
Tableau proof, 367
Tarski, A., 11, 12, 28, 83, 105, 175, 181, 182, 188, 191, 204, 216, 217, 234, 248
Tarski-style conditions, 175
Tautology, 113
Theorem, 107
Theory, 121
Theory, Scapegoat, 194
Trace (of an interpreter), 337
Tree, Derivation, 5
Tree, Formula, 40
Tree, Refutation, 314
Tree, Semantic, 303
Tree, SLD-resolution, 345
Tree, Syntactic, 5
Truth function, 78
Truth table, 77
Truth value, 77
Truth-preservation, 178

Index

Truth-value assignment, 81
Turing machine, 16
Turing machine, Universal, 23
Turing, A., 22, 223, 225, 227, 233
Turing-completeness, 155
Turing-decidability, 20
Turing-recognizability, 20
Turing's Theorem, 225

U

Ultrafilter theorem, 219
Unicity of decomposition, 35
Unification, 63
Unification problem, 64
Unifier, Most general (MGU), 63

V

Validity, 112
Validity, Analytical, 235
Validity, Refutation, 293
Valuation, 77
Venn diagram, 143
Venn, J., 9, 27

W

Wajsberg, M., 248
Whitehead, A. N., 28

www.ingramcontent.com/pod-product-compliance
Lightning Source LLC
Chambersburg PA
CBHW071646160426
43195CB00012B/1370